General Radiotelephone Operator License

FCC Commercial Radio License Preparation Element 1 and Element 3 Question Pools

BY
FRED MAIA, W5YI
GORDON WEST, WB6NOA

GENERAL RADIOTELEPHONE OPERATOR LICENSE

This book was developed and published by:
Master Publishing, Inc.
Richardson, Texas

Design and artwork by:
Plunk Design
Dallas, Texas

Editing by:
Gerald Luecke, KB5TZY
Charles Battle

Printing by:
Custom Printing Company
Owensville, Missouri

Acknowledgements:
All photographs that do not have a source identification are either courtesy of Radio Shack, authors or Master Publishing, Inc. originals.

The authors wish to thank Matt J. McCullar, KJ5BA, for his extensive research, text suggestions, contributions and reviews.

The authors wish to acknowledge the fine cooperation of the FCC staff at the Washington, DC and Gettysburg, PA offices.

A print of the Hoover photograph in Chapter 1 was obtained from the Herbert Hoover Presidential Library-Museum. Supposedly, Henry Miller News Picture Service was original copyright owner; however, no such listing is available from the copyright office.

Trademarks:
Teflon® is a registered trademark of E.I. DuPont DeNemours Co., Inc.
Bakelite® is a registered trademark of Union Carbide Corporation.
Newmar® is a registered trademark of Newmar, Inc.

Copyright © 1994
Master Publishing, Inc.
14 Canyon Creek Village MS 31
Richardson, Texas 75080
(214) 907-8938
All Rights Reserved

REGARDING THESE BOOK MATERIALS

Reproduction, publication, or duplication of this book, or any part thereof, in any manner, mechanically, electronically, or photographically is prohibited without the express written permission of the publisher.

The Author, Publisher or Seller assume no liability with respect to the use of the information contained herein.

For permission and other rights under this copyright, write
Master Publishing, Inc.

9 8 7 6 5 4 3 2 1

Table of Contents

Preface	iv
CHAPTER 1. History of Radio Regulation	1
CHAPTER 2. Commercial Radio Operator's Licenses – Then and Now	9
CHAPTER 3. Commercial License Examinations – Then and Now	25
CHAPTER 4. Question Pool – Element 1 *(Radio Law, Operating Practice)*	33
CHAPTER 5. Question Pool – Element 3 *(Electronic Fundamentals)*	79
CHAPTER 6. Taking the License Examinations	269
Appendix	
List of COLEMs	275
FCC Field Offices	276
FCC Rules and Regulations	
Part 13 – Commercial Radio Operators	277
Part 23 – International Fixed Public Radio Services	284
Part 73 – Broadcast Radio Services	285
Part 80 – Maritime Radio Services	286
Part 87 – Aviation Radio Services	310
Summary of Question Pool Formulas	317
Filled-out Examples of Forms 753, 755 and 756	320
Glossary	323
Index	329

QUESTION POOL RELEASE AND NOMENCLATURE

The Element 1 and Element 3 question pools are as released by the FCC on September 2, 1993 with corrections approved by the FCC Aviation and Marine Branch on March 3, 1994.

    Our interpretation and implementation may be different from other publishers. We are sure that any slight differences will not contribute to improper understanding of the question or its answer.

Preface

Welcome and congratulations on taking a very important step—your decision to prepare for a commercial radio operator's license. Gaining a commercial radio license is certification by the federal government that you have what it takes to operate and/or maintain commercial radio equipment.

Gaining a commercial license is not easy. The examinations are long and difficult. You will need to work hard to learn about operating rules and regulations, electronic devices and their use in electronic circuits, and maintaining electronic equipment. But when you finish, official documentation from the Federal Communications Commission—the FCC—will verify to the world that you know your stuff.

With your license, you can participate in the rewards—financial security, a wise addition to your resume, increased technical knowledge, personal satisfaction, and, if you're already an amateur, another ticket for the wall of the shack.

General Radiotelephone Operator License contains Element 1 and Element 3 question pools. Everything you need to pass the written examinations required for a Marine Radio Operator Permit (MROP) or a General Radiotelephone Operator License (GROL) is included. Passing the Element 1 and Element 3 examinations not only gains you your MROP or GROL, but puts you in position to earn additional commercial licenses. For example, an Element 1 examination is a partial requirement for the Global Maritime Distress and Safety System Radio Operator (GMDSS/O) and Radio Maintainer (GMDSS/M) licenses, and the three Radiotelegraph Operator Certificates, T-1, T-2 and T-3. An Element 3 examination is a partial requirement for the GMDSS/M license.

General Radiotelephone Operator License begins by providing a brief history of commercial radio license regulation in Chapter 1. Chapter 2 lists the commercial licenses available today, spells out when you do or do not need a commercial radio license, and defines which permit, certificate, license, or endorsement is required. Chapter 3 details which examinations must be passed to earn a particular license. The Element 1 question pool of 169 questions is Chapter 4, and the Element 3 question pool of 720 questions is Chapter 5. The book proper concludes in Chapter 6 with details about taking your examination—where to locate an examination, how the examination will be administered, and hints on passing the examination. To aid in your study, especially of Element 1, and to have important FCC Rules and Regulations on hand, the Appendix contains the complete FCC Part 13, and excerpts from Parts 23, 73, 80 and 87.

So step up, study hard, and congratulations will come your way when you pass your examination(s) and have your commercial license in hand.

                                          Fred Maia W5YI
                                          Gordon West WB6NOA

History of Radio Regulation

INTRODUCTION
Communications by radio has been called the most regulated business in the land. It is easy to understand why. Progress cannot continue without reliable communication, and this has been understood since biblical times. Many people choose to live without electronic communication, but no one would choose to live with unregulated electronic communication.

IN THE BEGINNING
Table 1-1 lists some significant events and important actions between 1835 and 1910 that furthered electronic communications, which started with the invention of the telegraph. *Figure 1-1* shows a typical system.

Table 1-1. Electronic Communications 1835-1910

Year	Event and Action
1835	Electronic communications began with the invention of the telegraph by Samuel F.B. Morse, a professor at New York University. Morse code is named after him, and is still the international CW code used today.
1849	Two European countries were linked by telegraph. This caused initial international agreements on rules and regulations governing sharing of information.
1850	Other treaties followed.
1865	25 European nations met in Paris and formed the International Telegraphic Convention. Later it became the International Telecommunications Union (ITU).
1865	Guglielmo Marconi, an Italian inventor, proved the feasibility of radio communications for which he received a patent in 1897.
1899	Marconi demonstrated the first transatlantic radio transmission from England to Newfoundland.
1901	Marine radio was born when the U.S. Navy adopted a wireless system.
1903	First international conference on governing radio communications was held in Berlin. Nine nations agreed that public safety took precedence over squabbles between commercial ventures.
1906	International conference in Berlin agreed to require ships to be properly equipped with wireless transmitters and receivers, and to set the first international distress frequency as 500 kHz for ships to use to call for help. Wireless radio was added to ITU's responsibilities.
1909	Steamships *Republic* and *Florida* collided off the coast of New York and 1500 lives were saved by a distress call sent by radio operator Jack Binns. Later in the year, the *S.S. Arapahoe* brought help with "SOS", which was adopted this year as the international radiotelegraph distress call. It is still in use today. "Mayday" was adopted in 1927 as the international distress call for radiotelephony.

1 GENERAL RADIOTELEPHONE OPERATOR LICENSE

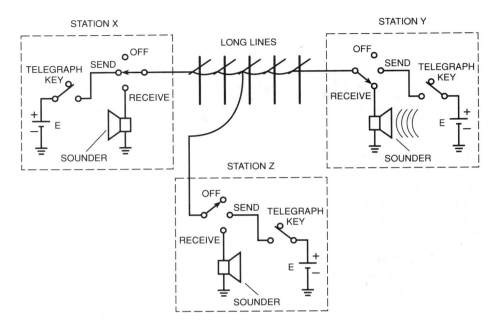

Figure 1-1. Early Long-Line Telegraph (Station X Sending to Station Y) System

AN ACCIDENT BRINGS CHANGE

In the early years of the 20th century, wireless communications remained confined primarily to the sea. Yet not many ships operated radio equipment, and those that did saw no logic in staffing it 24 hours a day. To do so was considered a luxury. That changed after the terrible night in April 1912 when the *Titanic* was ripped open by an iceberg in the North Atlantic and sent to the bottom of the sea three hours later.

The radio operator on board that massive ship frantically called for help over wireless, and the *Carpathia,* 58 miles away, responded and managed to rescue 700 survivors. Some 1,500 passengers perished! Later inquiries learned that another ship, called the *California,* had been only 20 miles away from the *Titanic* at the time of the accident and could have rushed to the scene much faster. Why didn't it? Because the *California's* radio operator had gone to bed, and there was no one to relieve him.

In the United States, *The Wireless Ship Act of 1910* — passed two years before the *Titanic* disaster — should have prevented the loss of so many lives. But not until after 1912 was this Act amended to require at least two operators and a constant watch, with emergency backup power supplies.

THE BEGINNINGS OF RADIO LEGISLATION

Partially as a result of the *Titanic* disaster, the United States Congress enacted the first law for the domestic control of general radio communications — the *Radio Act of 1912*. It regulated the character of emissions, as well as the transmission of distress calls, set aside certain frequencies for government use, and placed the control of all wireless stations under the jurisdiction of the Department of Commerce. The Act was the beginning of government radio licensing. It made access to the electromagnetic spectrum a privilege granted only by government approval.

The *Radio Act of 1912* was praised by many, but not by all. Radio was so new that very few understood its potential. Some barely knew what "wireless" meant. Lee de Forest, the inventor of the Audion three-element vacuum tube amplifier—a very essential ingredient in the advancement of both wired and wireless communications—was prosecuted for using the mails to defraud. The prosecutor accused de Forest of, "...willfully and deliberately misleading the public by stating that soon it will be possible to transmit the human voice across the Atlantic Ocean!"

At the time the *Radio Act of 1912* was passed, the radio spectrum was so spacious and so unoccupied that no one thought that channels would ever have to be assigned, much less share frequencies. If you wanted to operate a transmitter, there was plenty of room in which to do it. All you had to do was apply for a license. The fact that the man in charge of issuing licenses had no legal right to turn down anyone would become an important point years later in court.

WORLD WAR I: CLOSED FOR THE DURATION

When the United States entered the Great War, the Woodrow Wilson administration released two edicts with regard to communications. First, all commercial radio stations would fall under direct government operation for the duration of the war. Second, all amateur radio operators must not only sign off the air until further notice, but also dismantle all of their equipment under penalty of imprisonment. In addition, major radio manufacturers pooled their knowledge and expertise into getting the war won.

THE GREAT RADIO RUSH

World War I ended in November, 1918 and the massive military market for radio receivers came to an end. When the Radio Corporation of America (RCA) acquired the American Marconi company in 1919, David Sarnoff became its manager and convinced the company to go into the radio business. Sarnoff, a young wireless operator, even before World War I had contemplated a "radio music box" that would receive programs broadcast for public information and entertainment. At about the same time, General Electric and Westinghouse also began making radio receivers. At first, the public demand for receivers was small because there were few stations to listen to. That changed quickly as licensed stations went on the air and began regular broadcasting.

At the end of 1920, only 30 radio stations in the U.S. offered regular broadcasts, and all of them used amplitude modulation (AM). Pittsburgh's 8XK (an amateur radiotelephone station set up in 1916 by Westinghouse engineer Dr. Frank Conrad) became KDKA, the nation's first commercial broadcast station working on a wavelength of 360 meters. It signed on the air on election night, November 2, 1920, and today is the nation's oldest radio station still in operation. Licensing of broadcast stations on a regular basis began in 1921 with WBZ, Springfield, Mass., being the first station licensed. By the end of 1922 there were 382 stations, and by 1927 there were 733 stations! Most were operated by radio manufacturers, dealers and department stores selling receivers. The 1920s brought forth a virtual explosion of growth in the radio industry.

It seems almost impossible to believe that there was once a time when millions of citizens lived their entire lives without ever once hearing the voices of the leaders of their own country. But with a radio, there was so much that could be brought right into the home. Lonely and isolated people eagerly snapped them up to hear what was going on in the world around them. Thousands of musical novices could hear operas on the radio almost every night. Getting the news faster than newspapers could report it was truly exciting, and getting election results in hours instead of days was almost unbelievable.

WHEN BEDLAM REIGNS

The constant flood of radio stations joining the airwaves proved to be overwhelming by the middle of the 1920s. From 1923 to 1926, Secretary of Commerce Herbert Hoover (an engineer in his own right) submitted and resubmitted bills before Congress to straighten out the radio legislative process. Hoover gathered radio conferences each year from 1922 to 1925 in an effort to entice voluntary cooperation from the radio industry. He even listened in on a special receiver at his home as shown in *Figure 1-2,* to better understand the needs of the listening public.

Secretary Hoover (who later became President Hoover) understood that the radio world could no longer be adequately controlled by the *Radio Act of 1912.* As the Federal government's man in charge, he used his power to grant or deny licenses, assign frequencies, and dictate the time of day when a station could operate.

Although radio stations held licenses, they began to bend the rules. Stations lengthened their working hours, changed frequencies, and increased output power without authorization. When one of these station's owners, Zenith, was taken to court, Zenith charged that the Secretary of Commerce had no legal authority to tell them when or

HISTORY OF RADIO REGULATION 1

Figure 1-2. Commerce Secretary Herbert Hoover listening to radio receiver installed in his home so he could better understand complaints received by his department.
(Print Received from Herbert Hoover Presidential Library-Museum)

where to transmit a radio signal. On July 8, 1926, the acting Attorney General of the United States decreed that the Secretary of Commerce had no legal authority to assign wavelengths, power levels, or hours of operation, or to restrict the length of a station's license. Officially, no government agency controlled radio for the next six months.

The predictable happened. Without regulations, hundreds of stations had a field day. They cranked up power levels and changed frequencies whenever they desired. New stations, completely unlicensed, went on the air. The result was chaos. There was so much interference that millions of listeners all across America switched off their receivers in disgust, and the sale of new receivers slowed to a trickle.

A NEW AGENCY — THE FRC

It was the lack of income from sales of radio receivers that finally convinced the radio industry that some form of legislation and government control was the answer. This time they were glad to meet with Secretary Hoover. In February, 1927, Congress enacted the *Radio Act of 1927*. The new law created an agency known as the Federal Radio Commission (FRC). Led by a five-man panel, the FRC had power, and used it.

The FRC was granted legal authority to decide how much of the radio spectrum each service would get, and to change it if necessary. License applications could now be legally refused to be granted or renewed. Engineers were placed in charge to keep up with the latest scientific developments, and incorporate them into the rules and regulations as necessary. So sweeping were the changes, and so well were they received, that the *Radio Act of 1927* has been called radio's "Magna Charta."

Also taking place in 1927 was the International Radiotelegraph Convention in Washington. Practically every nation in the world joined in, deciding as a whole who would use what portions of the shortwave bands for different purposes.

THE FINAL REWRITE — THE FCC

By the beginning of the 1930s, it became clear that the FRC, while a good beginning, needed to be expanded to include regulation of other forms of electrical communication. The result was the famous *Communications Act of 1934*; it contained much of the 1927 Act, with several major additions. The FRC was abolished, and in its place the Federal Communications Commission (FCC) was born. Not only would it regulate radio, but also telegraph and telephone services.

Among the duties of the FCC are allocating radio frequency bands along international guidelines, assigning frequencies for various radio services and individual stations, regulating common carriers involved in foreign and interstate commerce, and determining the operational and technical qualifications of radio operators.

The FCC was created from the start to be a powerful agency. Standards were explicitly laid out, and stations that failed to follow them were subject to criminal prosecution. The *Communications Act of 1934* itself is only about 150 pages; however, regulations set up under the law currently occupy several thousand pages. To this day, no radio or television station in the United States can be sold, moved, shut down, or change its operating hours or power level without express permission from the FCC. Because of licensing, the FCC's word is law on whether or not a station can legally exist. In the first 50 years of FCC operation, only 149 stations lost their broadcast licenses. These days the FCC imposes massive fines on stations that break the rules (such as the broadcasting of hoaxes).

FREQUENCY MODULATION

Even though the principle of frequency modulation (FM) was known previously, a patent on FM was not issued until 1902. In addition, its advantages for broadcasting were not developed until much later — shortly before World War II — largely as a result of developmental work by Edwin H. Armstrong. It was not until 1940 that the FCC authorized commercial FM broadcasting to start January 1, 1941. The first licensed commercial FM station was WSM-FM in Nashville, Tennessee. It operated from May 29, 1941 until 1951.

HISTORY OF RADIO REGULATION 1

LANDMARK FCC DECISIONS
- Frequency-modulation was first presented to the FCC in 1935.
- Spectrum space was set aside for television in 1939.
- The FCC created the National Television System Committee (NTSC) to develop standards for monochrome television transmission.
- VHF channels 1–13 were assigned. Later, channel 1 was reassigned to land-mobile communication. Remaining channels, 2-13, are in use today.
- Full NTSC monochrome television operation began in 1941.
- 70 new UHF channels were assigned for TV broadcast in 1952.
- First UHF station signed on in 1952.
- Color TV standards were adopted in 1953.
- Stereo FM was approved in 1961.
- It was deceed in 1964 that all TV sets hence forth would be equipped with UHF tuning.
- First maximum-power (5 megawatts) UHF station came on the air in 1974.

KEEPING UP WITH THE TIMES
The President of the United States appoints the five FCC members. No more than three of them may be from the same political party. The five-year terms are staggered so that no one can replace all of them at once. FCC operations are continually monitored by Congress, and, because the FCC is no longer a permanent agency, it must be reauthorized by Congress every two years.

As technology changes, the FCC works steadily to keep up. Cable television, digital data transmission over the telephone lines, satellite broadcasting, high-definition television (HDTV)—all fall under their jurisdiction. Whenever a new format is invented and brought to the FCC's attention, the FCC examines it and makes rules accordingly.

INTERNATIONAL REGULATION
Almost every nation in the world belongs to the International Telecommunication Union (ITU). This is a United Nations agency, headquartered in Geneva, Switzerland. International standards are agreed upon here. All radio and TV station call sign prefixes for individual countries are set aside in groups by the ITU and only allowed in regions specified by ITU. For example, in the United States, which is in ITU Region 2 (see *Figure 1-3*), all station call signs begin with K, W, N or certain A prefixes.

World Administrative Radio Conferences (WARCs) are held about every 10 years to hammer out shares of the broadcast spectrum. The latest developments in technology help determine who gets to use which frequencies, and for what purposes, agreed upon through the WARC. During the most recent WARC, delegates set aside a range of microwave frequencies that will be used exclusively for the first manned mission to Mars!

1 *GENERAL RADIOTELEPHONE OPERATOR LICENSE*

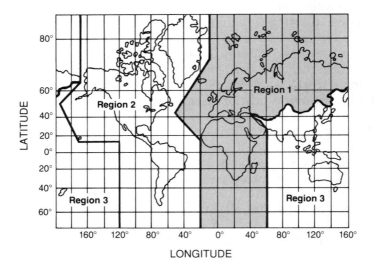

Figure 1-3. ITU Regions

The FCC and the ITU govern the use of all telecommunications equipment and standards in the U.S.A. Any new mode of information exchange will be examined by them, and will be agreed on with other countries through a WARC.

IN CONCLUSION

If everyone tried to communicate following one's own set of rules while ignoring everyone else's, it would be like a metropolitan area without traffic lights. It is for the benefit of the world community that international standards be agreed upon, and that engineers and technicians follow them. Lives are saved and bettered every day because of this. Anyone studying to further the progress of technical communication should be applauded. For the world to march forward, we must listen to our engineers. And we must listen to each other.

Commercial Radio Operator's Licenses – Then and Now

LICENSES AND EXAMINATIONS

There are two basic classes of licensed radio operators: amateur and commercial. Both licensing programs exist for the same reasons: to reduce interference to other radio stations, to insure technician and operator qualifications, and to bring order to the radio spectrum. As a general rule, the FCC issues a radio license when an applicant passes an examination demonstrating adequate knowledge of radio rules, operating procedures and electronics knowledge. An amateur license is required when the transmissions are for hobby purposes. A commercial radio license is needed when the radio equipment is used in connection with a business venture.

Besides being able to pass the necessary radio law and technical examinations, applicants for commercial radio licenses must be citizens of the United States (or eligible for employment in the United States) and be able to transmit and receive messages in English. Radiotelephone licenses permit operation, maintenance and repair of any type of transmitting equipment not transmitting Morse code. Up until ten years ago, all radiotelephone and radiotelegraph licenses were issued for a term of five years. *Figure 2-1* shows the line up of the different commercial radio licenses that *existed prior to 1985*. The elements required at that time for the examinations are identified in *Figure 2-2*.

License	Examination Elements
Radiotelephone – First Class License	1, 2, 3 and 4
Radiotelephone – Second Class License	1, 2 and 3
Radiotelephone – Third Class Permit	1 and 2
Radiotelegraph – First Class License	1, 2, 5 and 6 plus 20 CG[1] and 25 PL[3] code
Radiotelegraph – Second Class License	1, 2, 5 and 6 plus 16 CG[2] and 20 PL[4] code
Radiotelegraph – Third Class Permit	1 and 2 plus 16 CG[2] and 20 PL[4] code
Aircraft Radiotelegraph Endorsement	7
Ship Radar Endorsement	8
Broadcast Endorsement	9
Restricted Radiotelephone Permit	No oral or written examination

1. CG means 20-wpm* Code Groups
2. CG means 16-wpm Code Groups
3. PL means 25-wpm Plain Language
4. PL means 20-wpm Plain Language

* wpm means words per minute. Five characters equals one word.

Figure 2-1. Discontinued Commercial Radio Operator Licenses That Existed Prior to 1985

2 GENERAL RADIOTELEPHONE OPERATOR LICENSE

Element	Description	Element	Description
1	Basic Law	5	Radiotelegraph Law
2	Basic Operating Practice (Law)	6	Advanced Radiotelegraph (Theory)
3	Basic Radiotelephone (Theory)	7	Aircraft Radiotelegraph
4	Advanced Radiotelephone (Theory)	8	Ship Radar Techniques
		9	Broadcast Techniques

Figure 2-2. Elements Required for License Examinations Prior to 1985

MAJOR CHANGES MADE IN 1984!

On February 24, 1984, the Federal Communications Commission voted to make major changes to its commercial radio operator licensing program. On June 15, 1984, the following amendments were made:

1. The First Class and Second Class Radiotelephone Operator Licenses were discontinued and replaced with a special lifetime General Radiotelephone Operator License. Most First Class holders were *very unhappy* with this handling by the FCC, since they had worked hard to obtain the "First Phone" license, which was being reduced (as of 1985) to the level of a permanent Second Class Radiotelephone license. The Third Class Radiotelephone Operator Permit, Aircraft Radiotelegraph Endorsement and the Broadcast Endorsement were eliminated and not replaced.

2. All domestic operating license requirements for the private two-way radio services were abolished. That meant that anyone — commercially licensed or not — could now install, maintain or repair transmitting equipment in any of the private two-way radio services. The licensee of the station is responsible for the proper operation of the station. The radio services affected are
    a. Private Land Mobile Radio Service – Part 90
    b. Private Operational Fixed Microwave Service – Part 94
    c. General Mobile Radio Service (GMRS) – Part 95 – Subpart A
    d. Radio Control Services (R/C) – Part 95 – Subpart C
    e. Citizens Band Radio Service (CB) – Part 95 – Subpart D

3. Commercial radio licenses still required by international law will be indefinitely continued. These radio services are
    a. Aviation Services – Part 87
    b. Maritime Services – Part 80
    c. International Fixed Service – Part 23

4. A new Marine Radio Operator Permit (MROP) was established to be required by operators:
    a. aboard certain vessels navigating the Great Lakes, any tidewater or in the open sea,
    b. certain aviation radiotelephone stations, and
    c. certain maritime coast radiotelephone stations.

WHO NEEDS A COMMERCIAL RADIO OPERATOR LICENSE?
The answer depends on whether you wish to operate only, or to also repair and maintain radio stations.

Radio Operations
You *need* a commercial radio operator license to operate the following:
1. Ship radio stations if:
    a. the vessel carries more than six passengers for hire; or
    b. the radio operates on medium or high frequencies (300 kHz to 30 MHz); or
    c. the ship sails to foreign ports; or
    d. the ship is larger than 300 gross tons and is required to carry a radio station for safety purposes.
2. Coast (land) stations which operate on medium or high frequencies, or operate with more than 1500 watts of peak envelope power.
3. Aircraft radio stations, except those that use only VHF frequencies (higher than 30 MHz) on domestic flights;
4. Civil Air Patrol stations on other than VHF frequencies;
5. AM, FM or TV broadcast stations and international broadcast stations;
6. Auxiliary broadcast stations in the following services:
    a. Developmental broadcast
    b. Low-power TV
    c. TV translator
    d. FM broadcast translator
    e. FM booster;
7. International fixed public radiotelephone and radiotelegraph stations; and
8. Coast and ship stations transmitting radiotelegraphy.

   You *do not need* a commercial radio operator license to operate the following:
1. Coast stations operating on VHF frequencies with 250 watts or less of carrier power.
2. Ship stations operating only on VHF frequencies while sailing on domestic voyages.
3. Aircraft stations which operate only on VHF frequencies and do not make foreign flights.

Radio Maintenance and Repair
You *need* a commercial radio operator license to repair and maintain the following:
1. All ship radio and radar stations,
2. All coast radio stations,
3. All hand-carried units used to communicate with ships and coast stations on marine frequencies.

4. All aircraft stations and aeronautical ground stations (including hand-carried portable units) used to communicate with aircraft.
5. All AM, FM, TV, and international broadcast stations
6. Auxiliary broadcast stations operated in the following services:
    a. Experimental Broadcast
    b. Low-Power Television
    c. Television Broadcast Translator
    d. FM Broadcast Translator
    e. FM Broadcast Booster
7. International fixed public radiotelephone and radiotelegraph stations.

You *do not need* a commercial radio operator license to operate, repair or maintain any of the following types of stations:
1. Two-way land mobile radio equipment such as that used by police and fire departments; taxicabs and truckers; businesses and industries; ambulances and rescue squads; local, state and federal government agencies.
2. Personal radio equipment used in the Citizens Band, Radio Control, and General Mobile Radio Services.
3. Auxiliary broadcast stations such as remote pickup stations and others (except those listed in item 6 above)
4. Domestic public fixed and mobile radio systems such as mobile telephone systems, cellular systems, rural radio systems, point-to-point microwave systems, multipoint distribution systems, etc.
5. Stations operated in the Cable Television Relay Service.
6. Satellite stations, both uplink and downlink, of all types.

Important Caution!
These listings only describe when radio ***operator*** licenses are necessary. *Before you operate any radio **station**, make certain that the **station** is licensed as required.* CB and Radio Control radio stations do not require individual station licenses.

The holder of a commercial radio operator license or permit is not authorized to operate amateur radio stations. Only a person holding an amateur radio operator license may operate an amateur radio station.

TYPES OF LICENSES, PERMITS AND ENDORSEMENTS
The FCC now issues nine types of commercial radio operator licenses and permits, and two types of endorsements. They are listed in *Table 2-1* through *Table 2-10*. Commercial Radio Operator licenses normally carry a term of five years. The exceptions are the General Radiotelephone Operator License (GROL) and the Restricted Radiotelephone Operator Permits (RP), which are valid for the lifetime of the holder.

Table 2-1. Restricted Radiotelephone Operator Permit (RP)

Restricted Radiotelephone Operator Permit (RP) holders are authorized to operate most aircraft and aeronautical ground stations. They can also operate marine radiotelephone stations aboard pleasure craft (other than those carrying more than six passengers for hire on the Great Lakes or bays or tidewaters or in the open sea) when operator licensing is required. RP holders may also operate, repair and maintain any kind of AM, FM, TV or international broadcast station. Most licensees of broadcast stations, however, require their technicians and engineers to hold a General Radiotelephone Operator license.

Residents
An RP is issued without any examination. To qualify for an RP, you *must meet all four* of the following requirements:
1. Be either a legal resident of (or otherwise eligible for employment in) the United States, or hold an aircraft pilot certificate valid in the United States, or hold an FCC radio station license in your own name (see limitation on validity below.)
2. Be able to speak or hear
3. Be able to keep at least a rough written log
4. Be familiar with provisions of applicable treaties, laws and rules which govern the radio station you will operate.

You *do not need* an RP to operate the following:
1. A voluntarily-equipped ship or aircraft station (including a Civil Air Patrol (CAP) station) which operates only on VHF frequencies and does not make foreign voyages or flights
2. An aeronautical ground or coast station which operates only on VHF frequencies
3. On-board stations
4. A marine utility station, unless it is taken aboard a vessel which makes a foreign voyage
5. A survival craft station, when using telephony or an emergency position-indicating radio beacon (EPIRB)
6. A ship radar station, if the operating frequency is determined by a fixed-tuned device and if the radar is capable of being operated by only external controls.
7. Shore radar, shore radiolocation, maritime support or shore radio-navigation stations.

Non-Residents
If you are a non-resident alien, you must hold *at least one* of the following three documents to be eligible for an RP:
1. A valid United States pilot certificate issued by the Federal Aviation Administration
2. A foreign aircraft pilot certificate which is valid in the United States on the basis of reciprocal agreements with foreign governments.
3. A valid radio station license issued by the FCC in your own name. (An RP issued on this basis will authorize you to operate *only* your own station.)

2 GENERAL RADIOTELEPHONE OPERATOR LICENSE

Table 2-2. Marine Radio Operator Permit (MROP)

A Marine Radio Operator Permit (MROP) is required to operate radiotelephone stations aboard certain vessels that sail the Great Lakes. It is also required to operate radiotelephone stations aboard vessels of more than 300 gross tons and vessels which carry more than six passengers for hire in the open sea or any tidewater area of the United States. It is also required to operate certain aviation radiotelephone stations and certain coast radiotelephone stations. It does not authorize the operation of any AM, FM, or TV broadcast stations.

A MROP is valid for a five-year term. It may be renewed from anytime in the last year of the license term to five years following expiration. (An expired MROP is not valid for any radio operation.)

You *must meet all three* of the following requirements to be eligible for a MROP:
1. Be a legal resident of (eligible for employment in) the United States
2. Be able to receive and transmit spoken messages in English
3. Pass a written examination covering basic radio law and operating procedures (Element 1).

Table 2-3. General Radiotelephone Operator License (GROL)

A General Radiotelephone Operator License (GROL) is required to adjust, maintain or internally repair FCC-licensed radiotelephone transmitters in the aviation, maritime and international fixed public radio services. It conveys all of the operating authority of the Restricted Permit (RP) and Marine Radio Operator Permit (MROP). A GROL is issued for the lifetime of the holder.

A GROL is also required to operate the following:
1. Any maritime land radio station or compulsorily equipped ship radiotelephone station operating with more than 1500 watts of peak envelope power.
2. Voluntarily-equipped (pleasure) ship and aeronautical (including aircraft) stations with more than 1000 watts of peak envelope power.

You *must meet all three* of the following requirements to be eligible for a GROL:
1. Be a legal resident of (or otherwise eligible for employment in) the United States
2. Be able to receive and transmit spoken messages in English
3. Pass a written examination covering basic radio laws and operating procedures (Element 1), and electronics fundamentals and techniques required to repair and maintain radio transmitters and receivers (Element 3).

Table 2-4. Third Class Radiotelegraph Operator's Certificate (T-3)

The Third Class Radiotelegraph Operator's Certificate (T-3) authorizes operation of certain coast radiotelegraph stations. It also confers the operating authority of both the Restricted Radiotelephone Operator Permit and the Marine Radio Operator Permit.

You *must meet all four* of the following requirements to be eligible for a T-3:
1. Be a legal resident of (or otherwise eligible for employment in) the United States
2. Be able to receive and transmit spoken messages in English
3. Pass Morse code examinations at 16 code groups per minute (16 CG, Telegraphy Element 1) and 20 words-per-minute plain language (20 PL, Telegraphy Element 2) — both receiving and transmitting by hand*
4. Pass written examinations covering basic radio law and operating procedures for telephony (Element 1), and basic operating procedures for telegraphy (Element 5).

---

* A Morse code hand-sending examination probably will not be required. This is because the FCC has taken the position that applicants who can receive telegraphy by ear can also send code manually.

COMMERCIAL RADIO OPERATOR'S LICENSES – THEN AND NOW 2

Table 2-5. Second Class Radiotelegraph Operator's Certificate (T-2)

The Second Class Radiotelegraph Operator's Certificate (T-2) authorizes the holder to operate, repair and maintain ship and coast radiotelegraph stations in the maritime services. It also confers all of the operating authority of the Restricted Radiotelephone Operator Permit, Marine Radio Operator Permit, Third Class Radiotelephone Operator's Certificate and the General Radiotelephone Operator License. It is a very desirable commercial radio operator license.

You *must meet all four* of the following requirements to be eligible for a T-2:
1. Be a legal resident of (or otherwise eligible for employment in) the United States
2. Be able to receive and transmit spoken messages in English
3. Pass Morse code examinations at 16 code groups per minute (16 CG, Telegraphy Element 1) and 20 words-per-minute plain language (20 PL, Telegraphy Element 2) — both receiving and transmitting by hand*
4. Pass written examinations covering basic radio law and operating procedures for radiotelephony (Element 1), basic operating procedures for radiotelegraphy (Element 5), and electronics technology applicable to radiotelegraph stations (Element 6).

Amateur Extra Class licensees (or those within the grace period) receive Telegraphy Element 1 (16 Code Groups per minute) and Telegraphy Element 2 (20 Plain Language words per minute) examination credit toward the Third and Second Class Radiotelegraph Operator's Certificate. A photocopy of the Amateur Extra Class license must be attached to the FCC Form 756 application. See §13.9(c)(2)(ii)

Table 2-6. First Class Radiotelegraph Operator's Certificate (T-1)

The First Class Radiotelegraph Operator's Certificate (T-1) is required only for those who serve as a chief radio operator on a U.S. passenger ship. The privileges granted are the same as the Second Class Radiotelegraph Operator's Certificate; i.e., it authorizes the holder to operate, repair and maintain ship and coast radiotelegraph stations in the maritime services. It also confers all of the operating authority of the Restricted Radiotelephone Operator Permit, Marine Radio Operator Permit, Third Class Radiotelephone Operator's Certificate, Second Class Radiotelegraph Operator's Certificate, and the General Radiotelephone Operator License.

You *must meet all six* of the following requirements to be eligible for a T-1:
1. Be at least 21 years old
2. Have at least one year of experience in sending and receiving public correspondence by radiotelegraph at ship stations, public coast stations or both
3. Be a legal resident of (or otherwise eligible for employment in) the United States
4. Be able to receive and transmit spoken messages in English
5. Pass Morse code examinations at 20 code groups per minute (20 CG, Telegraphy Element 3) and 25 words-per-minute plain language (25 PL, Telegraphy Element 4) — both receiving and transmitting by hand*
6. Pass written examinations covering basic radio law and operating procedures for radiotelephony (Element 1), basic operating procedures for radiotelegraphy (Element 5), and electronics technology applicable to radiotelegraph stations (Element 6).

* A Morse code hand-sending examination probably will not be required. This is because the FCC has taken the position that applicants who can receive telegraphy by ear can also send code manually.

2 GENERAL RADIOTELEPHONE OPERATOR LICENSE

First, Second and Third Class Radiotelegraph Operator's Certificates are valid for a five-year term. They may be renewed from anytime in the last year of their term up to five years following expiration.

No person may hold more than one unexpired radiotelegraph operator's certificate at the same time — nor may a person hold any class of radiotelegraph operator's certificate and a Marine Radio Operator Permit (MROP) or a Restricted Radiotelephone Operator Permit (RP).

While the First and Second Class Radiotelegraph Operator's Certificates confer all of the operating authority of a GROL, they *do not* grant Element 3 examination credit toward the GMDSS Radio Maintainer's License. See §13.9(c)(2)(i).

Table 2-7. GMDSS Radio Operator's License (GMDSS/O)

The Global Maritime Distress and Safety System Operator's License (GMDSS/O) is required for operation of the new satellite-based marine emergency subsystems and equipment. GMDSS licenses will eventually replace all radiotelegraph operator certificates as Morse code is phased out on the high seas.

Effective August 1, 1993, all United States Coast Guard communications stations and cutters discontinued watchkeeping on the distress frequency (500 kHz) and have ceased all Morse code services in the medium-frequency radiotelegraphy band.

You *must meet all three* of the following requirements to be eligible for a GMDSS/O:
1. Be a legal resident of (or otherwise eligible for employment in) the United States
2. Be able to receive and transmit spoken messages in English
3. Pass written examinations covering basic radio law and maritime operating procedures (Element 1), and GMDSS radio operating practices (Element 7).

Table 2-8. GMDSS Radio Maintainer's License (GMDSS/M)

The Global Maritime Distress and Safety System Maintainer's License (GMDSS/M) is required to repair and maintain satellite-based marine emergency subsystems and equipment.

You *must meet all three* of the following requirements to be eligible for a GMDSS/M:
1. Be a legal resident of (or otherwise eligible for employment in) the United States
2. Be able to receive and transmit spoken messages in English
3. Pass written examinations covering basic radio law and maritime operating procedures (Element 1), technical examination on electronic fundamentals and techniques (Element 3), and GMDSS radio maintenance practices and procedures (Element 9).

Table 2-9. Ship Radar Endorsement

The Ship Radar Endorsement may be placed only on General Radiotelephone Operator Licenses (GROL), GMDSS Radio Maintainer's licenses (GMDSS/M) or on First or Second Class Radiotelegraph Operator's Certificates (T-1 or T-2.) Only persons whose commercial radio operator license bears this endorsement may repair, maintain or internally adjust ship radar equipment.

COMMERCIAL RADIO OPERATOR'S LICENSES – THEN AND NOW 2

You *must meet both* of the following requirements to be eligible for a Ship Radar Endorsement:
1. Hold (or qualify for) a First or Second Class Radiotelegraph Operator's Certificate (T-1 or T-2), GMDSS Radio Maintainer's license (GMDSS/M) or a General Radiotelephone Operator License (GROL)
2. Pass a written examination (Element 8) covering special rules applicable to ship radar stations and the technical fundamentals of radar and radar maintenance procedures.

Table 2-10. Six-Months Service Endorsement

The Six-Months Service Endorsement is required on the radiotelegraph operator's certificate of anyone who serves as the sole radio operator aboard large U.S. cargo ships sailing the high seas.

You *must meet all five* of the following requirements to be eligible for a Six-Months Service Endorsement:
1. You have been employed as a radio operator on board ships of the United States for a period totaling at least six months.
2. The ships were equipped with radio stations complying with the provisions of Part II of Title III of the Communications Act, or the ships were owned and operated by the U.S. Government.
3. The ships were in service during the applicable six-months period and no portion of any single in-port period included in the qualifying six-months period exceeded seven days.
4. You held a First or Second Class Radiotelegraph Operator's Certificate (T-1 or T-2) issued by the FCC during this entire six-months qualifying period.
5. You hold a radio officer's license issued by the U.S. Coast Guard at the time the six-months service endorsement is requested.

For those applicants who are qualified by having at least 180 days of creditable service, the following is to be submitted to the Federal Communications Commission, 1270 Fairfield Road, Gettysburg, PA 17325-7245:
1. Certification letter signed by the vessel's master or owner/agent specifying the vessel name, vessel call sign, dates of service (shipment and discharge), total number of days served (minus any in-port periods exceeding seven days); and the name(s) and certificate number of the chief radio officer holding a Six-Months Service Endorsement on the vessel during (shipment) service
2. Completed FCC Form 756 Commercial Radio Operator License Application
3. Original T-1 or T-2 certificate
4. Valid copy of U.S. Coast Guard license
5. Certificate(s) of Discharge to Merchant Seaman.

PAPERWORK TO OBTAIN A COMMERCIAL RADIO OPERATOR LICENSE

FCC Form 753

Use this form to apply for a restricted Radiotelephone Operator Permit (RP) if you are *legally eligible* for employment in the United States. A temporary permit for immediate operating authority is included on the form. No examination is required. Be aware that only a Form 753 that has been issued on or after 10/88 (October, 1988) will be accepted. Others will be returned and you will have to refile. The latest form will be changed to include a 4-character fee type code. Until a new edition is available, write in **"Fee Code: PARR"** on the form. See *Table 2-11*. (See example in the Appendix.)

2 GENERAL RADIOTELEPHONE OPERATOR LICENSE

FCC Form 755
Use this form to apply for a Restricted Radiotelephone Operator Permit – Limited Use (RP) if you are *not legally eligible* for employment in the United States (non-resident alien, foreign aircraft pilot, etc.) A temporary permit for immediate operating authority is included on the form. No examination is required. Be aware that only a Form 755 that has been issued on or after 8/88 (August, 1988) will be accepted. Others will be returned and you will have to refile. See note on fee type code for Form 753. Write in **"Fee Code: PARR"** on this form also. See *Table 2-11*. (See example in the Appendix.)

FCC Form 756
Use this form to apply for all other new Commercial Radio Operator licenses, permits and endorsements; to renew your expiring license; or to apply for replacement of your lost, stolen or mutilated license, or if you change your name. Each application for a new Commercial Radio Operator license, permit or endorsement, or a change in class which requires an examination, must include an *original* PPC (Proof-of-Passing Certificate), issued by examiner(s), showing that the applicant has passed the necessary examination element(s). See § 13.9(c) and § 13.13(c). Only editions of forms on or after 10/93 are acceptable. Make sure you include the proper 4-character fee type code from *Table 2-11*. (See example in the Appendix.)

Applicants may obtain current editions of these three forms at no cost by contacting the FCC Forms Distribution Center, 2803 52nd Avenue, Hyattsville, Maryland 20781 – Tel: 202/632-FORM (3676.)

Photographs Required for Radiotelegraph Licenses
If you are applying for a new, renewed or replacement radiotelegraph operator's certificate, you must submit two identical, signed photographs of yourself with your completed application.
1. The photographs must be *signed in ink* by the applicant on the front along the left margin. This signature must be clearly visible and match the signature on the FCC Form 756 application.
2. The photographs must not be less than 2 x 2 inches nor more than 2.5 x 3 inches.
3. The photographs must be a front view showing head and shoulders only with clear, full face.
4. The photographs may be color or black and white with a light, plain background.
5. The photographs must have been taken within six months of the application date.
6. Newspaper, magazine or photocopied photographs are not acceptable.
7. Enclose your photos in a plain envelope. Write the word "photos" and your name on the outside. Staple the envelope to your application.

EXAMINATION, REGULATORY AND APPLICATION FEES
There are three separate fees that may be charged to applicants for Commercial Radio Operator licenses, but an applicant only pays a maximum of two of them, depending on the type of license and action required. The three fees are examination fee, regulatory fee, and application fee.

Examination Fee
The examination fee is charged by a COLEM (Commercial Operator License Examination Manager.) This fee can vary from $25.00 to $120.00, depending upon which COLEM administers the examination, and the number and type of test elements administered. This fee is paid to the examiners.

Regulatory Fee *(Section 9 of Ominbus Budget Reconciliation Act of 1993)*
The regulatory fee is levied on the applicant by the Congress to assist the FCC in recovering the costs associated with Government regulation and enforcement of the radio service. The amount of this fee may be adjusted annually by the FCC to reflect changes in its appropriations. The beginning *Annual Regulatory Fee Schedule* for Commercial Radio Operator (47 CFR Part 13) licenses and permits is $7.00 annually which is collected in advance based on a full license term. Lifetime commercial radio operator licenses are based on fifteen times the annual rate or $105.00. Regulatory fees do not apply to amateur radio licensees – or to government and non–profit public safety entities.

Application Fee *(Section 8 of Ominbus Budget Reconciliation Act of 1993)*
The application fee (also called a processing fee) is a $45.00 charge that applicants pay to reimburse the Government for the costs involved in the administrative processing of radio licenses. The application fee is paid only for duplicate, renewal and replacement licenses. Make check payable to the FCC.

WHERE TO SEND THE APPLICATION AND FEES
Although there are three separate fees, as mentioned previously, no Commercial Radio Operator applicant pays more than two different charges. Between September 2, 1993 and July 1, 1994, there was only one cost to *new Commercial Radio Operator applicants* – that being the examination fee collected by the COLEM.

New Licenses
Beginning July 1, 1994 (see footnote in *Table 2-11*), candidates for new Commercial Radio Operator licenses requiring an examination must pay both a COLEM (examination) *and* a regulatory fee. You must attach a check payable to the FCC for the regulatory fee portion on all applications for a new license. Send to the appropriate following address (see *Table 2-11* for more detail):

2 GENERAL RADIOTELEPHONE OPERATOR LICENSE

1. GENERAL RADIOTELEPHONE OPERATOR LICENSE (New)

Federal Communications Commission
Radio Operator Permits
P.O. Box 358725
Pittsburgh, PA 15251–5725

Fee type code that must be entered on application: PACQ

2. ALL OTHER COMMERCIAL RADIO OPERATOR LICENSES (New)

Federal Communications Commission
Radio Operator Permits
P.O. Box 358800
Pittsburgh, PA 15251–5800

Fee type code that must be entered on RP application: PARR; on MROP, GMDSS/O/M and all Radiotelegraph certificates: PACR

3. SHIP RADAR AND SIX-MONTHS SERVICE ENDORSEMENTS (New)

Federal Communications Commission
1270 Fairfield Road
Gettysburg, PA 17325–7245

Renewal, Replacement and Duplicate Licenses

Commercial Radio Operator applicants not being tested for a new license will pay a combined regulatory and application fee. A single check for the combined regulatory fee and application fee amounts should be made payable to the FCC. All applications for renewal, replacement, duplicate or modified (includes name change) Commercial Radio Operator licenses must be sent to the appropriate following address (see *Table 2-11* for more detail):

4. ALL LICENSE RENEWALS

Federal Communications Commission
Radio Operator Permits
P.O. Box 358805
Pittsburgh, PA 15251–5805

Fee type code that must be entered on application: PACS

5. ALL REPLACEMENT OR DUPLICATE LICENSES

Federal Communications Commission
Feeable Correspondence
P.O. Box 358305
Pittsburgh, PA 15251–5305

Fee type code that must be entered on application: PADM

EXAMINATION, REGULATORY AND APPLICATION FEES BY LICENSE TYPE

Table 2-11 details all of the required fees associated with Commercial Radio Operator licenses. The COLEM examination fee is paid to the examiners; the regulatory and application fee to the FCC. Examples of the proper handling of fees and the required paperwork to be submitted to the FCC follow in paragraphs after *Table 2-11*.

COMMERCIAL RADIO OPERATOR'S LICENSES – THEN AND NOW

Table 2–11. Required Fees for Commercial Radio Operator Licenses

LICENSE	Exam	NEW LICENSE Regulatory Section 9	Send To:[1]	RENEWAL LICENSE Regulatory Section 9	Application Section 8	Send To:[1]	REPLACEMENT OR DUPLICATE[2] Regulatory Section 9	Application Section 8	Send To:[1]
RP	None PARR[3]	$105.00	2	(Lifetime)	(Lifetime)		None PADM[3]	$45.00	5
MROP	COLEM PACR[3]	$35.00	2	$35.00 PACS[3]	$45.00	4	None PADM[3]	$45.00	5
GROL	COLEM PACQ[3]	$105.00	1	(Lifetime)	(Lifetime)		None PADM[3]	$45.00	5
T–1, T–2, T–3	COLEM PACR[3]	$35.00	2	$35.00[4] PACS[3]	$45.00[4]	4	None PADM[3]	$45.00	5
GMDSS/ O/M	COLEM PACR[3]	$35.00	2	$35.00 PACS[3]	$45.00	4	None PADM[3]	$45.00	5
Radar	COLEM	None	3	None	None		None	None	
6-Months	None	None	3	None	None		None	None	

Important footnote: Section 8 fees are under Congressional review and are subject to change. Effective date is tentatively scheduled for July 1, 1994. Call FCC at (202) 632–0923 to confirm the effective date. Call FCC at (202) 632-3337 if you have a question on the amount of the fee.

[1] Refer to item number on page 20 for address.
[2] Includes Change of Name.
[3] Fee type code
[4] Be sure to include the two small signed photographs.

FEES AND HANDLING OF COMMERCIAL RADIO LICENSE APPLICATIONS

Here are some examples of the required fees and application handling associated with Commercial Radio Operator licenses.

1. ***New* General Radiotelephone Operator License: Examination fee – plus $105.00 lifetime license regulatory fee.**
    a. New examination fee assessed by and paid to the COLEM. Examination fees vary, so check with the COLEM handling the examination. ($35.00 if National Radio Examiners administer the Element 1 and 3 exams.)
    b. Regulatory (Section 9) fee for a lifetime license: **$105.00**
    c. Form 756 application package, PPC and check or money order covering the $105.00 regulatory fee payable to FCC and sent to FCC address listed as **item 1** under *New Licenses* paragraph. Application may be filed by the applicant or his agent – most likely the COLEM that administered the examination. The fee type code to be entered on the application is **PACQ**.

2. ***Routine Renewal*** **of Second Class Radiotelegraph Operator's Certificate: $80.00**
   a. Regulatory fee: **$35.00** – New 5-year term at **$7.00** per year payable in advance.
   b. Application fee: **$45.00**. This amount was determined by Congress.
   c. FCC Form 756 application package, expiring license (or copy), two signed photographs, and single check or money order for $80.00 for the combined regulatory and application fees made payable to FCC. Licensee sends to FCC address listed as **item 4** under *Renewal, Replacement and Duplicate Licenses* paragraph. The fee type code is **PACS**.
3. ***Replacement*** **Marine Radio Operator Permit (MROL): $45.00**
   a. Regulatory fee: None.
   b. Application fee: **$45.00**.
   c. FCC Form 756 application package and check or money order for $45.00 covering the application fee made payable to FCC. Licensee sends to FCC address listed as **item 5** under *Renewal, Replacement and Duplicate Licenses* paragraph. The fee type code is **PADM**.
4. ***Name Change*** **on GMDSS Maintainer's License $45.00**
   a. Regulatory fee: None
   b. Application fee: **$45.00**.
   c. FCC Form 756 application package and check or money order for $45.00 covering the application fee made payable to FCC. Licensee sends to same FCC address as example 3. Same fee type code.
5. ***Endorsement*** **for Six-Months Service on Radiotelegraph Operator's Certificate: No Cost.**
   a. Regulatory fee: None. (Licensee will not be granted a new full license term.)
   b. Application fee: None
   c. FCC Form 756 application package sent to FCC, address listed as **item 3** under *New Licenses* paragraph. No fee type code.
6. ***Endorsement*** **for Ship Radar on existing GROL or GMDSS/M or Radiotelegraph license: Examination fee by COLEM only.**
   a. Regulatory fee: None. (Licensee will not be granted a new full license term.)
   b. Application fee: None
   c. FCC Form 756 application package sent, with appropriate documentation, to same FCC address as example 5.

OPERATING WHILE RENEWAL APPLICATION PENDING

If your application to renew your license is received by the Commission before it expires, you may continue to operate under the authority of your license while the FCC processes your renewal application. However, if you forgot to renew your license before it expired, you cannot operate equipment that requires that license until it is received. You may file to renew your expired license any time during the five-year grace period after your license expires. If you fail to renew your license within the grace period, you must apply for a new license and take the required examination(s.) *The FCC <u>does not notify you</u> when it is time to renew your license. The expiration date is on your license (except, of course, in the case of a lifetime license or permit.)*

LOST, STOLEN, MUTILATED, OR DESTROYED LICENSES

If your license is lost, stolen, mutilated or destroyed, you may apply on a Form 756 application for a duplicate license from the FCC at the same address as for replacement or duplicate licenses. Explain what happened to the original license and include the appropriate application fee.

NAME CHANGE

If you change your name and wish a license with your new legal name, you may apply for a new replacement license; however, it is not required by the FCC that you replace a license when you change your name. If you apply for a replacement license, be sure to indicate the reason for your application and give both your former and new legal names. Return your current license to the FCC along with the FCC Form 756 application and the $45.00 application fee. Send the package to the FCC address for replacement and duplicate licenses. Only the application fee applies and applicants requesting a replacement license will not be granted a new full license term.

POSTING APPLICATION INSTEAD OF LICENSE

If you file an application for a renewal with the FCC, you are required to send in your original or a photocopy of your commercial radio operator license along with the application. If you are employed at a station where your operator license must be posted, you may post a signed copy of your renewal application and a photocopy (or the original) of your current license while your renewal application is being processed.

WHERE TO GET HELP ON COMMERCIAL RADIO OPERATOR APPLICATIONS

If you need assistance with your application, contact the FCC at:
      Federal Communications Commission
      Consumer Assistance Branch
      1270 Fairfield Road
      Gettysburg, PA 17325-7245
Telephone: (717) 337-1212 (during normal business hours 8:00 a.m. to 4:30 p.m. Monday through Friday.)

FEE FILING GUIDE

We have discussed fees and fee type codes several times. Each of the FCC's five bureaus publishes a multi-page Fee Filing Guide that notifies the public of the charges they must pay to the government for their license. The guide does *not* include the examination fee which is due a COLEM, only the regulatory and application fees. Every fee has a three or four letter fee type code with the first letter indicating the appropriate bureau. As pointed out, the four-letter fee type codes are shown in *Table 2-11*. The letter P stands for Private Radio Bureau. The balance of the fee type code refers to a specific license fee. (Other bureaus that have different codes are: C=Common Carrier Bureau, E=Office of Engineering and Technology, F=Field Operations Bureau, and M=Mass Media Bureau.) For the Private Radio Bureau, the letters PACR are the fee type code used for the $35.00 regulatory fee assessed on a new five-year commercial radio operator license other than a lifetime GROL or RP. The code changes to PACS when this license is renewed, or PADM if a duplicate or replacement license is requested.

SUMMARY

In this chapter we have summarized when a person does or does not need a commercial radio license; when each permit, license, certificate and endorsement are required; and the forms and paperwork required to make an application for a license. An examination fee is paid to the COLEM examiners when an examination is needed. There will be fees required by the FCC for new, renewal, replacement and duplicate licenses – sometimes an application fee and sometimes a regulatory fee and an application fee. *Table 2-11* summarizes these fees. Be aware that the fees in *Table 2-11* are under congressional review and are subject to change. Call the FCC at the telephone number listed in the footnote of *Table 2-11* in case you are in doubt about a fee.

Remember, that when you submit your application, you need to include a fee type code on your application form. The fee type codes are also included on *Table 2-11* and next to the appropriate FCC address for the application.

When you send in an application for a new license, remember that you must include the *original* PPC (Proof-of Passing Certificate) that was signed by your COLEM examiners, and any necessary credit documents. Keep a copy for yourself so that you have a record of passing your examination. Include a single check for the regulatory fee and the application fee, if both are required. The check should be stapled to the top side of your application. Recheck the addresses given in "WHERE TO SEND THE APPLICATION AND FEES" and make sure you have the current address.

Commercial License Examinations – Then and Now

FCC CHANGES HAM OPERATOR TESTING

In the early 1980s, in order to conserve resources, a general trend developed at the FCC to privatize many FCC administrative functions. The government turned their commercial radio technician certification program over to industry groups and de-emphasized some of its commercial radio licenses. In 1982, President Ronald Reagan signed legislation (Public Law 97-259) authorizing Amateur Radio examinations to be prepared and administered by volunteer ham radio organizations.

During 1983, the FCC developed new guidelines for Amateur Radio operator license testing which were patterned after a system used by the FAA, where all examination questions are in the public domain and widely published.

The FCC, with help from the amateur community, developed and released to the public examination question pools that contained all possible verbatim questions that could be asked of amateur operators. These question pools, which contained ten times the required number of questions for an examination, were released to the public in the form of PR (for Private Radio) 1035 bulletins.

VEC SYSTEM ADOPTED

The Commission then began a search for testing system administrators who would recruit volunteer examiners (VEs) and further develop a program to test amateur radio operator applicants. The heart of the volunteer examination program in the amateur radio service is the volunteer examiner coordinator (VEC). A VEC acts as the link between the FCC and the VE. They provide their VE teams with license testing materials; collect, screen, approve and forward amateur radio license applications to the FCC; and maintain records of all test sessions. The FCC acts only in an oversight capacity and provides guidance to the VECs. The "VEC System," as it came to be called, is funded by a small testing fee paid by the examinee to cover the testing VEs out-of-pocket costs. In 1986, maintenance of the various question pools was turned over to a committee of VECs who are responsible for keeping all ham examination question pools up to date.

During calendar year 1993, more than 100,000 ham radio operator applicants were administered over 200,000 examinations at some 10,000 VEC System test sessions around the world! The questions are current; fraud and abuse are rare; and there are neighborhood VE test teams everywhere — hardly a city is without one.

COMMERCIAL RADIO TESTING DILEMMA

A couple of years later, the FCC recognized that the VEC System was indeed working as hoped. The program was conserving dwindling government resources while at the same time increasing examination opportunities and improving the quality of operator license testing. The FCC believed the same mechanism could be, and should be, implemented in the Commercial Radio Services as well.

All through the 1980s, applicants wishing to take a commercial radio examination had to travel to the nearest of 25 FCC field offices located across the nation where commercial radio examinations were only administered quarterly. Later, the examination schedule was reduced to only twice a year.

Nevertheless, the demand for commercial radio operator licenses continued, and the demand for commercial radio license testing and the need to update examination questions far outstripped the FCC's fiscal and manpower capability. Public comments concluded that a more efficient commercial radio examination system was needed, and, finally in 1990, the FCC received legislative authority from Congress to delegate the examination of commercial radio operators to private groups.

FCC SEEKS TO PRIVATIZE COMMERCIAL TESTING

In August 1992, a *Notice of Proposed Rulemaking* (FO Docket 92-206), which looked toward implementing privatized commercial radio operator examinations, was adopted by the FCC. Then in October 1992, the FCC transferred the responsibility for handling their commercial radio operator license testing program to the Private Radio Bureau — the same department that handles Amateur Radio operator examinations.

COMMERCIAL RADIO OPERATOR TESTING PRIVATIZED

The final *Report and Order* to privatize commercial radio license testing was released on February 12, 1993. Commercial radio license testing would operate very much like the amateur service's successful VEC System, after which it was patterned. In the process, the FCC completely rewrote Part 13 of its Rules, which covers commercial radio operator examination requirements. (See Appendix for the complete text of Part 13.)

There are now nine types of commercial radio operator licenses, certificates, permits or endorsements that are either required by international radio law or private industry. They are listed in *Table 3-1* and *Table 3-2* along with the examinations required. They are granted after passing examinations selected from seven written and four telegraphy examinations on the subjects shown in *Table 3-3*.

Table 3-1. Commercial Radio Licenses — Examination Requirements

License Class	Test Element	Type of Examination	Pass[3]
Marine Radio Operator Permit (MROP)	Written Element 1	24-question written examination	18
General Radiotelephone Operator License (GROL)	Written Element 1 Written Element 3	24-question written examination 76-question written examination	18 57
T-1 – First Class Radiotelegraph Operator Certificate - 1st R/T (Must be 21 years old and have 1 year of experience)	Telegraphy Element 3[2] Telegraphy Element 4[2] Written Element 1 Written Element 5 Written Element 6	20 code groups (CG) per minute 25 plain language (PL) words per minute 24-question written examination 50-question written examination 100-question written examination	 18 38 75
T-2 – Second Class Radiotelegraph Operator Certificate - 2nd R/T	Telegraphy Element 1[2] Telegraphy Element 2[2] Written Element 1 Written Element 5 Written Element 6	16 code groups (CG) per minute[4] 20 plain language (PL) words per minute[4] 24-question written examination 50-question written examination 100-question written examination	 18 38 75
T-3 – Third Class Radiotelegraph Operator Certificate - 3rd R/T	Telegraphy Element 1[2] Telegraphy Element 2[2] Written Element 1 Written Element 5	16 code groups (CG) per minute[4] 20 plain language (PL) words per minute[4] 24-question written examination 50-question written examination	 18 38
GMDSS Radio Operator[1] (GMDSS/O)	Written Element 1 Written Element 7	24-question written examination 76-question written examination	18 57
GMDSS Radio Maintainer[1] (GMDSS/M)	Written Element 1 Written Element 3 Written Element 9	24-question written examination 76-question written examination 50-question written examination	18 57 38
Ship Radar Endorsement (RADAR)	Written Element 8	50-question written examination	38
Six Months Endorsement (Must hold First or Second Class Radiotelegraph Operator's Certificate)		No examination requirement, but requires a minimum of six months service as a radio operator aboard a U.S. ship.	
Restricted Radiotelephone Operator Permit		No examination requirement, but holder must be eligible for employment in the U.S., be familiar with radio law, be able to keep a log and be able to speak and understand English.	

Notes:
1. The Global Maritime Distress and Safety System (GMDSS) is an automated ship-to-shore distress alerting system using satellite and advanced terrestrial communications systems. GMDSS basically replaces Morse code communications which is being phased out as a maritime distress communications medium.
2. Passing telegraphy examination requires copying CW (Morse code) for one minute without error.
3. Number of correct answers to pass examination.
4. Amateur Extra Class operators automatically receive credit for this element examination without testing.

3 GENERAL RADIOTELEPHONE OPERATOR LICENSE

Table 3-2. Summary of Elements Required for Commercial Licenses

License	Written Elements							Telegraphy Elements			
	1*	3	5	6	7	8	9	1	2	3	4
MROP	X										
GROL	X	X									
T-3 – 3rd R/T	X		X					X	X		
T-2 – 2nd R/T	X		X	X				X	X		
T-1 – 1st R/T	X		X	X						X	X
GMDSS/O	X				X						
GMDSS/M	X	X					X				
RADAR						X**					

*Formerly Elements 1 and 2   **Endorsement on GROL, T-1, T-2, and GMDSS/M

Table 3-3. Commercial Radio Operator Question Pool Element Subjects

Element	Element Subject
Telegraphy Element 1 (Formerly Elements 1 and 2) T-2 – 2nd R/T T-3 – 3rd R/T	16 Code Groups (CG) per minute. (A code group is random groups of three to six numerals, punctuation and prosigns.)
Telegraphy Element 2 T-2 – 2nd R/T T-3 – 3rd R/T	20 Plain Language (PL) words per minute text.
Telegraphy Element 3 T-1 – 1st R/T	20 Code Groups (CG) per minute.
Telegraphy Element 4 T-1 – 1st R/T	25 Plain Language (PL) words per minute text.
Written Element 1 GROL MROP T-1 – 1st R/T T-2 – 2nd R/T T-3 – 3rd R/T GMDSS/O GMDSS/M	Basic radio law and radiotelephone operating practice with which every maritime operator should be familiar.
Written Element 3 GROL GMDSS/M	Electronic fundamentals and techniques required to adjust, repair and maintain radio equipment in the aviation, maritime and international fixed public radio services.
Written Element 5 T-1 – 1st R/T T-2 – 2nd R/T T-3 – 3rd R/T	Radiotelegraph operating procedures and practices primarily required in other than the Maritime Mobile Services.
Written Element 6 T-1 – 1st R/T T-2 – 2nd R/T	Advanced radiotelegraph operating procedures and practices. Contains technical, legal and other matters, including those required in the Maritime Mobile Services.
Written Element 7 GMDSS/O	GMDSS operating practices and regulations.
Written Element 8 RADAR	Ship Radar Techniques. Specialized theory and practice concerning proper installation, servicing and maintenance of ship radar equipment.
Written Element 9 GMDSS/M	GMDSS radio maintenance practices and procedures.

COMMERCIAL OPERATOR LICENSE EXAMINATION MANAGERS

The FCC's new Commercial Radio Operator testing program is directed by nine private groups known as Commercial Operator License Examination Managers (COLEMs). The FCC said they believed "...a system with multiple entities managing operator examinations will encourage competition between the entities and result in good service, responsiveness, and lower prices to the applicants." The nine COLEMs were chosen by the FCC's Private Radio Bureau. One of those chosen was The W5YI Group, Inc., which is headed up by one of your authors, Fred Maia. The W5YI Group was not only the first organization to provide amateur radio operator examinations on a national basis, but, through its National Radio Examiners Division, nationwide commercial radio operator testing as well. The other eight COLEMs are listed below. A more detailed list is provided in the Appendix.

> Drake Training and Technologies
> Electronic Technicians Association International (ETAI)
> Elkins Institute, Inc.
> International Society of Certified Electronics Technicians (ISCET)
> National Association of Business and Educational Radio, Inc. (NABER)
> Sea School (a maritime training organization)
> Sylvan KEE Systems
> The National Association of Radio Telecommunications Engineers, Inc. (NARTE)

COLEMs are responsible for:
- Announcing examination sessions
- Verifying the identity of each examinee
- Preparing, administering and grading examinations
- Notifying examinees of examination results (pass/fail)
- Certifying that an applicant has passed the test elements required to qualify for a commercial operator license
- Issuing to the examinee a Proof-of-Passing Certificate (PPC) within 10 days of the examination
- Ensuring that no activity takes place that would compromise the examination and that no unauthorized material is permitted in the examination room
- Handling post-examination questions and problems
- Treating all applicants equally regarding fees or services rendered.

Examiners are prohibited from administering an examination to an employee, relative, or relative of an employee.

COMMERCIAL QUESTION POOLS

On September 2, 1993, the FCC released the first question pools — Element 1 and Element 3 — required for the Marine Radio Operator Permit and the General Radiotelephone Operator License. Both Element 1 and Element 3 question pools contained in this book are complete and revised from the release of September 2, 1993, updated March 3, 1994. The FCC will develop a common question pool for each of the other examination elements from which all test questions must be taken. The FCC's Chief of the Aviation and Marine Branch (Washington, DC) is in charge of all question pools. Each question pool is (or will be) made up of subelements or topics, and does (or will) contain at least five times the number of questions required for a single examination. All test questions and their multiple-choice answers will be released to the public by the FCC in much the same manner as they did for Element 1 and Element 3. It is anticipated that all seven question pools will be released by the end of 1994.

WRITTEN EXAMINATIONS — CONTENT AND PASS RATE

The pass rate for all written examinations is 75 percent. All examination questions must be taken from the respective question pools. *Table 3-4* shows the number of questions in each question pool, and how the examination questions will be selected from the question pool topics. *Table 3-1* identifies the number of correct answers needed to pass the written examinations.

Handicapped applicants must be accommodated and special examination procedures employed if necessary. A doctor's certification indicating the nature of the disability may be required by the examination manager. Applicants with uncorrected disabilities which adversely affect their performance as a commercial radio operator will be issued a license with a restrictive endorsement. (See § 13.7(c) (6).)

Table 3-4. General Radiotelephone Operator License Question Pools

Examination	Subelement	Topic	Page	Total Questions	Examination Questions
Element 1	A	Maritime Radio Law and Operating Practices	34	169	24
Element 1	Totals			169	24
Element 3	A	Operating Procedures	80	40	3
	B	Radio Wave Propagation	90	22	3
	C	Radio Practice	97	96	5
	D	Electrical Principles	122	115	16
	E	Circuit Components	164	75	13
	F	Practical Circuits	182	138	22
	G	Signals and Emissions	214	97	9
	H	Antennas and Feed Lines	237	137	5
Element 3	Totals			720	76
GROL	Totals			889	100

PASSING THE COMMERCIAL TELEGRAPHY CODE TESTS

Applicants are required to copy a telegraphy test message by ear for a period of one minute. As in the amateur service, the sending examination need not be administered since the FCC has taken the position that "Passing a telegraphy-receiving examination is adequate proof of an examinee's ability to send and receive telegraphy." Each Morse code test message must contain all letters of the alphabet, numerals 0-9, the period, comma, question mark, slant mark and prosigns $\overline{AR}$, $\overline{BT}$, and $\overline{SK}$. All numerals, punctuation and prosigns count as two characters. The examination manager is responsible for determining the correctness of the examinee's answers.

CREDIT FOR AMATEUR OPERATORS

Of particular interest to amateur radio operators is that examination credit toward commercial telegraphy Elements 1 and 2 will be routinely allowed to Amateur Extra Class operators who have already passed the amateur Element 1(C) 20 words-per-minute code test. This means that Extra Class amateurs can qualify for the Second or Third Class Radiotelegraph Operator's Certificate by passing only the associated written examinations. Examinees who merely pass amateur examination Element 1(C) without fulfilling all requirements for the Extra Class license are not granted credit for commercial telegraphy Elements 1 and 2. And an additional advantage is that two-thirds of the Element 3 question pool questions are based on those found in the amateur radio Element 4(A) and 4(B) question pools for Advanced Class and Extra Class examinations.

FCC FEES AND FORMS

The FCC fees and forms to use for particular examinations are detailed in Chapter 2, and information for filling out the forms is covered in Chapter 6.

QUESTION CODING

In *Figure 3-1*, a number from the Element 3 pool is used as an example to explain the coding used in the question pools. 3A2 means Element 3; A means subelement A, Operating Procedures; and 2 means it is the second question in subelement A.

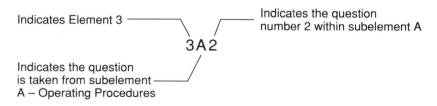

Figure 3-1. Examination Question Coding

EXAMINATION QUESTION POOLS

The complete Element 1 question pool of 169 questions is contained in Chapter 4, and the complete Element 3 question pool of 720 questions is contained in Chapter 5. These question pools are as issued by the FCC into the public domain on September 2, 1993, except FCC corrections have been updated on March 3, 1994. In this book, each question in the pools has the question, its four multiple-choice answers, the right answer identified, and a detailed explanation of why the right answer is correct.

HINTS FOR STUDYING FOR THE EXAMINATIONS

The Element 1 Radio Law questions are based on the *Code of Federal Regulations,* Title 47 *(Telecommunication)* Part 80 *(Maritime Service Rules.)* Part 80 is more than 200 pages long, but answers to the Element 1 questions can be found by reading only about 24 pages. The important Part 80 pages are in the Appendix. All Element 1 questions have been cross indexed to the exact Part 80 section that applies.

Go through all the questions and study the easy ones first. Mark the ones to which you already know the answers. Then go back and concentrate on the harder questions (or topics). Memorize the formulas and learn how to use them. Obtain reference texts from Radio Shack, or your library, or ask your teacher, amateur friends, or work companions for recommendations. Study with the confidence you are only going to miss two or three questions. You can miss up to 6 questions on Element 1 and 19 on Element 3 and still pass! *NEVER* leave an answer blank. You have a 25% chance of getting it right just by taking a wild guess! And usually you can identify two of the choices as being incorrect, which increases your chances to 50%!

ADDITIONAL STUDY MATERIALS

There are computer-aided software programs—your author's National Radio Examiners has one especially for GROL—that run on an IBM-PC (or compatible) that allow you to take properly constructed (sample) tests right at your keyboard—or you can print out examinations and take them later. The software lets you know when you have attained the skill level to pass either (or both) Element 1 and 3.

CREDIT FOR PASSING ELEMENTS

If you are striving for a GROL, you get a credit document (called a PPC —for *Proof-of-Passing Certificate)* if you pass only one element and not the other. This allows you a whole year to complete the failed element before you lose examination credit. Also, if you pass Element 1 and fail Element 3, you still qualify for the Marine Radio Operator Permit. While you do not have to accept the MROP, you can obtain additional time credit for Element 1 if you do. This is because the MROP grants Element 1 credit as long as the license is valid. (MROPs are issued for a 5-year term.)

Question Pool – Element 1

The questions asked in each Element 1 examination must be taken from the following question pool. The release date of this pool is *September 2, 1993, updated March 3, 1994*. This pool will remain in effect until superseded by a release of an updated pool from the FCC.

An Element 1 examination is used to prove that the examinee possesses the qualifications to operate licensed radio stations required of a person holding a Marine Radio Operator Permit (MROP). An examination on Element 1 is also necessary to properly perform the duties required of a person holding a General Radiotelephone Operator License (GROL). To pass, an examinee must answer correctly at least 18 out of 24 questions from this pool. Element 1 is also a partial requirement for the Global Maritime Distress and Safety System Maintainer's License and the three Radiotelegraph Operator Certificates, T-1, T-2, and T-3.

An MROP must be held by a person required to operate radiotelephone stations (1) aboard certain vessels that sail the Great Lakes; (2) aboard vessels of more than 300 gross tons; (3) aboard vessels that carry more than six passengers for hire; and (4) in certain aviation and coast stations.

Each Element 1 examination is administered by a Commercial Operator License Examination Manager (COLEM). The COLEM must construct the examination by selecting 24 questions from this 169-question pool. You must have 18 or more answers correct from the 24 in order to pass the examination.

The COLEM may change the order of the answer and distractors (incorrect choices) of questions from the order that appears in this pool.

Suggestions concerning improvement to the questions in the pool or submittal of new questions for consideration in updated pools may be sent to the Federal Communications Commission, Aviation and Marine Branch, Room 5322, 2025 M Street N.W., Washington DC 20554.

4 GENERAL RADIOTELEPHONE OPERATOR LICENSE

Element 1A – Radio Law and Operating Practice   (24 questions)

1A1 What is the Global Maritime Distress and Safety System (GMDSS)?
A. An automated ship-to-shore distress alerting system using satellite and advanced terrestrial communications systems
B. An emergency radio service employing analog and manual safety apparatus
C. An association of radio officers trained in emergency procedures
D. The international organization charged with the safety of ocean-going vessels

ANSWER A: This system, coordinated worldwide by the International Maritime Organization (IMO), provides rapid transfer of a ship's distress message to units best suited for providing assistance. With the GMDSS, certain frequency bands are set aside and each station is assigned a unique call sign. See FCC Rule § 80.5.

1A2 What authority does the Marine Radio Operator Permit confer?
A. Grants authority to operate commercial broadcast stations and repair associated equipment
B. Allows the radio operator to maintain equipment in the Business Radio Service
C. Confers authority to operate licensed radio stations in the Aviation, Marine and International Fixed Public Radio Services
D. The non-transferable right to install, operate and maintain any type-accepted radio transmitter

ANSWER C: The MROP confers authority to operate licensed radio stations. It is earned by passing an exam containing 24 questions from this question pool. 18 correct answers pass. Element 1 question pool concentrates on operating techniques and practices in the maritime service. It is not concerned with maintenance and repair. See *Table 2-4* of Chapter 4.

1A3 Which of the following persons are ineligible to be issued a commercial radio operator license?
A. Individuals who are unable to send and receive correctly by telephone spoken messages in English
B. Handicapped persons with uncorrected disabilities which affect their ability to perform all duties required of commercial radio operators
C. Foreign maritime radio operators unless they are certified by the International Maritime Organization (IMO)
D. U.S. Military radio operators who are still on active duty

ANSWER A: By FCC Rule § 13.9, persons who cannot transmit or receive correctly voice messages in English are not eligible for a commercial operator license. This includes the mute and deaf. Persons afflicted with other physical handicaps may be issued a commercial radio license, if found qualified, with certain restrictive endorsements.

1A4 Who is required to make entries on a required service or maintenance log?
A. The licensed operator or a person whom he or she designates
B. The operator responsible for the station operation or maintenance

1A – RADIO LAW AND OPERATING PRACTICE

C. Any commercial radio operator holding at least a Restricted Radiotelephone Operator Permit
D. The technician who actually makes the adjustments to the equipment

ANSWER B: Since the person servicing the radio equipment knows better than anyone else what was required to fix a problem, that person is best qualified to state the problem and the steps taken to repair it in the maintenance log. See FCC Rule § 80.409.

1A5 What is a requirement of every commercial operator on duty and in charge of a transmitting system?

A. A copy of the Proof-of-Passing Certificate (PPC) must be on display at the transmitter location.
B. The original license or a photocopy must be posted or in the operator's personal possession and available for inspection.
C. The FCC Form 756 certifying the operator's qualifications must be readily available at the transmitting system site.
D. A copy of the operator's license must be supplied to the radio station's supervisor as evidence of technical qualification.

ANSWER B: The person in charge of maintaining correct operation of a commercial radio station must be able to provide proof of technical expertise and competence by prominently displaying his or her FCC commercial radio license while on duty. See FCC Rule § 13.19(c).

1A6 What is distress traffic?

A. In radiotelegraphy, SOS sent as a single character; in radiotelephony, the speaking of the word, "Mayday"
B. Health and welfare messages concerning the immediate protection of property and the safety of human life
C. Internationally recognized communications relating to emergency situations
D. All messages relative to the immediate assistance required by a ship, aircraft or other vehicle in imminent danger

ANSWER D: While choice A is technically correct, choice D covers the entire spectrum of possible situations in which distress traffic may occur. See definitions in § 80.5 and § 80.314.

1A7 What is a maritime mobile repeater station?

A. A fixed land station used to extend the communications range of ship and coast stations
B. An automatic on-board radio station which facilitates the transmissions of safety communications aboard ship
C. A mobile radio station which links two or more public coast stations
D. A one-way, low-power communications system used in the maneuvering of vessels

ANSWER A: A repeater automatically retransmits everything it hears, adding strength to the original signal, and extending its range. Public correspondence telephone service is a repeater station. See definitions in § 80.5.

4 GENERAL RADIOTELEPHONE OPERATOR LICENSE

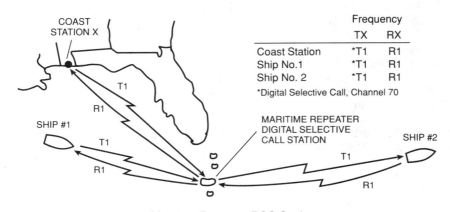

Maritime Repeater DSC Station

1A8 What is an urgency transmission?
  A. A radio distress transmission affecting the security of humans or property
  B. Health and welfare traffic which impacts the protection of on-board personnel
  C. A communications alert that important personal messages must be transmitted
  D. A communications transmission concerning the safety of a ship, aircraft or other vehicle, or of some person on board or within sight
ANSWER D: Urgency traffic covers situations which potentially could worsen into distress traffic. A person on board who is ill or near death could require an urgency message. See definitions in § 80.5, also § 80.327.

1A9 What is a ship earth station?
  A. A maritime mobile-satellite station located at a coast station
  B. A mobile satellite location located on board a vessel
  C. A communications system which provides line-of-sight communications between vessels at sea and coast stations
  D. An automated ship-to-shore distress alerting system
ANSWER B: A ship earth station allows messages to be exchanged between a ship and a satellite orbiting the earth. Many navigation systems use such equipment. While an earth station may be located anywhere on the planet, a ship earth station resides only on a ship. See definitions in § 80.5.

1A10 What is the internationally recognized urgency signal?
  A. The letters "TTT" transmitted three times by radiotelegraphy
  B. Three oral repetitions of the word "safety" sent before the call
  C. The word "PAN PAN" spoken three times before the urgent call
  D. The pronouncement of the word "Mayday"
ANSWER C: Remember that "Mayday" is reserved for distress traffic only; "PAN PAN" is for urgency. See § 80.327.

1A11 What is a safety transmission?
  A. A radiotelephony warning preceded by the words "PAN"
  B. Health and welfare traffic concerning the protection of human life

1A – RADIO LAW AND OPERATING PRACTICE 4

    C. A communications transmission which indicates that a station is preparing to transmit an important navigation or weather warning
    D. A radiotelegraphy alert preceded by the letters "XXX" sent three times

ANSWER C: Choices A and B are incorrect because they pertain to urgency traffic. Safety traffic concerns only navigation and weather information. See definitions in § 80.5, also § 80.329.

1A12 What is a requirement of all marine transmitting apparatus used aboard United States vessels?
    A. Only equipment that has been type-accepted by the FCC for Part 80 operations is authorized.
    B. Equipment must be approved by the U.S. Coast Guard for maritime mobile use.
    C. Certification is required by the International Maritime Organization (IMO).
    D. Programming of all maritime channels must be performed by a licensed Marine Radio Operator.

ANSWER A: All the rules and regulations covered in this question pool come directly from Part 80 — Stations in the Maritime Services. Technical standards for radio equipment are covered in Subpart E. See § 80.43 and § 80.203(a).

1A13 Where do you submit an application for inspection of a ship radio station?
    A. To a Commercial Operator Licensing Examination Manager (COLE Manager)
    B. To the Federal Communications Commission, Washington, DC 20554
    C. To the Engineer-in-Charge of the FCC District Office nearest the proposed place of inspection
    D. To the nearest International Maritime Organization (IMO) review facility

ANSWER C: Per FCC Rule § 80.59, an application for inspection and certification must be submitted to the engineer in charge of the FCC District Office at least three days before the proposed inspection date.

1A14 What are the antenna requirements of a VHF telephony coast, maritime utility or ship station?
    A. The shore or on-board antenna must be vertically polarized.
    B. The antenna array must be type-accepted for 30-200 MHz operation by the FCC
    C. The horizontally-polarized antenna must be positioned so as not to cause excessive interference to other stations.
    D. The antenna must be capable of being energized by an output in excess of 100 watts.

ANSWER A: A vertically-polarized antenna takes up less space on board a ship and avoids the potential problem of cross-polarization. By requiring all antennas to be polarized the same way, the chance of receiving radio signals properly is greatly increased. Also, a non-directional antenna, such as an omni-directional one, reduces the chance that a distress call could be missed simply because the ship's antenna is pointing the wrong way. See § 80.72 and § 80.81.

4 GENERAL RADIOTELEPHONE OPERATOR LICENSE

1A15 What regulations govern the use and operation of FCC-licensed ship stations in international waters?
   A. The regulations of the International Maritime Organization (IMO) and Radio Officers Union
   B. Part 80 of the FCC Rules plus the international Radio Regulations and agreements to which the United States is a party
   C. The Maritime Mobile Directives of the International Telecommunication Union
   D. Those of the FCC's Aviation and Marine Branch, PRB, Washington, DC 20554

ANSWER B: Part 80 specifies rules for operating procedures, technical standards, safety watch requirements, emission classes, transmitter power, transmitter licensing, frequencies and tolerances, and more.

1A16 Which of the following transmissions are not authorized in the Maritime Service?
   A. Communications from vessels in dry dock undergoing repairs
   B. Message handling on behalf of third parties for which a charge is rendered
   C. Needless or superfluous radiocommunications
   D. Transmissions to test the operating performance of on- board station equipment

ANSWER C: With so many vessels on the water, spending time on the air exchanging trivial information could prevent a station from getting through with an important message. Keep all messages short and to the point. See § 80.89.

1A17 What are the highest-priority communications from ships at sea?
   A. All critical message traffic authorized by the ship's master
   B. Navigation and meteorological warnings
   C. Distress calls, and communications preceded by the international urgency and safety signals
   D. Authorized government communications for which priority right has been claimed

ANSWER C: The ultimate priority in radio traffic goes to those messages whose reception could mean the difference between life and death. In such cases, rules and regulations pertaining to the legality of frequency use and duration are temporarily suspended. See § 80.312 and § 80.91.

1A18 What is the best way for a radio operator to minimize or prevent interference to other stations?
   A. By using an omni-directional antenna pointed away from other stations
   B. Reducing power to a level that will not affect other on-frequency communications
   C. By changing frequency when notified that a radiocommunication causes interference
   D. Determine that a frequency is not in use by monitoring the frequency before transmitting

ANSWER D: Answer A is not correct because an omni-directional antenna transmits equally well in all directions; you can't point it away from other stations.

1A – RADIO LAW AND OPERATING PRACTICE **4**

Answers B and C try to solve a problem after causing it. Even a few watts of power used on board a ship can be heard quite clearly thousands of miles away. This can disrupt other radio services, perhaps even masking a distress call from another ship. You may be able to hear only one of the two stations, so listen for a few minutes before transmitting—it's not only polite, but it's also the law. See § 80.92 and § 80.87.

1A19 Under what circumstances may a ship or aircraft station interfere with a public coast station?
- A. Under no circumstances during on going radiocommunications
- B. During periods of government priority traffic handling
- C. When it is necessary to transmit a message concerning the safety of navigation or important meteorological warnings
- D. In cases of distress

ANSWER D: Since distress messages have the ultimate priority, operators may break in on any station at any time. If the operator of the distress station feels that the best way to attract attention is by disrupting another station, then (and only then!) it is legal to do so. All other messages must be sent in such a manner as to not disrupt other stations. See § 80.312.

1A20 Who determines when a ship station may transmit routine traffic destined for a coast or government station in the maritime mobile service?
- A. Shipboard radio officers may transmit traffic when it will not interfere with ongoing radiocommunications.
- B. The order and time of transmission and permissible type of message traffic is decided by the licensed on-duty operator.
- C. Ship stations must comply with instructions given by the coast or government station.
- D. The precedence of conventional radiocommunications is determined by FCC and international regulation.

ANSWER C: Ensure the smooth flow of information by following a coast or government station's requests to the letter. Making sudden demands on the coast station could upset their methods of handling paperwork, with the possible result that your message gets damaged or even lost. See § 80.116(g).

1A21 Who is responsible for payment of all charges accruing to other facilities for the handling or forwarding of messages?
- A. The licensee of the ship station transmitting the messages
- B. The third party for whom the message traffic was originated
- C. The master of the ship, jointly with the station licensee
- D. The licensed commercial radio operator transmitting the radiocommunication

ANSWER A: The person who owns the station license is held responsible for paying any charges that another station might present for handling message traffic. Most communications between a ship and a land station involve scheduling of vessel movements, obtaining supplies, and arranging for repairs. No charge is made for a distress, urgency, or safety message. See § 80.95.

1A22 Ordinarily, how often would a station using a telephony emission identify?
  A. At least every 10 minutes
  B. At 15-minute intervals, unless public correspondence is in progress
  C. At the beginning and end of each transmission and at 15-minute intervals
  D. At 20-minute intervals

ANSWER C: Stations engaged in conversation must identify their stations periodically. This not only lets listeners know who is communicating, but it also satisfies the FCC's rule that only licensed operators are allowed to use the airwaves. Operators must give their station call sign in English. See § 80.102.

1A23 When does a maritime radar transmitter identify its station?
  A. By radiotelegraphy at the onset and termination of operation
  B. At 20-minute intervals, using an automatic transmitter identification system
  C. Radar transmitters must not transmit station identification.
  D. By a transmitter identification label (TIL) secured to the transmitter

ANSWER C: Per FCC Rule § 80.104, a radar station is a self-contained transmitter and receiver that is designed to respond only to its own emissions and is not meant to be received by any other stations. See § 8.159 and § 80.165.

1A24 What is the general obligation of a coast or marine-utility station?
  A. To accept and dispatch messages without charge, which are necessary for the business and operational needs of ships
  B. To acknowledge and receive all calls directed to it by ship or aircraft stations
  C. To transmit lists of call signs of all fixed and mobile stations for which they have traffic
  D. To broadcast warnings and other information for the general benefit of all mariners

ANSWER B: The purpose of coast stations and marine-utility stations is to exchange messages. They cannot legally ignore any call. A coast station is simply a maritime radio station on land. A marine-utility station is simply a handheld radiotelephone unit, usually used while aboard a vessel. See § 80.105.

1A25 How does a coast station notify a ship that it has a message for the ship?
  A. By making a directed transmission on 2182 kHz or 156.800 MHz
  B. The coast station changes to the vessel's known working frequency
  C. By establishing communications using the eight-digit maritime mobile service identification
  D. The coast station may transmit, at intervals, lists of call signs in alphabetical order for which they have traffic.

ANSWER D: This "bulletin-board" method of traffic notification works well. It allows ships to find out if they have messages without querying the coast station. You should listen several times during the day, as the message queue constantly changes. See § 80.108.

1A26 Under what circumstances may a coast station using telephony transmit a general call to a group of vessels?
A. Under no circumstances
B. When announcing or preceding the transmission of distress, urgency, safety or other important messages
C. When the vessels are located in international waters beyond 12 miles
D. When identical traffic is destined for multiple mobile stations within range

ANSWER B: Messages concerning health and welfare are of importance to everyone. This timely information flows much more smoothly when offered to all stations simultaneously, rather than one ship at a time. See § 80.111(a).

1A27 Who has ultimate control of service at a ship's radio station?
A. The master of the ship
B. A holder of a First Class Radiotelegraph Certificate with a six months' service endorsement
C. The Radio Officer-in-Charge authorized by the captain of the vessel
D. An appointed licensed radio operator who agrees to comply with all Radio Regulations in force

ANSWER A: The master of the ship (usually the owner or "captain" of the vessel) owns the station license and enjoys ultimate authority on the air. He is permitted to designate a qualified operator to handle radio messages. See § 80.114.

1A28 What is the power limitation of associated ship stations operating under the authority of a ship station license?
A. The power level authorized to the parent ship station
B. Associated vessels are prohibited from operating under the authority granted to another station licensee.
C. The minimum power necessary to complete the radiocommunications
D. Power is limited to one watt.

ANSWER D: An associated ship station is a VHF transmitter, usually a hand-held transceiver. It may be used only in the vicinity of the ship station with which it is associated. It must not communicate with any other stations, so its power is limited. It may not be used from shore. See definitions in § 80.5 and § 80.115(a).

1A29 How is an associated vessel operating under the authority of another ship station license identified?
A. All vessels are required to have a unique call sign issued by the Federal Communications Commission.
B. With any station call sign self-assigned by the operator of the associated vessel
C. By the call sign of the station with which it is connected and an appropriate unit designator
D. Client vessels use the call sign of their parent plus the appropriate ITU regional indicator.

ANSWER C: More than one associated ship station may be in use at one time. A unique suffix added to the ship station's call sign identifies individual stations. See § 80.115(a).

1A30 On what frequency should a ship station normally call a coast station when using a radiotelephony emission?
A. On a vacant radio channel determined by the licensed radio officer
B. Calls should be initiated on the appropriate ship-to-shore working frequency of the coast station.
C. On any calling frequency internationally approved for use within ITU Region 2
D. On 2182 kHz or 156.800 MHz at any time

ANSWER B: A few VHF channel frequencies are set aside strictly for this type of message traffic. Commercial pilots use them for the movement and docking of vessels near ports, locks and waterways. See § 80.116(a).

1A31 On what frequency would a vessel normally call another ship station when using a radiotelephony emission?
A. Only on 2182 kHz in ITU Region 2
B. On the appropriate calling channel of the ship station at 15 minutes past the hour
C. On 2182 kHz or 156.800 MHz, unless the station knows the called vessel maintains a simultaneous watch on another intership working frequency
D. On the vessel's unique working radio channel assigned by the Federal Communications Commission

ANSWER C: Not only are 2182 kHz and 156.800 MHz distress frequencies, but they are also calling frequencies. If you do not know the intership working frequency in advance, make a call on the calling frequency. When you get a response, ask about the intership frequency and move to it. A radio "watch" is the act of listening on a designated frequency for any possible distress messages. See definitions in § 80.5, and § 80.116(b).

1A32 What is required of a ship station which has established initial contact with another station on 2182 kHz or 156.800 MHz?
A. The stations must check the radio channel for distress, urgency and safety calls at least once every ten minutes.
B. The stations must change to an authorized working frequency for the transmission of messages.
C. Radiated power must be minimized so as not to interfere with other stations needing to use the channel
D. To expedite safety communications, the vessels must observe radio silence for two out of every fifteen minutes.

ANSWER B: After making contact on the calling frequency, move to another frequency to keep the distress channel clear. See § 80.116(c).

1A33 What type of communications may be exchanged by radioprinter between authorized private coast stations and ships of less than 1600 gross tons?
A. Public correspondence service may be provided on voyages of more than 24 hours.
B. All communications, providing they do not exceed 3 minutes after the stations have established contact

C. Only those communications which concern the business and operational needs of vessels
D. There are no restrictions.

ANSWER C: Per FCC Rule § 80.1155, radioprinter and facsimile communications are also allowed between ships to accomplish the business and operation of the ships.

1A34 What are the service requirements of all ship stations?
A. Each ship station must receive and acknowledge all communications with any station in the maritime mobile service.
B. Public correspondence services must be offered for any person during the hours the radio operator is normally on duty.
C. All ship stations must maintain watch on 500 kHz, 2182 kHz and 156.800 MHz.
D. Reserve antennas, emergency power sources and alternate communications installations must be available.

ANSWER A: All maritime radio operators must exchange information without delay or concealment. See § 80.141(b).

1A35 When may the operator of a ship radio station allow an unlicensed person to speak over the transmitter?
A. At no time. Only commercially-licensed radio operators may modulate the transmitting apparatus.
B. When the station power does not exceed 200 watts peak envelope power
C. When under the supervision of the licensed operator
D. During the hours that the radio officer is normally off duty

ANSWER C: In some cases a telephone call must be made from ship to shore; e.g., a passenger may want to call home. The passenger does not own the station license, therefore, the station operator supervises the communication to keep it legal. A marine operator is contacted over a VHF frequency and the call is placed just like any other phone call. The holder of the ship station license is charged for the call. See § 80.156.

1A36 What are the radio operator requirements of a cargo ship equipped with a 1000 watt peak-envelope-power radiotelephone station?
A. The operator must hold a General Radiotelephone Operator License or higher-class license.
B. The operator must hold a Restricted Radiotelephone Operator Permit or higher-class license.
C. The operator must hold a Marine Radio Operator Permit or higher-class license.
D. The operator must hold a GMDSS Radio Maintainer's License.

ANSWER C: Per FCC Rule § 80.159, operating a radiotelephone station aboard a cargo ship requires an MROP as long as the transmitter's output power is less than 1500 watts peak envelope power. If it exceeds 1500 watts, then a General Radiotelephone Operator License is required.

4 GENERAL RADIOTELEPHONE OPERATOR LICENSE

1A37 What are the radio operator requirements of a small passenger ship carrying more than six passengers equipped with a 1000-watt carrier power radiotelephone station?
- A. The operator must hold a General Radiotelephone Operator or higher-class license.
- B. The operator must hold a Marine Radio Operator Permit or higher-class license.
- C. The operator must hold a Restricted Radiotelephone Operator Permit or higher-class license.
- D. The operator must hold a GMDSS Radio Operator's License.

ANSWER A: FCC Rule § 80.159 also covers passenger ships. As long as the carrier power of the transmitter is less than 250 watts, an MROP is all that is required. In this case, however, the carrier power is over 250 watts, so a General Radiotelephone Operator License is needed.

1A38 Which commercial radio operator license is required to operate a fixed tuned ship radar station with external controls?
- A. A radio operator certificate containing a Ship Radar Endorsement
- B. A Marine Radio Operator Permit or higher
- C. Either a First or Second Class Radiotelegraph certificate or a General Radiotelephone Operator License
- D. No radio operator authorization is required.

ANSWER D: FCC Rule § 80.177 authorizes operation without a radio operator license.

1A39 Which commercial radio operator license is required to install a VHF transmitter in a voluntarily-equipped ship station?
- A. A Marine Radio Operator Permit or higher class of license
- B. None, if installed by, or under the supervision of, the licensee of the ship station and no modifications are made to any circuits
- C. A Restricted Radiotelephone Operator Permit or higher class of license
- D. A General Radiotelephone Operator License

ANSWER B: FCC Rule § 80.5 says that a "voluntary ship" is "any ship which is not required by treaty or statute to be equipped with radiotelecommunication equipment." No license is required to own or operate its radio equipment; however, the ship must have a station license.

1A40 What transmitting equipment is authorized for use by a station in the maritime services?
- A. Transmitters that have been certified by the manufacturer for maritime use
- B. Unless specifically excepted, only transmitters type-accepted by the Federal Communications Commission for Part 80 operations
- C. Equipment that has been inspected and approved by the U.S. Coast Guard
- D. Transceivers and transmitters that meet all ITU specifications for use in maritime mobile service

ANSWER B: If any equipment does not meet Part 80 requirements, it could cause interference to other radio services. Refer to questions 1A12 and 1A15.

1A41 The FCC deleted this question from the pool.

1A – RADIO LAW AND OPERATING PRACTICE

1A42 What is a distress communication?
A. An internationally recognized communication indicating that the sender is threatened by grave and imminent danger and requests immediate assistance
B. Communications indicating that the calling station has a very urgent message concerning safety
C. Radiocommunications which, if delayed, will adversely affect the safety of life or property
D. An official radiocommunications notification of approaching navigational or meteorological hazards

ANSWER A: *Grave and imminent danger* defines a distress communication; it has top priority over messages of urgency and safety. See § 80.5 and § 80.314.

1A43 Who may be granted a ship station license in the maritime service?
A. Anyone, including foreign governments
B. Only FCC-licensed operators holding a First or Second Class Radiotelegraph Operator's Certificate or the General Radiotelephone Operator License
C. Vessels that have been inspected and approved by the U.S. Coast Guard and Federal Communications Commission
D. The owner or operator of a vessel, or their subsidiaries

ANSWER D: Per FCC Rule § 80.15, "A ship station license may only be granted to the owner or operator of the vessel, a subsidiary communications corporation of the owner or operator of the vessel, a state or local government subdivision, or any agency of the U.S. Government" as the FCC sees fit.

The same rule also specifies who may not receive a station license. "A station license cannot be granted to or held by a foreign government or its representative." An alien must request a station license from the communications bureau of his or her own country.

1A44 Who is responsible for the proper maintenance of station logs?
A. The station licensee and the radio operator in charge of the station
B. The station licensee
C. The commercially-licensed radio operator in charge of the station
D. The ship's master and the station licensee

ANSWER A: FCC Rule § 80.409. "The station licensee and the radio operator in charge of the station are responsible for the maintenance of station logs. ... These persons must keep the log in an orderly manner. Key letters or abbreviations may be used if their proper meaning or explanation is contained elsewhere in the same log." Erasures are not allowed. See question 1A71.

1A45 How long should station logs be retained when there are entries relating to distress or disaster situations?
A. Until authorized by the Commission in writing to destroy them
B. Indefinitely, or until destruction is specifically authorized by the U.S. Coast Guard
C. For a period of three years from the date of entry, unless notified by the FCC
D. For a period of one year from the date of entry

ANSWER C: Logs are always required to be retained by the station licensee for at least one year, but for at least three years if a distress message was received or transmitted. See § 80.409(b).

4 *GENERAL RADIOTELEPHONE OPERATOR LICENSE*

1A46 Where must ship station logs be kept during a voyage?
A. At the principal radiotelephone operating position
B. They must be secured in the vessel's strongbox for safekeeping.
C. In the personal custody of the licensed commercial radio operator
D. All logs are turned over to the ship's master when the radio operator goes off duty.

ANSWER A: Per FCC Rule § 80.409, station logs (both radiotelephone and radiotelegraph) must be kept at the principal radio operating room during the entire voyage.

1A47 What is the antenna requirement of a radiotelephone installation aboard a passenger vessel?
A. The antenna must be located a minimum of 15 meters from the radiotelegraph antenna.
B. An emergency reserve antenna system must be provided for communications on 156.8 MHz.
C. The antenna must be vertically polarized and as non-directional and efficient as is practicable for the transmission and reception of ground waves over seawater.
D. All antennas must be tested and the operational results logged at least once during each voyage.

ANSWER C: See explanation at question 1A14.

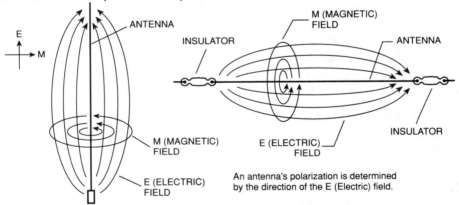

a. Vertically-Polarized Antenna   b. Horizontally-Polarized Antenna
Polarized Antennas

1A48 Where must the principal radiotelephone operating position be installed in a ship station?
A. At the principal radio operating position of the vessel
B. In the room or an adjoining room from which the ship is normally steered while at sea
C. In the chart room, master's quarters or wheel house
D. At the level of the main wheel house or at least one deck above the ship's main deck

ANSWER B: To get the person in command correct and timely information, it makes sense to keep the radio close to the steering room at all times, particularly in a busy shipping channel or at sea during a rough storm. See § 80.853(d).

1A – RADIO LAW AND OPERATING PRACTICE

1A49 What are the technical requirements of a VHF antenna system aboard a vessel?
A. The antenna must provide an amplification factor of at least 2.1 dbi.
B. The antenna must be vertically polarized and non-directional.
C. The antenna must be capable of radiating a signal a minimum of 150 nautical miles on 156.8 MHz.
D. The antenna must be constructed of corrosion-proof aluminum and capable of proper operation during an emergency.
ANSWER B: See explanation at question 1A14.

1A50 How often must the radiotelephone installation aboard a small passenger boat be inspected?
A. Equipment inspections are required at least once every 12 months.
B. When the vessel is first placed in service and every 2 years thereafter
C. At least once every five years
D. A minimum of every 3 years, and when the ship is within 75 statute miles of an FCC field office
ANSWER C: Radio systems in large ships must be inspected once a year. Because small passenger boats have much smaller radio systems with fewer parts, inspection is required only every five years. Also, the station licensee is in a better position to detect equipment failure in a small passenger boat. See § 80.903.

1A51 How far from land may a small passenger vessel operate when equipped only with a VHF radiotelephone installation?
A. No more than 20 nautical miles from the nearest land if within the range of a VHF public coast or U.S. Coast Guard station
B. No more than 100 nautical miles from the nearest land
C. No more than 20 nautical miles unless equipped with a reserve power supply
D. The vessel must remain within the communications range of the nearest coast station at all times.
ANSWER A: VHF signals travel by line-of-sight, and, therefore, cannot be heard reliably by any station more than 20 miles away. Without radio equipment that operates at lower frequencies (which offer longer propagation), it makes sense to stay close to shore. See § 80.905(a).

1A52 What is the minimum transmitter power level required by the FCC for a medium-frequency transmitter aboard a compulsorily fitted vessel?
A. At least 100 watts, single-sideband, suppressed-carrier power
B. At least 60 watts PEP
C. The power predictably needed to communicate with the nearest public coast station operating on 2182 kHz
D. At least 25 watts delivered into 50 ohms effective resistance when operated with a primary voltage of 13.6 volts DC
ANSWER B: An international radiotelephone distress frequency, 2182 kHz, is classified in the medium-frequency range. Since all compulsorily fitted vessels must carry radio equipment capable of transmitting on this frequency, 60 watts peak envelope power (PEP) is considered the minimum power level necessary to ensure contact with another station. See § 80.807(c)(2) and § 80.855(d)(2).

4 GENERAL RADIOTELEPHONE OPERATOR LICENSE

1A53 What is a Class "A" EPIRB?
  A. An alerting device notifying mariners of imminent danger
  B. A satellite-based maritime distress and safety alerting system
  C. An automatic, battery-operated emergency position-indicating radiobeacon that floats free of a sinking ship
  D. A high-efficiency audio amplifier

ANSWER C: Per FCC Rule § 80.1053, "EPIRB" means Emergency Position Indicating Radio Beacon. It is designed to attract attention by broadcasting a distress signal. A Class "A" EPIRB must be able to ballast itself into upright position in less than one second; its antenna must deploy automatically; it must contain a visual indicator that shows when it is operating; and it must be waterproof.

1A54 What are the radio watch requirements of a voluntary ship?
  A. While licensees are not required to operate the ship radio station, general-purpose watches must be maintained if they do.
  B. Radio watches must be maintained on 500 kHz, 2182 kHz and 156.800 MHz, but no station logs are required.
  C. Radio watches are optional but logs must be maintained of all medium-, high-frequency and VHF radio operation.
  D. Radio watches must be maintained on the 156-158 MHz, 1600-4000 kHz and 4000-23000 kHz bands

ANSWER A: Per Rule § 80.310, "Voluntary vessels must maintain a watch on 156.800 MHz whenever the radio is operating and is not being used to communicate."

1A55 What is the Automated Mutual-Assistance Vessel Rescue System?
  A. A voluntary organization of mariners who maintain radio watch on 500 kHz, 2182 kHz and 156.800 MHz
  B. An international system operated by the Coast Guard, providing coordination of search and rescue efforts
  C. A coordinated radio direction-finding effort between the Federal Communications Commission and U.S. Coast Guard to assist ships in distress
  D. A satellite-based distress and safety-alerting program operated by the U.S. Coast Guard

ANSWER B: The Automated Mutual-Assistance Vessel Rescue System, known as AMVER for short, has saved many lives over the years. Information is available to ships or search-and-rescue agencies of any country to increase marine safety. It is truly an international effort. See definitions in § 80.5.

1A56 What is a bridge-to-bridge station?
  A. An internal communications system linking the wheel house with the ship's primary radio operating position and other integral ship control points
  B. An inland waterways and coastal radio station serving ship stations operating within the United States
  C. A portable ship station necessary to eliminate frequent application to operate a ship station on board different vessels
  D. A VHF radio station located on a ship's navigational bridge or main control station that is used only for navigational communications

1A – RADIO LAW AND OPERATING PRACTICE

ANSWER D: A bridge-to-bridge station operates on designated frequency Channel 13 in the 156-162 MHz band. These operations take place strictly between the bridges of two ships, as a means of letting each other know exactly who they are and where they are going. See definitions in § 80.5.

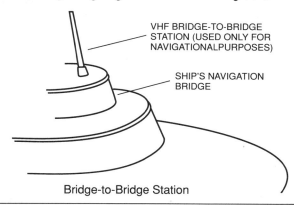

Bridge-to-Bridge Station

1A57 Which of the following statements is true as to ships subject to the Safety Convention?
 A. A cargo ship participates in international commerce by transporting goods between harbors.
 B. Passenger ships carry six or more passengers for hire as opposed to transporting merchandise.
 C. A cargo ship is any ship that is not licensed or certificated to carry more than 12 passengers.
 D. Cargo ships are FCC-inspected on an annual basis, while passenger ships undergo U.S. Coast Guard inspections every six months.
ANSWER C: Per FCC Rule § 80.5, a passenger ship carries, or is licensed or certificated to carry, more than twelve passengers. A cargo ship is any ship that is not a passenger ship.

1A58 What is a "passenger-carrying vessel" when used in reference to the Great Lakes Radio Agreement?
 A. A vessel that is licensed or certificated to carry more than 12 passengers
 B. Any ship carrying more than six passengers for hire
 C. Any ship, the principal purpose of which is to ferry persons on the Great Lakes and other inland waterways
 D. A ship which is used primarily for transporting persons and goods to and from domestic harbors or ports
ANSWER B: FCC Rule § 80.5 states that a "...'passenger-carrying vessel', when used in reference to the Great Lakes Radio Agreement, means any ship transporting more than six passengers for hire." Note the difference in "passenger-carrying vessel" here and "passenger ship" in question 1A57.

1A59 How do the FCC's rules define a power-driven vessel?
 A. A ship that is not manually propelled or under sail
 B. Any ship propelled by machinery
 C. A watercraft containing a motor with a power rating of at least 3 HP
 D. A vessel moved by mechanical equipment at a rate of 5 knots or more

ANSWER B: Rule § 80.5 mentions this definition specifically. If a ship is propelled by engines, jets, paddlewheels, or any other machinery, it is considered a power-driven vessel.

1A60 How do the rules define "navigational communications"?
A. Safety communications pertaining to the maneuvering or directing of vessels' movements
B. Important communications concerning the routing of vessels during periods of meteorological crisis
C. Telecommunications pertaining to the guidance of maritime vessels in hazardous waters
D. Radio signals consisting of weather, sea conditions, notices to mariners and potential dangers

ANSWER A: At first glance, the other three answers appear to be legitimate, but they cover dangerous situations. Answer A covers the movements of ships in all kinds of weather. See definitions in § 80.5.

1A61 What traffic management service is operated by the U.S. Coast Guard in certain designated water areas to prevent ship collisions, groundings and environmental harm?
A. Water Safety Management Bureau (WSMB)
B. Vessel Traffic Service (VTS)
C. Ship Movement and Safety Agency (SMSA)
D. Interdepartmental Harbor and Port Patrol (IHPP)

ANSWER B: The VTS operates in the 156-MHz range and protects, among other regions, the Seattle (Puget Sound), New York, New Orleans and Houston shipping channels. Other regions use VTS when there is no interference to the regions just described. See definitions in § 80.5.

1A62 What action must be taken by the owner of a vessel who changes its name?
A. A Request for Ship License Modification (RSLM) must be submitted to the FCC's licensing facility.
B. The Engineer-in-Charge of the nearest FCC field office must be informed.
C. The Federal Communications Commission in Gettysburg, PA, must be notified in writing.
D. Written confirmation must be obtained from the U.S. Coast Guard.

ANSWER C: Per FCC Rule § 80.29, a station licensee must notify the FCC *in writing* whenever any of the following is changed: mailing address, licensee name, or vessel name. Be as complete and specific as possible with all information.

1A63 When may a shipboard radio operator make a transmission in the maritime services not addressed to a particular station or stations?
A. General CQ calls may only be made when the operator is off duty and another operator is on watch.
B. Only during the transmission of distress, urgency or safety signals or messages, or to test the station
C. Only when specifically authorized by the master of the ship
D. When the radio officer is more than 12 miles from shore and the nearest ship or coast station is unknown

1A – RADIO LAW AND OPERATING PRACTICE 4

ANSWER B: A station may attempt to reach any other station by making a general call during distress, when any contact is desired because immediate assistance is needed, or when testing equipment. When testing equipment, stating over the air that you are doing so avoids confusion, and all stations listening understand what you are doing and do not respond unless you ask for a radio check. See § 80.89.

1A64 What is the order of priority of radiotelephone communications in the maritime services?
  A. Distress calls and signals, followed by communications preceded by urgency and safety signals
  B. Alarm, radio direction-finding, and health and welfare communications
  C. Navigation hazards, meteorological warnings, priority traffic
  D. Government precedence, messages concerning safety of life and protection of property, and traffic concerning grave and imminent danger

ANSWER A: Distress calls always take precedence over all other types of communication. Distress means lives are in danger and immediate assistance is required. Urgency calls concern the safety of a ship or person. Safety traffic concerns navigation problems and weather warnings. See § 80.91.

1A65 What should a station operator do before making a transmission?
  A. Transmit a general notification that the operator wishes to utilize the channel
  B. Except for the transmission of distress calls, determine that the frequency is not in use by monitoring the frequency before transmitting
  C. Check transmitting equipment to be certain it is properly calibrated
  D. Ask if the frequency is in use

ANSWER B: Always listen before transmitting; however, it is legal to transmit a distress call immediately. See question 1A18 for consequences of interference on the airwaves.

1A66 What is the proper procedure for testing a radiotelephone installation?
  A. Transmit the station's call sign, followed by the word "test" on the radio channel being used for the test.
  B. A dummy antenna must be used to insure the test will not interfere with ongoing communications.
  C. Permission for the voice test must be requested and received from the nearest public coast station.
  D. Short tests must be confined to a single frequency and must never be conducted in port.

ANSWER A: While answers B and C sound plausible, it is perfectly legal to test a radiotelephone installation — provided that it is done legally. Using the word "test" informs all stations listening that you are simply testing the radio equipment and do not require any emergency assistance. Using the ship's antenna during this test (instead of a dummy load) also verifies the antenna system. See § 80.101(a)(2).

4 GENERAL RADIOTELEPHONE OPERATOR LICENSE

1A67 What is the minimum radio operator requirement for ships subject to the Great Lakes Radio Agreement?
A. Third Class Radiotelegraph Operator's Certificate
B. General Radiotelephone Operator License
C. Marine Radio Operator Permit
D. Restricted Radiotelephone Operator Permit
ANSWER C: Interestingly, the FCC gives this information in a rule (§ 80.161) by itself: "Each ship subject to the Great Lakes Radio Agreement must have on board an officer or member of the crew who holds a Marine Radio Operator Permit or higher class license." The Great Lakes carry heavy marine traffic, and effective radio communications help prevent navigation mishaps.

1A68 What FCC authorization is required to operate a VHF transmitter on board a vessel voluntarily equipped with radio and sailing on a domestic voyage?
A. No radio operator license or permit is required.
B. Marine Radio Operator Permit
C. Restricted Radiotelephone Operator Permit
D. General Radiotelephone Operator License
ANSWER A: The key word is *voluntarily*. Voluntarily equipped vessels do not require operator permits for radio operation, but a station license is still needed. See definitions in § 80.177(a).

1A69 On what frequencies does the Communications Act require radio watches by compulsory radiotelephone stations?
A. Watches are required on 500 kHz and 2182 kHz.
B. Continuous watch is required on 2182 kHz only.
C. On all frequencies between 405-535 kHz, 1605-3500 kHz and 156-162 MHz
D. Watches are required on 2182 kHz and 156.800 MHz.
ANSWER D: The *watch* frequencies for *radiotelephone* stations are 2182 kHz and 156.800 MHz. Watches must be on these frequencies at all times. This is a common question on the examination, so be ready for it. See § 80.147 and § 80.148.

1A70 What is the purpose of the international radiotelephone alarm signal?
A. To notify nearby ships of the loss of a person or persons overboard
B. To call attention to the upcoming transmission of an important meteorological warning
C. To alert radio officers monitoring watch frequencies of a forthcoming distress, urgency or safety message
D. To actuate automatic devices giving an aural alarm to attract the attention of the operator where there is no listening watch on the distress frequency
ANSWER D: While a constant watch is required on distress frequencies, it may not always be possible to keep an operator on duty around the clock. In these cases, an automatic alarm signal, triggered when a message with an encoded alarm signal is received on a distress frequency, alerts the operator. See § 80.305(a).

1A – RADIO LAW AND OPERATING PRACTICE

1A71 What is the proper procedure for making a correction in the station log?
A. The ship's master must be notified, approve and initial all changes to the station log.
B. The mistake may be erased and the correction made and initialized only by the radio operator making the original error.
C. The original person making the entry must strike out the error, initial the correction and indicate the date of the correction.
D. Rewrite the new entry in its entirety directly below the incorrect notation and initial the change.

ANSWER C: Per FCC Rule § 80.409, "Erasures, obliterations or willful destruction within the retention period are prohibited. Corrections may be made only by the person originating the entry by striking out the error, initialing the correction and indicating the date of correction." Though not required, it is a good idea to keep station logs in ink.

1A72 What authorization is required to operate a 350-watt, PEP maritime voice station on frequencies below 30 MHz aboard a small, non-commercial pleasure vessel?
A. Third Class Radiotelegraph Operator's Certificate
B. General Radiotelephone Operator License
C. Restricted Radiotelephone Operator Permit
D. Marine Radio Operator Permit

ANSWER C: The Restricted Radiotelephone Operator Permit is often abbreviated as RP. A chart showing what licenses are required to operate various types of transmitters is in FCC Rule § 80.165, which is specifically for voluntary stations.

1A73 What is selective calling?
A. A coded transmission directed to a particular ship station
B. A radiotelephony communication directed at a particular ship station
C. An electronic device which uses a discriminator circuit to filter out unwanted signals
D. A telegraphy transmission directed only to another specific radiotelegraph station

ANSWER A: Selective calling operates very much like dialing a telephone. Without it, a station operator must continually monitor the airwaves, listening for a specific call sign. A digital coding system permits the radio to be squelched until a message containing the proper digital code is received. The selection code activates the radio and normal communication continues. See definitions in § 80.5.

1A74 In the International Phonetic Alphabet, the letters D, N, and O are represented by the words:
A. Delta, November, Oscar
B. Denmark, Neptune, Oscar
C. December, Nebraska, Olive
D. Delta, Neptune, Olive

ANSWER A: The following is the International Phonetic Alphabet. All radio operators around the world use it:

GENERAL RADIOTELEPHONE OPERATOR LICENSE

ITU/ICAO	Alternates	ITU/ICAO	Alternates
A - Alpha	(Able, Adam)	N - November	(Norway, Nancy)
B - Bravo	(Baker)	O - Oscar	(Ocean, Otto)
C - Charlie	(Cocoa)	P - Papa	(Peter, Pacific)
D - Delta	(Dog, David)	Q - Quebec	(Queen)
E - Echo	(Easy, Edward)	R - Romeo	(Roger, Radio)
F - Foxtrot	(Fox, Frank)	S - Sierra	(Sugar, Susan)
G - Golf	(George, Gulf)	T - Tango	(Thomas, Tokyo)
H - Hotel	(Henry)	U - Uniform	(United, Union)
I - India	(Item, Ida, Italy)	V - Victor	(Victory, Victoria)
J - Juliette	(Japan, Jig)	W - Whiskey	(Willie, William)
K - Kilo	(King, Kilowatt)	X - X-Ray	(Xylophone)
L - Lima	(Love, Lewis)	Y - Yankee	(Yoke, Young)
M - Mike	(Mexico, Mary)	Z - Zulu	(Zebra, Zed, Zanzibar)

Phonetics in parentheses are sometimes used on the ham bands, especially by older, long-term amateurs, but the ITU/ICAO phonetics are preferable.

ITU/ICAO Phonetic Alphabet
Adopted by the International Telecommunication Union and International Civil Aviation Organization
Source: *Amateur Radio Logbook,* Contributions by F. Maia, ©1993 Master Publishing, Inc.

English is the closest we have to a universal language. Even so, it can be troublesome for some radio operators from other countries. That is why the International Telecommunication Union (ITU) created this internationally recognized standard phonetic alphabet. It helps all radio operators to identify their stations and receive information when interference makes communication difficult. Whenever you cannot understand the other operator, ask him or her to spell the word phonetically. Station call signs, upon initial contact, are always spelled out this way to lessen confusion. This is especially important during an emergency.

Learning this phonetic alphabet is easier than it looks. Try to spell your own name with it. Practice spelling ordinary words with it. Very soon it will become second nature to you.

1A75 When is it legal to transmit high power on channel 13?
A. Failure of vessel being called to respond
B. In a blind situation such as rounding a bend in a river
C. During an emergency
D. All of the above
ANSWER D: Channel 13 messages must be about navigation; for example, passing or meeting other vessels. Your power must not be more than one watt unless you are declaring an emergency, the other vessel fails to respond, or if an obstruction prevents low-power communication. In such cases, higher output power is allowed. See § 80.331(c).

1A76 What must be in operation when no operator is standing watch on a compulsory radio-equipped vessel while out at sea?
A. An auto alarm
B. Indicating Radio Beacon signals
C. Distress-Alert signal device
D. Radiotelegraph transceiver set to 2182 kHz
ANSWER A: An auto alarm activates the receiver's speaker only when a special alarm signal is received. Such a signal is transmitted by a station in distress to

1A77 When may a bridge-to-bridge transmission be more than 1 watt?
  A. When broadcasting a distress message
  B. When rounding a bend in a river or traveling in a blind spot
  C. When calling the Coast Guard
  D. Both A and B above

ANSWER D: Bridge-to-bridge communications are usually on Channel 13 using only 1 watt. See also question 1A75.

1A78 When are EPIRB batteries changed?
  A. After emergency use; after battery life expires
  B. After emergency use; as per manufacturer's instructions marked on outside of transmitter within month and year replacement date
  C. After emergency use; every 12 months when not used
  D. Whenever voltage drops to less than 50% of full charge

ANSWER B: All companies that make EPIRB batteries must stamp on the outside of the battery case the month and year of the battery's manufacture and the month and year when 50% of its useful life is due to expire. When the expiration date is reached, the battery must be replaced; but if the EPIRB is actually used before that date, the battery must be replaced immediately afterward. Refer to question 1A53 for EPIRB definition, and see also § 80.1053(e).

1A79 The radiotelephone distress message consists of:
  A. MAYDAY spoken three times, call sign and name of vessel in distress
  B. Particulars of its position, latitude and longitude, and other information which might facilitate rescue, such as length, color and type of vessel, and number of persons on board
  C. Nature of distress and kind of assistance required
  D. All of the above

ANSWER D: Many people become so frightened during an emergency that they forget how to call for help. It is important to stay calm and help the authorities help you. Provide them with as much information as possible to make their job easier. "MAYDAY MAYDAY MAYDAY" captures their attention; your call sign and vessel name tell them who you are; your latitude and longitude tell them where you are. State the nature of your distress so the authorities can bring along the necessary equipment. Give the number of persons aboard and conditions of any injured. Briefly describe your boat's length and color of the hull. If a ship is swamped, count heads immediately. If someone is missing, tell the authorities. See § 80.315(b).

1A80 If a ship sinks, what device is designed to float free of the mother ship, is turned on automatically and transmits a distress signal?
  A. EPIRB on 121.5 MHz/243 MHz or 406.025 MHz
  B. EPIRB on 2182 kHz and 405.025 kHz
  C. Bridge-to-bridge transmitter on 2182 kHz
  D. Auto alarm keyer on any frequency

ANSWER A: An EPIRB automatically transmits a distress call on a specially assigned frequency when a vessel sinks. A water-activated battery is the central power source. Note that these frequencies are in the VHF and UHF ranges, and that an EPIRB is meant to be tracked by orbiting satellites, part of the SARSAT COSPAS system. Direction-finding techniques help home in on the signal. (Refer to question 1A53.)

1A81 International laws and regulations require a silent period on 2182 kHz:
A. For three minutes immediately after the hour
B. For three minutes immediately after the half-hour
C. For the first minute of every quarter-hour
D. Both A and B above

ANSWER D: Each ship with a radiotelephone station must listen on 2182 kHz for three minutes immediately after the hour, and for three minutes immediately after the half-hour, Coordinated Universal Time (UTC). UTC is a worldwide standard of timekeeping. Since all ship clocks are synchronized to one universal time, all ships will listen during the same silent period. Silent periods give low-power stations and those with emergencies a chance to be heard above the din of other stations.

There is a different silent period for radiotelegraphy stations. For 500 kHz, the silent periods run for three minutes immediately after the 15-minute mark, and for three minutes immediately after the 45-minute mark. Don't get them mixed up. See § 80.304(b).

1A82 How should the 2182-kHz auto-alarm be tested?
A. On a different frequency into antenna
B. On a different frequency into dummy load
C. On 2182 kHz into antenna
D. Only under U.S. Coast Guard authorization

ANSWER B: While it is important to test emergency communications equipment to make sure it functions properly, doing so must not allow a false signal to be received by any other station and possibly trigger a needless rescue mission. Testing the auto-alarm on a different frequency ensures that the signal won't activate any other equipment, and the dummy load ensures that the radio waves emitted travel no farther than the interior of the ship's radio room. See § 80.855(h).

1A83 What is the average range of VHF marine transmissions?
A. 150 miles
B. 50 miles
C. 20 miles
D. 10 miles

ANSWER C: VHF radio waves cannot travel much farther than you can see. Beyond 20 miles, VHF radio waves cannot bring help if needed. That is why vessels equipped only with VHF radio equipment may not travel more than 20 miles from shore. See also question 1A51.

1A84 A ship station using VHF bridge-to-bridge Channel 13:
A. May be identified by call sign and country of origin
B. Must be identified by call sign and name of vessel
C. May be identified by the name of the ship in lieu of call sign
D. Does not need to identify itself within 100 miles from shore

1A – RADIO LAW AND OPERATING PRACTICE **4**

ANSWER C: Calling another ship station by that ship's name is perfectly legal because it is often impossible to know another ship's call sign in advance. When you make contact, the other ship station will identify itself with its own call sign. See § 80.102(c).

1A85 When using a SSB station on 2182 kHz or VHF-FM on Channel 16:
   A. Preliminary call must not exceed 30 seconds
   B. If contact is not made, you must wait at least 2 minutes before repeating the call
   C. Once contact is established, you must switch to a working frequency
   D. All of the above

ANSWER D: The idea is to take up as little time on the distress or calling frequency as possible. Keep in mind that some vessel in distress may need the frequency at any time. See § 80.116.

1A86 By international agreement, which ships must carry radio equipment for the safety of life at sea?
   A. Cargo ships of more than 300 gross tons and vessels carrying more than 12 passengers
   B. All ships traveling more than 100 miles out to sea
   C. Cargo ships of more than 100 gross tons and passenger vessels on international deep-sea voyages
   D. All cargo ships of more than 100 gross tons

ANSWER A: Cargo ships travel all over the world, most of the time far away from shore, and may require immediate assistance. Passenger ships can be in danger at a moment's notice. For these reasons, these ships must carry radio equipment. See § 80.851 and § 80.901.

1A87 What is the most important practice that a radio operator must learn?
   A. Monitor the channel before transmitting
   B. Operate with the lowest power necessary
   C. Test a radiotelephone transmitter daily
   D. Always listen to 121.5 MHz

ANSWER A: Always listen first! See explanation at question 1A18.

1A88 Portable ship radio transceivers operated as associated ship units:
   A. Must be operated on the safety and calling frequency 156.8 MHz (Channel 16) or a VHF intership frequency
   B. May not be used from shore without a separate license
   C. Must only communicate with the ship station with which it is associated or with associated portable ship units
   D. All of the above

ANSWER D: Associated ship stations operating under the authority of the ship station license are restricted to only one watt of output power. The portable station must be identified by the call sign of the ship station with which it is associated, followed by an appropriate unit designator. See questions 1A28 and 1A29 for clarification.

1A89 Which is a radiotelephony calling and distress frequency?
   A. 500 kHz
   B. 2182 kHz
   C. 156.3 MHz
   D. 3113 kHz
ANSWER B: This is an important frequency to remember. The FCC requires all ship stations to maintain a watch on 2182 kHz. See also questions 1A69 and 1A81.

1A90 What is the priority of communications?
   A. Distress, urgency, safety and radio direction-finding
   B. Safety, distress, urgency and radio direction-finding
   C. Distress, safety, radio direction-finding, search and rescue
   D. Radio direction-finding, distress and safety
ANSWER A: Distress calls always take precedence over all other messages. An urgency message warns other stations of dangerous situations. Safety messages warn stations of navigation warnings or meteorological warnings. See also questions 1A17 and 1A64.

1A91 Cargo ships of 300 to 1600 gross tons should be able to transmit a minimum range of:
   A. 75 miles
   B. 150 miles
   C. 200 miles
   D. 300 miles
ANSWER B: Per FCC Rule § 80.855, the minimum range is 150 *nautical* miles. Since cargo ships must travel throughout international waters, they must carry radio equipment capable of reaching across great distances.

1A92 Radiotelephone stations required to keep logs of their transmissions must include:
   A. Station, date and time
   B. Name of operator on duty
   C. Station call signs with which communication took place
   D. All of the above
ANSWER D: And that's not all. Logs must contain the following and the time of their occurrence: a summary of all distress, urgency and safety traffic; the position of the ship at least once a day; the time the watch is discontinued, the reason for it, and the time the watch resumes; the times when storage batteries are placed on charge and taken off charge; results of required equipment tests; and a daily statement of the condition of the required radiotelephone equipment. See § 80.409.

1A93 Each cargo ship of the United States which is equipped with a radiotelephone station for compliance with Part II of Title III of the Communications Act shall, while being navigated outside of a harbor or port, keep a continuous and efficient watch on:
   A. 2182 kHz
   B. 156.8 MHz
   C. Both A and B
   D. Monitor all frequencies within the 2000 kHz to 27500 kHz band used for communications

1A – RADIO LAW AND OPERATING PRACTICE 4

ANSWER C: Per FCC Rule § 80.305, each cargo ship equipped with a radiotelephone station must maintain a continuous watch on 2182 kHz and 156.800 MHz. You will almost certainly have a question about this topic on your examination.

1A94 What call should you transmit on Channel 16 if your ship is sinking?
- A. SOS three times
- B. MAYDAY three times
- C. PAN three times
- D. URGENCY three times

ANSWER B: "SOS" is reserved for telegraphy only; it can be sent in a very short time. "MAYDAY" is easily and universally understood when spoken, so it's used in telephony. See § 80.315.

Message Type	Telephony	Telegraphy
Distress	MAYDAY MAYDAY MAYDAY	SOS SOS SOS
Urgency	PAN PAN PAN PAN PAN PAN	XXX XXX XXX
Safety	SECURITY SECURITY SECURITY	TTT TTT TTT

International Distress, Urgency, and Safety Signals

1A95 Under normal circumstances, what do you do if the transmitter aboard your ship is operating off-frequency, overmodulating or distorting?
- A. Reduce to low power
- B. Stop transmitting
- C. Reduce audio volume level
- D. Make a notation in station operating log

ANSWER B: Operating a transmitter out of its specifications is not only illegal, it is also dangerous. It could interfere with other radio services. The operator must constantly be aware of the radio's behavior and repair any problems as soon as possible. See § 80.90.

1A96 The urgency signal has lower priority than:
- A. Direction-finding
- B. Distress
- C. Safety
- D. Security

ANSWER B: See explanations at questions 1A8, 1A10, 1A17 and 1A64.

1A97 The primary purpose of bridge-to-bridge communications is:
A. Search and rescue emergency calls only
B. All short-range transmission aboard ship
C. Transmission of Captain's orders from the bridge
D. Navigational communications

ANSWER D: Bridge-to-bridge communications are only for navigation. VHF Channel 13 is known as the bridge-to-bridge channel. It is available to all ships. See also question 1A56.

1A98 What is the international VHF digital selective calling channel?
A. 2182 kHz
B. 156.35 MHz
C. 156.525 MHz
D. 500 kHz

ANSWER C: There are several international digital selective calling (DSC) frequencies, but the only one in the VHF band is 156.525 MHz. This is Channel 70. See questions 1A73 and 1A105.

During normal communications, a ship and a coast station talk to each other over digital selective calling on two different frequencies: one to transmit and one to receive. When calling another station or making a distress call, however, the same frequency is used for both transmitting and receiving.

1A99 When your transmission is ended and you expect no response, say:
A. BREAK
B. OVER
C. ROGER
D. CLEAR

ANSWER D: Radio operators say "over and out" only in the movies. "CLEAR" means "I am ceasing transmission." The station operator may continue to listen for other stations, however.

1A100 When attempting to contact other vessels on Channel 16:
A. Limit calling to 30 seconds
B. If no answer is received, wait 2 minutes before calling vessel again
C. Channel 16 is used for emergency calls only
D. Both A and B

ANSWER D: Answer C is incorrect because Channel 16 is used not only for distress calls, but also for a calling frequency. See also question 1A85, and remember that Channel 16 is VHF-FM.

1A101 When a message has been received and will be complied with, say:
A. MAYDAY
B. OVER
C. ROGER
D. WILCO

ANSWER D: This one is easy to remember because "WILCO" is short for "will comply." Remember never to say "MAYDAY" unless you are in a distress situation!

1A – RADIO LAW AND OPERATING PRACTICE

CLEAR	I am ceasing transmission.
WILCO	I have received your message and will comply.
BREAK	Do you acknowledge receipt?
OVER	I am awaiting your response.
ROGER	I understand your entire message.

Common Over-the-Air Protocol

1A102 The FCC may suspend an operator license upon proof that the operator:
A. Has assisted another to obtain a license by fraudulent means
B. Has willfully damaged transmitter equipment
C. Has transmitted obscene language
D. Any of the above

ANSWER D: If the FCC can prove that an operator is involved in any of the above, that person is in serious trouble. Radio communication, particularly at sea, is serious business. Sabotaging radio equipment or helping an unqualified operator obtain a license can bring forth the wrath of the FCC, and the consequences can range from a severe fine to jail time. Good station operators take pride in their work. See § 13.9(e).

1A103 What channel must compulsorily-equipped vessels monitor at all times in the open sea?
A. Channel 8, 156.4 MHz
B. Channel 16, 156.8 MHz
C. Channel 22A, 157.1 MHz
D. Channel 6, 156.3 MHz

ANSWER B: Per FCC Rule § 80.148, Channel 16, 156.800 MHz is an international radiotelephone distress frequency in the VHF range. All vessels using radio equipment are required by law to maintain a watch on this frequency. VHF Channel 16 is also a calling channel.

1A104 When testing is conducted on 2182 kHz or 156.8 MHz, testing should not continue for more than _____ in any 5-minute period.
A. 10 seconds
B. 1 minute
C. 2 minutes
D. None of the above

ANSWER A: Per FCC Rule § 80.101, "Test signals must not exceed ten seconds, and must not be repeated until at least one minute has elapsed. On these distress frequencies, the time between tests must be a minimum of five minutes."

Testing radio equipment in this manner requires transmitting the station's call sign, followed by the word "test." If another station responds with "wait," then you must cease transmitting for at least 30 seconds.

1A105 Which VHF channel is used only for digital selective calling?
A. Channel 70
B. Channel 16
C. Channel 22A
D. Channel 6

ANSWER A: The international VHF digital selective calling frequency is 156.525 MHz, which is Channel 70. With this method of communication, digital codes trigger circuits on specially designed receivers. Ordinarily, once you make contact with another station set up for DSC, both of you can switch to another frequency. DSC keeps the conversation private, since no one else can hear it. See also questions 1A73 and 1A98.

A distress message on DSC is different because it triggers all Class A DSC radios. Ordinary messages trigger only those radios set up to receive messages that include a specific code.

1A106 VHF ship station transmitters must have the capability of reducing carrier power to:
 A. 1 watt
 B. 10 watts
 C. 25 watts
 D. 50 watts

ANSWER A: FCC rules require radio operators to use as little output power as possible to maintain reliable communications. Restricting VHF transmissions to only one watt is common when calling for another ship, because it is usually within eyeshot and one watt is more than powerful enough. If the station does not respond, then try calling with higher power. See § 80.873(c).

1A107 The system of substituting words for corresponding letters is called:
 A. International code system
 B. Phonetic system
 C. Mnemonic system
 D. 10 codes

ANSWER B: The International Phonetic Alphabet allows messages to be transmitted and received through noisy conditions. Do you remember them? Refer to question 1A74.

1A108 How long should station logs be retained when there are no entries relating to distress or disaster situations?
 A. For a period of three years from the date of entry, unless notified by the FCC
 B. Until authorized by the Commission in writing to destroy them
 C. Indefinitely, or until destruction is specifically authorized by the U.S. Coast Guard
 D. For a period of one year from the date of entry

ANSWER D: Logs in which no distress traffic was handled must be retained for at least one year. If distress traffic is handled, however, the logs must be retained for at least three years. See also question 1A45.

1A109 The auto alarm device for generating signals shall be:
 A. Tested monthly, using a dummy load
 B. Tested every three months, using a dummy load
 C. Tested weekly, using a dummy load
 D. None of the above

ANSWER C: It is imperative that a dummy load be used during this test. Refer to question 1A82.

1A – RADIO LAW AND OPERATING PRACTICE

1A110 Licensed radiotelephone operators are not required on board ships for:
 A. Voluntarily equipped ship stations on domestic voyages operating on VHF channels
 B. Ship radar, provided the equipment is non-tunable, pulse-type magnetron and can be operated by means of exclusively-external controls
 C. Installation of a VHF transmitter in a ship station where the work is performed by or under the immediate supervision of the licensee of the ship station
 D. Any of the above

ANSWER D: FCC Rule § 80.177 also states that no license is required to operate a survival craft station or an emergency position indicating radio beacon.

1A111 Under what license are hand-held transceivers covered when used on board a ship at sea?
 A. The ship station license
 B. Under the authority of the licensed operator
 C. Walkie-talkie radios are illegal to use at sea.
 D. No license is needed

ANSWER A: Per FCC Rule § 80.13, "One ship station license will be granted for operation of all maritime services transmitting equipment on board a vessel." Refer to question 1A88.

1A112 What should an operator do to prevent interference?
 A. Turn off transmitter when not in use
 B. Monitor channel before transmitting
 C. Transmissions should be as brief as possible
 D. Both B and C

ANSWER D: A good radio operator would never dream of turning on the power with the transmit button pushed! Always listen before transmitting. See also explanation at question 1A18.

1A113 Identify a ship station's radiotelephone transmissions by:
 A. Country of registration
 B. Call sign
 C. Name of the vessel
 D. Both B and C

ANSWER B: Per FCC Rule § 80.102, stations must give their call signs in English at the beginning and end of each communication with any other station, and at least every 15 minutes in between. See also question 1A22.

1A114 Maritime emergency radios should be tested:
 A. Before each voyage
 B. Weekly while the ship is at sea
 C. Every 24 hours
 D. Both A and B

ANSWER D: Emergency equipment must be inspected and tested, before a voyage and weekly while underway, to make sure it will be operational when you need it. The sea is a hazardous environment for electrical and electronic circuits. See § 80.832(a).

1A115 The URGENCY signal concerning the safety of a ship, aircraft or person shall be sent only on the authority of:
   A. Master of ship
   B. Person responsible for mobile station
   C. Either A or B above
   D. An FCC-licensed operator

ANSWER C: The FCC rules state this clearly in § 80.327 a): "The urgency signal must be sent only on the authority of the master or person responsible for the mobile station." The person sending urgency traffic has a "very urgent message to transmit concerning the safety of a ship, aircraft, or other vehicle, or the safety of a person." It is important that such a message be made by a dependable person. If the master of the ship is busy (as often happens in urgency situations), then the responsibility falls upon the person so designated.

1A116 Survival craft emergency transmitter tests may NOT be made:
   A. For more than 10 seconds
   B. Without using station call sign, followed by the word "test"
   C. Within 5 minutes of a previous test
   D. All of the above

ANSWER D: Follow these rules to the letter. If anyone monitoring the airwaves picks up a signal that does not conform to the rules of testing, they will assume that it is a genuine distress call. Make survival radio equipment tests as quickly as possible, and take pains to restrict their RF waves to within the testing area. Remember to use the word "test." See § 80.101(a)(3).

1A117 International laws and regulations require a silent period on 2182 kHz:
   A. For three minutes immediately after the hour
   B. For three minutes immediately after the half-hour
   C. For the first minute of every quarter-hour
   D. Both A and B above

ANSWER D: See the explanation at question 1A81.

1A118 How should the 2182-kHz auto-alarm be tested?
   A. On a different frequency into antenna
   B. On a different frequency into dummy load
   C. On 2182 kHz into dummy load
   D. On 2182 kHz into antenna

ANSWER B: Never test any equipment on the international distress frequency! Not even when using a dummy load. Clarify by referring to question 1A82.

1A119 Each cargo ship of the United States which is equipped with a radiotelephone station for compliance with the Safety Convention shall, while at sea:
   A. Not transmit on 2182 kHz during emergency conditions
   B. Keep the radiotelephone transmitter operating at full 100% carrier power for maximum reception on 2182 kHz
   C. Reduce peak envelope power on 156.8 MHz during emergencies
   D. Keep continuous watch on 2182 kHz using a watch receiver having a loudspeaker and auto-alarm distress-frequency watch receiver

ANSWER D: Auto alarms allow a watch to be carried out without actually stationing someone to monitor it 24 hours a day. See also questions 1A70 and 1A76.

1A – RADIO LAW AND OPERATING PRACTICE 4

1A120 What is the procedure for testing a 2182-kHz ship radiotelephone transmitter with full carrier power while out at sea?
 A. Reduce to low power, then transmit test tone
 B. Switch transmitter to another frequency before testing
 C. Simply say: "This is (call letters) testing." If all meters indicate normal values, it is assumed transmitter is operating properly.
 D. It is not permitted to test on the air.

ANSWER C: Remember to use the word "test" and your call letters. If your test equipment reveals any erroneous readings, shut down the transmitter immediately and correct the problem. Take as little time with this kind of test as possible. See § 80.101(a).

1A121 If your transmitter is producing spurious harmonics or is operating at a deviation from the technical requirements of the station authorization:
 A. Continue operating until returning to port
 B. Repair problem within 24 hours
 C. Cease transmission
 D. Reduce power immediately

ANSWER C: Any malfunctions of the ship station transmitter must be noted by the station operator, and the transmitter must be taken off the air immediately. Faulty equipment is more dangerous than no equipment at all. See also question 1A95.

1A122 As an alternative to keeping watch on a working frequency in the band 1600-4000 kHz, an operator must tune station receiver to monitor 2182 kHz:
 A. At all times
 B. During distress calls only
 C. During daytime hours of service
 D. During the silence periods each hour

ANSWER A: This is law. The international maritime radiotelephone distress frequency is 2182 kHz. It must be monitored constantly for any possible distress calls. Radiotelegraph stations must maintain a watch on 500 kHz. Ships equipped with VHF radio equipment must listen on 156.800 MHz for any possible distress calls. See § 80.301(b).

1A123 An operator or maintainer must hold a General Radiotelephone Operator License to:
 A. Adjust or repair FCC-licensed transmitters in the aviation, maritime and international fixed public radio services
 B. Operate voluntarily-equipped ship maritime mobile or aircraft transmitters with more than 1,000 watts of peak envelope power
 C. Operate radiotelephone equipment with more than 1,500 watts of peak envelope power on cargo ships over 300 gross tons
 D. All of the above

ANSWER D: Chapter 2 details when a GROL is needed. To earn a General Radiotelephone Operator License, you must pass two written exams which are taken from the question pools for Element 1 and Element 3, both of which are contained in this book.

1A124 What is the radiotelephony calling and distress frequency?
A. 500 kHz
B. 500R122JA
C. 2182 kHz
D. 2182R2647

ANSWER C: Since frequencies are always given in kHz or MHz, answers B and D are obviously wrong. 500 kHz is the radiotelegraphy calling and distress frequency; since this is telephony, answer C is correct. See § 80.313.

1A125 If a ship radio transmitter signal becomes distorted:
A. Cease operations
B. Reduce transmitter power
C. Use minimum modulation
D. Reduce audio amplitude

ANSWER A: Defective radio equipment must be taken off the air immediately. This topic will almost certainly appear on your examination. Faulty transmitters can really mess up the airwaves. See § 80.90.

1A126 Tests of survival craft radio equipment, EXCEPT EPIRBs and two-way radiotelephone equipment, must be conducted:
A. At weekly intervals while the ship is at sea
B. Within 24 hours prior to departure when a test has not been conducted within a week of departure
C. Both A and B above
D. When required by the Commission

ANSWER C: Better to be sure, than to think, you're safe. Confirm that the survival craft transmitters work properly before you leave. During the voyage, make certain that the batteries maintain a healthy charge. They could save your life. See § 80.832(a).

1A127 Each cargo ship of the United States which is equipped with a radiotelephone station for compliance with Part II of Title III of the Communications Act shall, while being navigated outside of a harbor or port, keep a continuous watch on:
A. 2182 kHz
B. 156.8 MHz
C. Both A and B
D. Cargo ships are exempt from radio watch regulations

ANSWER C: Keep radio watches on both frequencies at all times. Question 1A93 is almost identical!

1A128 When may you test a radiotelephone transmitter on the air?
A. Between midnight and 6:00 AM local time
B. Only when authorized by the Commission
C. At any time as necessary to assure proper operation
D. After reducing transmitter power to 1 watt

ANSWER C: Rather than risk operating with broken equipment, it is better to test it on the air and find out for sure. Don't take any more time than is necessary. See § 80.101.

1A – RADIO LAW AND OPERATING PRACTICE 4

1A129 What is the required daytime range of a radiotelephone station aboard a 900-ton, ocean-going cargo vessel?
A. 25 miles
B. 50 miles
C. 150 miles
D. 500 miles
ANSWER C: Heavy cargo ships must be able to communicate across great distances, since they travel throughout the world's oceans. The minimum range is 150 *nautical* miles. See § 80.855(c).

1A130 What do you do if the transmitter aboard your ship is operating off-frequency, overmodulating or distorting?
A. Reduce to low power
B. Stop transmitting
C. Reduce audio volume level
D. Make a notation in the station operating log
ANSWER B: Defective radio equipment is dangerous. Remember to always monitor your transmissions and equipment. Chances are good that you will be tested on this. Refer to questions 1A95, 1A121, and 1A125.

1A131 What is the authorized frequency for an on-board ship repeater for use with a mobile transmitter operating at 467.750 MHz?
A. 457.525 MHz
B. 467.775 MHz
C. 467.800 MHz
D. 467.825 MHz
ANSWER A: A repeater receives on one frequency and transmits on another. The separation between TX and RX is 10.225 MHz. To avoid unnecessary propagation, an on-board ship repeater's antenna must not be more than 10 feet above the deck. In addition, an on-board ship repeater must be able to shut off the transmitter if it receives a constant signal for more than three minutes.

Channel	Mobile	Repeater
1	467.750 MHz	457.525 MHz
2	467.775 MHz	457.550 MHz
3	467.800 MHz	457.575 MHz
4	467.825 MHz	457.600 MHz

1A132 Survival craft EPIRBs are tested:
A. With a manually-activated test switch
B. With a dummy load having the equivalent impedance of the antenna affixed to the EPIRB
C. With radiation reduced to a level not to exceed 25 microvolts per meter
D. All of the above
ANSWER D: An EPIRB contains test switches that are covered with switch guards so they cannot be activated accidentally. When the switch is placed in the test position, the outside antenna is disabled and an equivalent dummy load absorbs the attenuated RF energy from the transmitter. Full power occurs only when in actual use. An indicator light shows that the unit is working properly. When the switch is released, the EPIRB shuts off and the outside antenna is automatically reconnected. See § 80.1053.

1A133 What safety signal call word is spoken three times, followed by the station call letters spoken three times, to announce a storm warning, danger to navigation, or special aid to navigation?
  A. PAN
  B. MAYDAY
  C. SECURITY
  D. SAFETY

ANSWER C: Distress messages, urgency messages and safety messages each have their own special call word. The word for safety messages is SECURITY. Be careful not to choose answer D. When you hear "SECURITY SECURITY SECURITY," you know that a message concerning a storm warning or danger to navigation is about to be sent. See definitions in § 80.5 and § 80.329.

1A134 When should both the call sign and the name of the ship be mentioned during radiotelephone transmissions?
  A. At all times
  B. During an emergency
  C. When transmitting on 2182 kHz
  D. Within 100 miles of any shore

ANSWER B: During ordinary radio communications, the ship station's call sign is enough for identification purposes. During an emergency, however, stating the name of the ship in addition to the call sign lessens confusion. The more information presented to those assisting, the more efficiently they can provide help. See § 80.316.

1A135 How often is the auto alarm tested?
  A. During the 5-minute silent period
  B. Monthly on 121.5 MHz, using a dummy load
  C. Weekly on frequencies other than the 2182-kHz distress frequency, using a dummy antenna
  D. Each day on 2182 kHz, using a dummy antenna

ANSWER C: Never test emergency alarm equipment on international distress frequencies! Ship station equipment must be tested weekly due to the highly corrosive nature of seawater. See also question 1A82, and § 80.855(h).

1A136 One nautical mile is approximately equal to how many statute miles?
  A. 1.61 statute miles
  B. 1.83 statute miles
  C. 1.15 statute miles
  D. 1.47 statute miles

ANSWER C: The nautical mile is recognized throughout the world as the fundamental unit of distance for navigation. It is defined as the length of 1 minute of arc on a great circle drawn on the surface of a sphere that has the same area as the earth. The nautical mile is slightly longer (6,080 feet) than the statute mile (5,280 feet). To convert nautical miles to statute miles, multiply by 1.15.

Miles		Statute mile = 5280 ft.
Nautical	Statute	*Nautical mile = 6080.20 ft.
1.0	1.152	Nautical mile = 1.1515 statue miles
5.0	5.758	
10.0	11.516	*A Nautical mile is equal to $\frac{1}{21600}$ of the
20.0	23.030	great circle of the earth.

Converting Nautical Miles to Statute Miles

1A137 A reserve power source must be able to power all radio equipment, plus an emergency light system, for how long?
   A. 24 hours
   B. 12 hours
   C. 8 hours
   D. 6 hours

ANSWER D: The rules governing a ship station's reserve power supply are quite strict. The reserve power supply is usually made up of batteries. The owner of the station license must be able to prove, upon request of an FCC inspector, that the reserve power supply can last for at least six hours. The reserve power source must be located as close to the backup transmitter and receiver as practicable, and illuminated if it is not located within the radio room. Batteries must remain fully charged every day while the ship is at sea. All of this equipment, plus emergency lighting, must be ready to activate within one minute at all times. Take shortcuts with reserve equipment at your peril. See § 80.808(a)(9).

1A138 Frequencies used for portable communications on-board ship:
   A. 9300-9500 MHz
   B. 1636.5-1644 MHz
   C. 2900-3100 MHz
   D. 457.525-467.825 MHz

ANSWER D: These are UHF frequencies authorized for stations on board ship and may also be used for portable ship stations. On-board repeater stations also use these frequencies. See also question 1A131.

1A139 In the FCC rules, the frequency band from 30 to 300 MHz is also known as:
   A. Very-High Frequency (VHF)
   B. Ultra-High Frequency (UHF)
   C. Medium Frequency (MF)
   D. High Frequency (HF)

ANSWER A: All ships with an FCC station license must carry VHF equipment. Bridge-to-bridge transmissions and the international distress frequency fall within this frequency range.

4 GENERAL RADIOTELEPHONE OPERATOR LICENSE

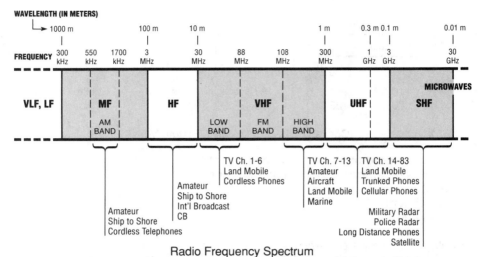

Radio Frequency Spectrum
Source: *Mobile 2-Way Radio Communications,* G. West, ©1993 Master Publishing, Inc.

1A140 What channel must VHF-FM-equipped vessels monitor at all times the station is operated?
 A. Channel 8; 156.4 MHz
 B. Channel 16; 156.8 MHz
 C. Channel 5A; 156.25 MHz
 D. Channel 1A; 156.07 MHz
ANSWER B: Per FCC Rule § 80.148, each VHF ship station during its hours of operation must maintain a watch on 156.800 MHz (Channel 16) whenever such station is not being used for exchanging communications. See also question 1A103.

1A141 When testing is conducted within the 2170-2194-kHz and 156.75-156.85-MHz bands, transmissions should not continue for more than _____ in any 5-minute period.
 A. 10 seconds
 B. 1 minute
 C. 5 minutes
 D. No limitation
ANSWER A: Conducting service tests too often on the distress frequency could prevent another station from being heard clearly. A wide interval between transmissions allows others to use the frequency. See § 80.101(a)(3).

1A142 What emergency radio testing is required for cargo ships?
 A. Tests must be conducted weekly while ship is at sea.
 B. Full-power carrier tests into dummy load
 C. Specific-gravity check in lead-acid batteries, or voltage under load for dry-cell batteries
 D. All of the above
ANSWER D: Take care of your radio and it will take care of you. Be especially careful about the batteries. Check all parameters constantly. Check for broken cables, broken antennas, loose connectors, etc. See § 80.855 (h) and § 80.811(a)(1).

1A – RADIO LAW AND OPERATING PRACTICE

1A143 The master or owner of a vessel must apply how many days in advance for an FCC ship inspection?
- A. 60 days
- B. 30 days
- C. 3 days
- D. 24 hours

ANSWER C: Three days is minimum; however, waiting until the last minute may result in some stiff charges for the ship's master or owner if FCC engineers have to work overtime to check your radio equipment. See § 80.59(d).

1A144 Marine transmitters should be modulated between:
- A. 75% - 100%
- B. 70% - 105%
- C. 85% - 100%
- D. 75% - 120%

ANSWER A: There are several different ways to modulate a radio signal, but all face the same percentage of modulation. Station radio equipment must include a modulation limiter to prevent modulation over 100% to prevent distortion. Always monitor your signal. See § 80.213(a).

1A145 What is a good practice when speaking into a microphone in a noisy location?
- A. Overmodulation
- B. Change phase in audio circuits
- C. Increase monitor audio gain
- D. Shield microphone with hands

ANSWER D: If noise enters a transmitter, it must be filtered out electronically, and no filter is perfect. It is best to prevent extraneous noise from entering the transmitter in the first place, and the best way to do so is to shield the microphone. Shielding the microphone with your hands narrows the pickup area to your voice only.

1A146 When pausing briefly for station copying message to acknowledge, say:
- A. BREAK
- B. OVER
- C. WILCO
- D. STOP

ANSWER A: Due to the unpredictability of radio communication, it is not unusual for a station operator to miss part of a message during copying. This is especially true when a proper name is spoken, which is why spelling words with the phonetic alphabet is required. Using "BREAK" periodically gives the other station a chance to either confirm reception or ask for a repeat.

1A147 Overmodulation is often caused by:
- A. Turning down audio gain control
- B. Station frequency drift
- C. Weather conditions
- D. Shouting into microphone

ANSWER D: Overmodulation occurs when the amplitude of an incoming audio signal extends beyond a given level. It is "clipped," which reduces signal quality. Shouting or talking too closely into the microphone can cause overmodulation.

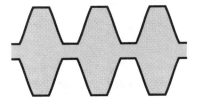

Overmodulation of audio in transmitter causes distortion of modulation envelope.

Oscilloscope Waveform Showing "Flattopping"

1A148 To indicate a response is expected, say:
A. WILCO
B. ROGER
C. OVER
D. BREAK

ANSWER C: "OVER" informs the other station operator that it is now his or her turn to speak. This is a universally understood phrase and eliminates confusion, particularly during heavy interference.

1A149 When all of a transmission has been received, say:
A. ATTENTION
B. ROGER
C. RECEIVED
D. WILCO

ANSWER B: Note that "ROGER" is spoken only when *ALL* of a transmission is understood. If you hear only part of a message, ask for a repeat.

1A150 What information must be included in a *DISTRESS* message?
A. Name of vessel
B. Location
C. Type of distress and specifics of help requested
D. All of the above

ANSWER D: People can become so scared during an emergency that they may panic and forget to include even the most basic information needed to receive assistance. On land, a 911 operator may be able to learn the location from which a phone call is received, but no such luxury exists at sea. The more information you provide over the air, the better your chances of receiving help. See § 80.316.

1A151 The maritime MF radiotelephone silence periods begin at _____ and _____ minutes past the UTC hour.
A. :15, :45
B. :00, :30
C. :20, :40
D. :05, :35

ANSWER B: Don't transmit on the distress frequency during silent periods unless you are declaring an emergency. See also question 1A81, and § 80.304(b).

1A – RADIO LAW AND OPERATING PRACTICE 4

1A152 A marine public coast station operator may not charge a fee for what type of communication?
A. Port Authority transmissions
B. Storm updates
C. Distress
D. All of the above

ANSWER D: You may call for help at any time, anywhere in the world, when you genuinely need assistance, without charge. See also question 1A21.

1A153 Which of the following represent the first three letters of the phonetic alphabet?
A. Alpha Bravo Charlie
B. Adam Baker Charlie
C. Alpha Baker Crystal
D. Adam Brown Chuck

ANSWER A: If you cannot remember the phonetic alphabet, refer to question 1A74 and keep practicing.

1A154 Two-way communications with both stations operating on the same frequency is:
A. Radiotelephone
B. Duplex
C. Simplex
D. Multiplex

ANSWER C: A simplex system allows both stations to use the same frequency for transmitting and receiving, but only one station can transmit at a time. A duplex system allows both stations to transmit and receive simultaneously, but not on the same frequency.

1A155 When a ship is sold:
A. New owner must apply for a new license
B. FCC inspection of equipment is required
C. Old license is valid until it expires
D. Continue to operate; license automatically transfers with ownership

ANSWER A: FCC Rule § 80.56 states this plainly: "Whenever the vessel ownership is transferred, the previous authorization must be forwarded to the Commission for cancellation. The new owner must file for a new authorization." In other words, when you purchase a boat, the radio license does not come with it automatically.

What happens if the boat is sold for scrap, to a collector, or for some other reason is not meant to be used any more? Then the radio equipment aboard permanently discontinues operation, and the new owner of the boat must return the station license to the Commission for cancellation. This is covered in FCC Rule § 80.31.

1A156 What is the second in order of priority?
A. URGENT
B. DISTRESS
C. SAFETY
D. MAYDAY

ANSWER A: An urgency message is second in importance only to distress traffic. An urgent situation could evolve into a distress situation unless assistance is given. See § 80.327(d).

1A157 Portable ship units, hand-helds or walkie-talkies used as an associated ship unit:
A. Must operate with 1 watt and be able to transmit on Channel 16
B. May communicate only with the mother ship and other portable units and small boats belonging to mother ship
C. Must not transmit from shore or to other vessels
D. All of the above

ANSWER D: Portable ship stations are meant to be used for low-power communications, using the same ship station license. See also questions 1A88, 1A111 and § 80.115.

1A158 The HF (High-Frequency) band is:
A. 3 - 30 MHz
B. 3 - 30 GHz
C. 30 - 300 MHz
D. 300 - 3000 MHz

ANSWER A: The High-Frequency band is much lower in the RF spectrum than VHF or UHF. HF signals propagate much farther than VHF signals. See the figure at question 1A139.

1A159 Omega operates in what frequency band?
A. Below 3 kHz
B. 3 - 30 kHz
C. 30 - 300 kHz
D. 300 - 3000 kHz

ANSWER B: Omega is a form of radio navigation using very low frequencies. It currently resides in the 10-14 kHz band. What makes Omega special is the way in which radio navigation is determined. Rather than measure the differences in time between signals, Omega measures the difference in phase between signals.

Fewer than 10 Omega transmitting stations are required for the entire world. While all Omega stations use the same frequency, only one is on at any time — they are turned on and off in sequence, approximately one second at a time. All Omega transmitters send out the same signal and they are phase-synchronized with atomic clocks. The Omega receiver on board a ship measures the difference in phase between one received signal and another. From these comparisons, a precise location of the ship can be extrapolated to within 10 nautical miles.

1A160 Shipboard transmitters using F3E emission (FM voice) may not exceed what carrier power?
A. 500 watts
B. 250 watts
C. 100 watts
D. 25 watts

ANSWER D: FM radio equipment aboard a vessel cannot exceed 25 watts carrier power. It also must be able to reduce output power to one watt (refer to question 1A106). Such a wide range ensures that ship-to-ship communications will not interfere with other stations, and that a distress message has a good chance of being received elsewhere. See § 80.215(g).

1A161 Loran C operates in what frequency band?
A. VHF; 30 - 300 MHz
B. HF; 3 - 30 MHz
C. MF; 300 - 3000 kHz
D. LF; 30 - 300 kHz

ANSWER D: Loran was invented before World War II, and has been undergoing refinement ever since. Loran-C has been around for over 20 years. It is being used by the Federal Aviation Administration, the U.S. Coast Guard, and the marine community for precise navigation assistance. It uses a low frequency band, 90 kHz to 110 kHz.

Loran-C consists of transmitters arranged in chains. At least three transmitters make up a chain. All transmitters emit bursts of digital information, with each station sending certain unique bits of data at different intervals. A Loran-C receiver detects all these bits at different times, depending on the receiver's location. Loran-C extrapolates the receiver's location this way with atomic-clock timing, allowing one-quarter mile accuracy.

Loran-C chains extend around the world, including over 20 million square miles around the United States and 600,000 users.

1A162 What has most priority:
A. URGENT
B. DISTRESS
C. SAFETY
D. SECURITY

ANSWER B: Distress traffic always has the ultimate priority. All other traffic must stand by while distress traffic is acknowledged. See § 80.312.

1A163 When and how may Class A and B EPIRBs be tested?
A. Within the first 5 minutes of the hour; tests not to exceed 3 audible sweeps or one second, whichever is longer
B. Within first 3 minutes of hour; tests not to exceed 30 seconds
C. Within first 1 minute of hour; test not to exceed 1 minute
D. At any time ship is at sea

ANSWER A: If at all possible, testing any EPIRB must be coordinated with the Coast Guard to prepare them for receiving such a signal. Otherwise, brief operational tests must be conducted very quickly and within certain time windows. If tests of these devices are to be expected within specified time frames, then the "blips" these tests make over the air are known by other stations to be tests. Testing an EPIRB at any other time, even for only an instant, could trigger a false rescue mission. See also explanation at question 1A53.

4 GENERAL RADIOTELEPHONE OPERATOR LICENSE

1A164 When is the Silent Period on 2182 kHz, when only emergency communications may occur?
  A. One minute at the beginning of every hour and half hour
  B. At all times
  C. No designated period; silence is maintained only when a distress call is received
  D. Three minutes at the beginning of every hour and half hour

ANSWER D: 2182 kHz is the radiotelephone distress frequency, so the silent periods are at the top of the hour and at every half-hour. See also questions 1A81, 1A151, and § 80.304(b).

1A165 What is the frequency range of UHF?
  A. 0.3 to 3 GHz
  B. 0.3 to 3 MHz
  C. 3 to 30 kHz
  D. 30 to 300 MHz

ANSWER A: UHF means Ultra-High Frequency. GHz means gigahertz; 1 GHz equals 1,000 MHz. UHF ranges from 300 MHz to 3,000 MHz; therefore, 0.3 to 3 GHz. UHF frequencies permit extremely high rates of data transmission, but the range on earth is limited, usually line-of-sight. However, these frequencies are often used in satellite communications.

1A166 A room temperature of +30.0 degrees Celsius is equivalent to how many degrees Fahrenheit?
  A. 104
  B. 83
  C. 95
  D. 86

ANSWER D: The Celsius temperature scale is much more widely accepted throughout the world than Fahrenheit. The mathematical formulas for converting between the two scales are:

$$°F = 9/5 °C + 32$$
$$°C = 5/9 (°F - 32)$$

The figure on page 77 shows the derivation of the formulas and a conversion chart for selected values.

1A167 Atmospheric noise or static is not a great problem:
  A. At frequencies below 20 MHz
  B. At frequencies below 5 MHz
  C. At frequencies above 1 MHz
  D. At frequencies above 30 MHz

ANSWER D: Most atmospheric noise and static takes place at rather low frequencies in the RF spectrum. Lightning, while rather broad-band, centers near 7 MHz, but its harmonics disrupt many other frequencies. The earth itself has its own geographic resonant frequency of only 7.6 Hz! Geomagnetic storms, sponsored by the sun, can wipe out radio communications all over the world in this low band.

1A – RADIO LAW AND OPERATING PRACTICE **4**

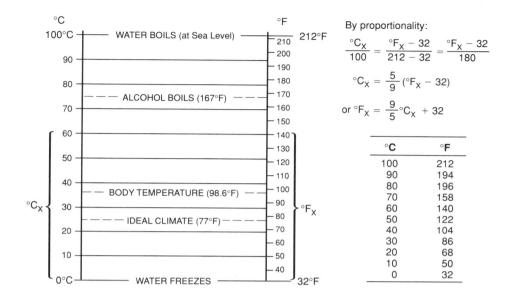

Relationship of Celsius (°C) to Fahrenheit (°F)

1A168 Frequencies which have substantially straight-line propagation characteristics similar to that of light waves are:
A. Frequencies below 500 kHz
B. Frequencies between 500 kHz and 1,000 kHz
C. Frequencies between 1,000 kHz and 3,000 kHz
D. Frequencies above 50,000 kHz

ANSWER D: Above 50,000 kHz, radio waves behave very much like light waves and do not propagate much farther than the visual horizon. Television transmitters operate in this range. They use extremely tall masts and towers for their transmitting antennas to extend their line of sight. Even so, most VHF television signals fade out after about 50 miles. Conversely, extremely low-frequency (below 500 kHz) radio waves can travel around the earth unimpeded several times, no matter what the height of the transmitting antenna.

1A169 In the International Phonetic Alphabet, the letters E, M, and S are represented by the words:
A. Echo, Michigan, Sonar
B. Equator, Mike, Sonar
C. Echo, Mike, Sierra
D. Element, Mister, Scooter

ANSWER C: Another phonetic alphabet question! If you have forgotten the International Phonetic Alphabet, refer to question 1A74. You will almost certainly receive one of these questions on your examination.

1A170 What is the international radiotelephone distress call?
- A. "SOS, SOS, SOS; THIS IS" followed by the call sign of the station (repeated 3 times)
- B. "MAYDAY, MAYDAY, MAYDAY; THIS IS" followed by the call sign (or name, if no call sign assigned) of the mobile station in distress, spoken three times
- C. For radiotelephone use, any words or message which will attract attention may be used
- D. The alternating two-tone signal produced by the radiotelephone alarm signal generator

ANSWER B: The key words are "radiotelephone" and "distress". "SOS" is meant to be sent only in radiotelegraphy distress traffic. "MAYDAY" is meant to be spoken in radiotelephony distress traffic. See § 80.315.

Question Pool – Element 3

The questions asked in each Element 3 examination must be taken from the following pool. The release date of this pool is *September 2, 1993, updated March 3, 1994*. This pool will remain in effect until superseded by a release of an updated pool from the FCC.

An Element 3 examination is used to prove that the examinee possesses the operational and technical qualifications necessary to properly perform the duties required of a person holding a General Radiotelephone Operator License (GROL). To pass, an examinee must answer correctly at least 57 out of 76 questions from this pool. Element 3 is also a partial requirement for the Global Maritime Distress and Safety System Maintainer's License.

A GROL must be held by any person who adjusts, maintains, or internally repairs a radiotelephone transmitter at a station licensed by the FCC in the aviation, maritime, or international fixed public radio services. A GROL must also be held by the operator of (1) certain aviation and maritime land radio stations; (2) compulsory equipped ship radiotelephone stations transmitting with more than 1,500 watts of peak envelope power output; and (3) voluntarily equipped ship stations transmitting with more than 1,000 watts peak envelope power.

Each Element 3 examination is administered by a Commercial Operator License Examination Manager (COLEM). The COLEM must construct the examination by selecting 76 questions from this 720-question pool using the following algorithm:

Subelement 3A — Operating Procedures (3 questions)
Subelement 3B — Radio Wave Propagation (3 questions)
Subelement 3C — Radio Practice (5 questions)
Subelement 3D — Electrical Principles (16 questions)
Subelement 3E — Circuit Components (13 questions)
Subelement 3F — Practical Circuits (22 questions)
Subelement 3G — Signals and Emissions (9 questions)
Subelement 3H — Antennas and Feed Lines (5 questions)

The COLEM may change the order of the answer and distractors (incorrect choices) of questions from the order that appears in this pool.

Suggestions concerning improvement to the questions in the pool or submittal of new questions for consideration in updated pools may be sent to the Federal Communications Commission, Aviation and Marine Branch, Room 5322, 2025 M Street N.W., Washington DC 20554.

5 GENERAL RADIOTELEPHONE OPERATOR LICENSE

Subelement 3A – Operating Procedures (3 questions)

3A1 What is facsimile?
- A. The transmission of characters by radioteletype that form a picture when printed
- B. The transmission of still pictures by slow-scan television
- C. The transmission of video by television
- D. The transmission of printed pictures for permanent display on paper

ANSWER D: In the marine and aviation radio services, the most common use of radio facsimile is for the printed pictures of weather charts. This is called "WEFAX", and is the term used to refer to the broadcast of facsimile images containing weather charts and satellite imagery. In the United States, these broadcasts are typically prepared by the National Weather Service (NWS), a branch of the National Oceanographic and Atmospheric Administration (NOAA).

Facsimile System
Courtesy of ALDEN ELECTRONICS

3A2 What is the modern standard scan rate for a facsimile picture transmitted by a radio station?
- A. The modern standard is 240 lines per minute.
- B. The modern standard is 50 lines per minute.
- C. The modern standard is 150 lines per second.
- D. The modern standard is 60 lines per second.

ANSWER A: 240 lpm requires 5 kHz of channel bandwidth, and this is what you would tune into when receiving marine and aviation weather products from GOES and some Russian APT weather satellites on VHF and UHF frequencies. Since 240 lpm requires 5-kHz channel bandwidth, it would not be legal for use in the marine bands on high frequency. High-frequency weather facsimile is at 120 lpm, a slower rate that stays within worldwide bandwidth limits.

3A3 What is the approximate transmission time for a facsimile picture transmitted by a radio station?
- A. Approximately 6 minutes per frame at 240 lines per minute
- B. Approximately 3.3 minutes per frame at 240 lines per minute
- C. Approximately 6 seconds per frame at 240 lines per minute
- D. 1/60 second per frame at 240 lines per minute

ANSWER B: When receiving geostationary weather facsimile signals on VHF and UHF frequencies at 240 lpm, it will take approximately 3.3 minutes per frame.

3A4 What is the term for the transmission of printed pictures by radio for the purpose of a permanent display?
A. Television
B. Facsimile
C. Xerography
D. ACSSB

ANSWER B: Printed pictures by facsimile may be permanently displayed on color or monochrome computer systems just as easily as they might be printed out on paper. In many cases, weather imagery displayed on a computer VGA monitor offers much more resolution and definition than what might come out of a typical dot matrix printer.

3A5 In facsimile, how are variations in picture brightness and darkness converted into voltage variations?
A. With an LED
B. With a Hall-effect transistor
C. With a photodetector
D. With an optoisolator

ANSWER C: The photodetector converts variations in picture brightness and darkness into voltage variations. This is one of the extremely sensitive components of weather satellites. It must never be aimed at the sun; to do so could cause permanent damage to the photodetector.

3A6 What is an *ascending pass* for a satellite?
A. A pass from west to east
B. A pass from east to west
C. A pass from south to north
D. A pass from north to south

ANSWER C: Ascending... going up! From south to north.

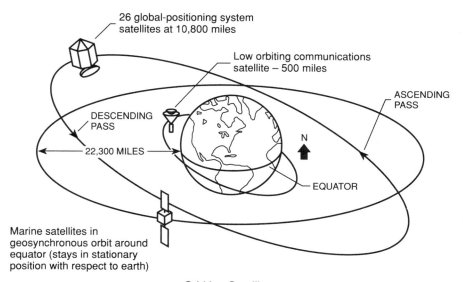

Orbiting Satellites

3A7 What is a *descending pass* for a satellite?
   A. A pass from north to south
   B. A pass from west to east
   C. A pass from east to west
   D. A pass from south to north
ANSWER A: Descending. . . going down! From north to south.

3A8 What is the *period* of a satellite?
   A. An orbital arc that extends from 60 degrees west longitude to 145 degrees west longitude
   B. The point on an orbit where satellite height is minimum
   C. The amount of time it takes for a satellite to complete one orbit
   D. The time it takes a satellite to travel from perigee to apogee
ANSWER C: Polar orbiting weather satellites complete two orbits, or two periods, every 24 hours. Four U.S. and two Russian polar orbiting satellites operate on the 137-MHz band and give us visible light imagery for the daytime pass, and infrared imagery for the nighttime pass. At 137 MHz, an inexpensive computer program in a laptop PC, a wide-band scanner, and a cross dipole is all that's necessary to easily pull in weather facsimile imagery when any one of these satellites is within view on its ascending or descending orbit.

3A9 What is a *linear transponder*?
   A. A repeater that passes only linear or binary signals
   B. A device that receives and retransmits signals of any mode in a certain passband
   C. An amplifier for SSB transmissions
   D. A device used to change a FM emission to an AM emission
ANSWER B: Ham radio operators have developed their own orbiting satellites carrying amateur radio (OSCARs), and their linear transponders will pass any mode when received by the satellite. FM is discouraged because this wide bandwidth occupies too much bandwidth and consumes too much satellite battery power. However, there *is some limited use* of FM over geostationary satellites by authorized land and marine radio users.

3A10 What are the two basic types of *linear transponders* used in satellites?
   A. Inverting and non-inverting
   B. Geostationary and elliptical
   C. Phase 2 and Phase 3
   D. Amplitude modulated and frequency modulated
ANSWER A: Some satellite linear transponders invert the uplink signal, others do not. The inverting transponder takes the upper sideband of the signal at the low end of the band and inverts it to a lower sideband at the top of the band. Non-inverting linear transponders will follow the input without inverting.

3A11 Why does the downlink frequency appear to vary by several kHz during a low-earth-orbit satellite pass?
   A. The distance between the satellite and ground station is changing, causing the Kepler effect
   B. The distance between the satellite and ground station is changing, causing the Bernoulli effect

C. The distance between the satellite and ground station is changing, causing the Boyles' law effect
D. The distance between the satellite and ground station is changing, causing the Doppler effect

ANSWER D: As a satellite is speeding toward you, the Doppler effect will make its frequencies several hundred hertz higher. As the satellite passes overhead and speeds away from you, the signals will begin to decrease in frequency.

3A12 Why does the received signal from a satellite stabilized by a computer-pulsed electromagnet exhibit a fairly rapid pulsed fading effect?
 A. Because the satellite is rotating
 B. Because of ionospheric absorption
 C. Because of the satellite's low orbital altitude
 D. Because of the Doppler effect

ANSWER A: Some satellites are continuously rotated to keep all sides of the "bird" equally bathed in sunlight. This distributes the heat buildup on the solar panels, and also allows the panels to receive energy from the sun throughout the rotation. A circularly-polarized antenna will help minimize the pulsed fading.

3A13 What type of antenna can be used to minimize the effects of *spin modulation* and *Faraday rotation*?
 A. A nonpolarized antenna
 B. A circularly-polarized antenna
 C. An isotropic antenna
 D. A log-periodic dipole array

ANSWER B: Many of you preparing for your commercial examination are already licensed amateur radio operators. If you have worked over the amateur satellites, you have encountered the problems of ham satellites spinning in space, as well as the problems of reflected signals off the surface of the moon, leading to Faraday rotation fluctuation of signal strengths. A circularly-polarized antenna will minimize this effect, and either right-hand or left-hand circular polarization may be required.

3A14 What is *blanking* in a video signal?
 A. Synchronization of the horizontal and vertical sync-pulses
 B. Turning off the scanning beam while it is traveling from right to left and from bottom to top
 C. Turning off the scanning beam at the conclusion of a transmission
 D. Transmitting a black and white test pattern

ANSWER B: The scanning beam is turned off when it completes its travel from left to right, and is returning from right to left to start another line. The scanning beam is also blanked when the beam reaches the bottom and is returning to the top to begin another scan from top to bottom.

3A15 What is the standard video voltage level between the sync tip and the whitest white at TV camera outputs and modulator inputs?
 A. 1 volt peak-to-peak
 B. 120 IEEE units
 C. 12 volts DC
 D. 5 volts RMS

ANSWER A: One volt peak-to-peak is the standard video voltage between sync tip and white.

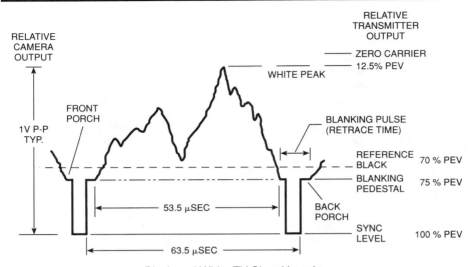

Black and White TV Signal Levels
Courtesy of *ARRL Handbook,* ©1987, The American Radio Relay League, Inc.

3A16 What is the standard video level, in percent PEV, for black?
 A. 0%
 B. 12.5%
 C. 70%
 D. 100%
ANSWER C: The standard video level voltage for black is 70 percent of the peak envelope voltage (PEV).

3A17 What is the standard video level, in percent PEV, for white?
 A. 0%
 B. 12.5%
 C. 70%
 D. 100%
ANSWER B: The standard video level voltage for white is 12.5 percent of the peak envelope voltage (PEV).

3A18 What is the standard video level, in percent PEV, for blanking?
 A. 0%
 B. 12.5%
 C. 75%
 D. 100%
ANSWER C: The standard video level voltage for blanking is very close to black, 75 percent of the peak envelope voltage (PEV).

3A19 A 25-MHz amplitude modulated transmitter's actual carrier frequency is 25.00025 MHz without modulation and is 24.99950 MHz when modulated. What statement is true?
 A. If the allowed frequency tolerance is 0.001%, this is an illegal transmission.
 B. If the allowed frequency tolerance is 0.002%, this is an illegal transmission.

3A – OPERATING PROCEDURES

C. Modulation should not change carrier frequency.
D. If the authorized frequency tolerance is 0.005% for the 25 MHz band this transmitter is operating legally.

ANSWER D: This question deals with an older AM transmitter assigned a carrier frequency of 25.000 MHz. Its actual carrier frequency is measured 250 Hz high. To calculate whether or not this transmitter is within the legal limits, take answer A, B, and D percentages of frequency tolerance, move the decimal point two places to the left for percentage, then six places to the right to calculate parts per million (ppm), and then multiply parts per million times the frequency in MHz. For answers:

A. 10 ppm × 25 MHz = 250 Hz — It's barely legal for 0.001%.
B. 20 ppm × 25 MHz = 500 Hz — well within for 0.002% permissible excursion
D. 50 ppm × 25 MHz = 1250 Hz — legal excursion, "This transmitter is operating legally" for the 0.005% tolerance allowed. Correct answer.

3A20 If you are listening to an FM radio station at 100.6 MHz on a car radio when an airplane in the vicinity is transmitting at 121.2 MHz and your car radio receives interference, the possible problem could be:
 A. Improper shielding in receive
 B. Poor "Q" receiver
 C. Image frequency
 D. Intermodulation or coupling

ANSWER C: Most FM receivers use an IF (intermediate frequency) tuned between 10.3 MHz and 10.7 MHz. Image interference is common at twice the IF frequency. The difference between 121.2 MHz and the desired frequency of 100.6 MHz is 20.6 MHz, and this is exactly two times an IF of 10.3 MHz. Because these numbers fall within the IF pass band, it can be safely assumed that the problem is an image frequency. However, if the interference saturated the entire FM broadcast band, chances are the other three answers would be correct.

3A21 One nautical mile is equal to how many statute miles?
 A. 1.5
 B. 8.3
 C. 1.73
 D. 1.15

ANSWER D: One nautical mile equals 1.15 statute miles. See Element 1, question 1A136.

3A22 1.73 nautical miles equals how many statute miles?
 A. 2
 B. 1.5
 C. 1.73
 D. 1

ANSWER A: Multiply 1.73 × 1.15 = 1.99, round off to 2 for answer A.

3A23 Solder is:
 A. 50% lead 50% tin
 B. 40% lead 60% tin
 C. 60% lead 40% tin
 D. 70% lead 30% tin

ANSWER B: You can remember the right percentages by remembering how solder *FLOWS*—F = four, L = Six. The percentages for solder are normally stated tin first, then lead; e.g., 60/40 for this solder.

3A24 What ferrite device can be used instead of a duplexer to isolate a microwave transmitter and receiver when both are connected to the same antenna?
A. Isolator
B. Circulator
C. Magnetron
D. Simplex

ANSWER B: A microwave circulator may be used to couple a transmitter and receiver to a common antenna through a phase shift in the incoming or outgoing signal. When the two ports are in phase, the signal transfers. The port out of phase will look like a reflective load, and will not accept the signal. An internal permanent magnet surrounds the ferrite device to allow only one out of many ports to accept the incoming and outgoing signal.

3A25 The ILS localizer measures what deviation of an aircraft?
A. Horizontal
B. Vertical
C. Ground speed
D. Distance between aircraft

ANSWER A: The instrument landing system (ILS) localizer measures horizontal deviation from the landing approach centerline; it is the glide slope that measures vertical deviation. Remember anything that is "local" is next to — side-to-side. This will help you memorize the correct answer.

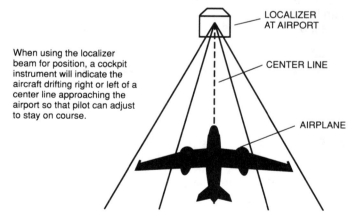

Flight Control with Localizer

3A26 3:00 PM Central Standard Time is:
A. 1000 UTC
B. 2100 UTC
C. 1800 UTC
D. 0300 UTC

ANSWER B: UTC stands for universal time coordinated, formerly called "GMT", and sometimes abbreviated as Zulu time, Z for the zero meridian.

From table below we see that:
 Eastern Standard Time plus 5 = UTC.
 Central Standard Time plus 6 = UTC.
 Mountain Standard Time plus 7 = UTC.
 Pacific Standard Time plus 8 = UTC.
In this question, 3:00 PM is 1500 hours in 24-hour time, and 1500 plus 600 (6 hours) gives 2100 hours UTC.

UTC	AST/EDT	EST/CDT	CST/MDT	MST/PDT	PST
0000	2000	1900	1800	1700	1600
0100	2100	2000	1900	1800	1700
0200	2200	2100	2000	1900	1800
0300	2300	2200	2100	2000	1900
0400	0000	2300	2200	2100	2000
0500	0100	0000	2300	2200	2100
0600	0200	0100	0000	2300	2200
0700	0300	0200	0100	0000	2300
0800	0400	0300	0200	0100	0000
0900	0500	0400	0300	0200	0100
1000	0600	0500	0400	0300	0200
1100	0700	0600	0500	0400	0300
1200	0800	0700	0600	0500	0400
1300	0900	0800	0700	0600	0500
1400	1000	0900	0800	0700	0600
1500	1100	1000	0900	0800	0700
1600	1200	1100	1000	0900	0800
1700	1300	1200	1100	1000	0900
1800	1400	1300	1200	1100	1000
1900	1500	1400	1300	1200	1100
2000	1600	1500	1400	1300	1200
2100	1700	1600	1500	1400	1300
2200	1800	1700	1600	1500	1400
2300	1900	1800	1700	1600	1500

Remember that the day changes at 0000 UTC. [6:00 p.m. Pacific Standard Time on Monday evening is actually 2:00 a.m. UTC the following day, Tuesday!]
To convert local time to UTC: For AST/EDT, add 4 hours. For EST/CDT, add 5 hours. For CST/MDT, add 6 hours. For MST/PDT, add 7 hours. For PST, add 8 hours.

Coordinated Universal Time (UTC) to Local Time Conversion Chart
Source: *Amateur Radio Logbook,* ©1993, Master Publishing, Inc.

3A27 10 statute miles per hour equals how many knots?
 A. 11.5
 B. 8.7
 C. 5
 D. 3

ANSWER B: Knots is short for "nautical miles per hour." One statute mile equals approximately 0.87 of a nautical mile. 10 × 0.87 = 8.7, answer B.

3A28 Which of the following is an acceptable method of solder removal from holes in a printed board?
 A. Compressed air
 B. Toothpick
 C. Soldering iron and a suction device
 D. Power drill

ANSWER C: Here at the workbench, this handy tool is affectionately called a "solder sucker."

3A29 100 statute miles equals how many nautical miles?
A. 87
B. 108
C. 173
D. 13

ANSWER A: Think of the "ST" in statute similar to the number "87". 100 × 0.87 = 87. Since 1 nm = 1.15 sm, then 1 sm = 1 ÷ 1.15 = 0.87 nm.

3A30 6:00 PM PST is equal to what time in UTC?
A. 0200
B. 1800
C. 2300
D. 1300

ANSWER A: Since PST is plus 8 and 6:00 PM is 1800 hours, 800 + 1800 = 2600, and 2600 − 2400 = 0200 UTC. See question 3A26.

3A31 An Auto Alarm signal consists of two sine wave audio tones transmitted alternately at what frequencies?
A. 121.5 and 243 MHz
B. 500 and 1000 kHz
C. 1300 and 220 kHz
D. None of the above

ANSWER D: None of the answers is correct because the auto alarm signal is 2200 and 1300 Hz (2.2 kHz and 1.3 kHz). Be careful—answer C looks correct at first glance.

3A32 What is *NOT* a good soldering practice in electronic circuits?
A. Use adequate heat
B. Clean parts sufficiently
C. Prevent corrosion by never using flux
D. Be certain parts do not move while solder is cooling

ANSWER C: Be careful—this question asks what is *NOT* a good practice. The use of flux with solder will help make a good solder joint and prevent corrosion. However, be sure to use electronic grade solder with rosin flux inside it. Do not use acid flux like plumbers use.

3A33 Waveguides are not utilized at VHF or UHF frequencies because:
A. Characteristic impedance
B. Resistance to high frequency waves
C. Large dimensions of waveguide are not practical
D. Grounding problems

ANSWER C: The lower the frequency, the longer the wavelength, and the larger the dimensions of a waveguide. Large waveguides are not practical and would be expensive.

3A34 2300 UTC time is:
A. 2 PM CST
B. 3 PM PST
C. 10 AM EST
D. 6 AM EST

ANSWER B: Refer to question 3A26 for conversion factors. All the answers using the conversion factors are calculated. B is correct answer.

A. 2 PM CST = 1400 + 600 = 2000 UTC
B. 3 PM PST = 1500 + 800 = 2300 UTC
C. 10 AM EST = 1000 + 500 = 1500 UTC
D. 6 AM EST = 0600 + 500 = 1100 UTC

3A35 One statute mile equals how many nautical miles?
A. 3.8
B. 1.5
C. 0.87
D. 0.7

ANSWER C: Did you spot that "ST" as a "87"? One statute mile is less than one nautical mile — 0.87 nautical mile.

3A36 When soldering electronic circuits be sure to:
A. Use sufficient heat
B. Use maximum heat
C. Heat wires until sweating begins
D. Use minimum solder

ANSWER C: Heat the wires only until "sweating" begins; you will see the solder wet onto the leads or printed circuit board traces. It flows and looks shiny.

3A37 2.3 statute miles equals how many nautical miles?
A. 2
B. 1.5
C. 1.73
D. 1

ANSWER A: Remember the "ST" means 0.87, and 0.87 × 2.3 = 2.

3A38 What is the purpose of flux?
A. Removes oxides from surfaces to be joined
B. Prevents oxidation during soldering
C. Acid cleans printed circuit connections
D. Both a and b

ANSWER D: Solder flux will clean the surfaces to be joined and prevent oxides from forming during soldering. Using acid flux will cause corrosion.

3A39 Which of these will be useful for insulation at UHF?
A. Rubber
B. Mica
C. Wax impregnated paper
D. Lead

ANSWER B: At UHF frequencies, Teflon® and mica are good insulators. Lead would be considered a conductor, and rubber is not appropriate as an insulator at UHF or microwave frequencies. Wax paper was used as insulation in capacitors back in the old days, but not at UHF frequencies!

3A40 The FCC originally omitted this question. 3A41 has been changed to 3A40.

3A40 The condition of a storage battery is determined with a(n):
A. Hygrometer
B. Manometer
C. FET
D. Hydrometer

ANSWER D: Watch out! There's only one letter difference in the answers for A and D. We use a hy**D**rometer to test the specific gravity of the electrolyte in a storage battery to determine the charge level. A hy**G**rometer is used to measure relative humidity of the atmosphere.

Subelement 3B – Radio Wave Propagation   (3 questions)

3B1 What is a *selective fading* effect?
A. A fading effect caused by small changes in beam heading at the receiving station
B. A fading effect caused by phase differences between radio wave components of the same transmission, as experienced at the receiving station
C. A fading effect caused by large changes in the height of the ionosphere, as experienced at the receiving station
D. A fading effect caused by time differences between the receiving and transmitting stations

ANSWER B: If you ever tried to watch television with a pair of rabbit ears, or an outside antenna, chances are you noticed what happened when a jet plane flew overhead. Your picture went in and out. This is caused by selective fading. As two signals arrive at your antenna, the reflection off the jet plane is slightly delayed from the direct signal, and this creates a phase difference. When two signals are out of phase, they tend to cancel each other. You can also hear selective fading at its best (at its worst?) on the 10-meter FM band at 29.6 MHz. While the signals are strong, they are many times distorted by their sidebands being out of phase. Worldwide AM broadcast stations are best received in upper or lower sideband, rather than AM double sideband, because AM double-sideband (DSB) reception brings in phase differences on the incoming DSB shortwave signal.

3B2 What is the *propagation effect* called when phase differences between radio wave components of the same transmission are experienced at the recovery station?
A. Faraday rotation
B. Diversity reception
C. Selective fading
D. Phase shift

ANSWER C: When received signals from the same original radio transmission come in out of phase, this is called selective fading. Watch out for answer D. The question reads "What is the effect of a phase difference," and the **effect** is "selective fading."

3B – RADIO WAVE PROPAGATION

3B3 What is the major cause of *selective fading*?
  A. Small changes in beam heading at the receiving station
  B. Large changes in the height of the ionosphere, as experienced at the receiving station
  C. Time differences between the receiving and transmitting stations
  D. Phase differences between radio wave components of the same transmission, as experienced at the receiving station
ANSWER D: The key words *phase differences* depict an incoming signal that will sound distorted due to selective fading of the sidebands.

3B4 Which emission modes suffer the most from *selective fading*?
  A. CW and SSB
  B. FM and double sideband AM
  C. SSB and image
  D. SSTV and CW
ANSWER B: Selective fading affects most wideband signals. Since FM is about 10 kHz wide, and AM double-sideband is 6 kHz wide, these two emissions are most influenced by selective fading.

3B5 How does the bandwidth of the transmitted signal affect *selective fading*?
  A. It is more pronounced at wide bandwidths.
  B. It is more pronounced at narrow bandwidths.
  C. It is equally pronounced at both narrow and wide bandwidths.
  D. The receiver bandwidth determines the selective fading effect.
ANSWER A: The wider the bandwidth, the more severely the signal will suffer from selective fading. Aeronautical and marine VHF stations may encounter selective fading when operating in the proximity of other nearby aircraft or metal-hull ship stations. Marine and aeronautical high-frequency SSB is a much narrower bandwidth signal, and is less susceptible to selective fading than older AM double-sideband equipment used years ago.

3B6 Why does the radio-path horizon distance exceed the geometric horizon?
  A. E-layer skip
  B. D-layer skip
  C. Auroral skip
  D. Radio waves may be bent.
ANSWER D: Radio waves bend slightly over the horizon because of the difference in the air's refractive index at higher altitudes. You can calculate your approximate VHF range to the horizon with the formula: D (miles) = 1.415 × √ H (in feet). Determine the height (in feet) of the station you want to communicate with, calculate its distance to the horizon, and add the two answers together for your rock-solid communications range. Depending on local weather conditions, a 15 to 30 percent range enhancement over the optical horizon will usually take place at VHF and UHF radio frequency bands. See chart at 3B7.

3B7 How much farther does the radio-path horizon distance exceed the geometric horizon?
  A. By approximately 15% of the distance
  B. By approximately twice the distance

C. By approximately one-half the distance
D. By approximately four times the distance

ANSWER A: This 15 percent increase in distance over the geometric horizon is enhanced during periods of stable high-pressure systems overriding a cool water mass. This could allow ship stations on marine VHF to intercommunicate in excess of 100 miles. It's the same phenomenon—tropospheric bending—that may allow aeronautical stations to communicate with ground stations hundreds, and sometimes, thousands of miles away.

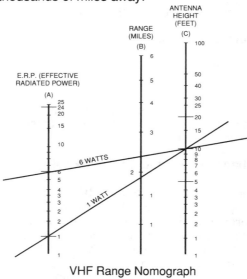

VHF Range Nomograph

3B8 To what distance is VHF propagation ordinarily limited?
A. Approximately 1000 miles
B. Approximately 500 miles
C. Approximately 1500 miles
D. Approximately 2000 miles

ANSWER B: About the furthest you could propagate a VHF signal on a regular basis would be 500 miles from a base station transmitter on a 10,000 foot peak. Although the calculated range to this land mobile base station might only be 100 miles, tropospheric bending could extend this three or four times its normal calculated line-of-sight coverage. Aeronautical stations up at 40,000 feet might enjoy 200-mile coverage, and amateur radio operators can normally intercommunicate with the space shuttle out to approximately 500 miles on VHF.

3B9 What propagation condition is usually indicated when a VHF signal is received from a station over 500 miles away?
A. D-layer absorption
B. Faraday rotation
C. Tropospheric ducting
D. Moonbounce

ANSWER C: Some incredible extra-long-range VHF and UHF propagation may take place in the presence of a high-pressure system. Sinking warm air, called a subsidence, overlays cool air just above the surface of the earth and over water. A well-defined inversion layer occurs when there is a sharp boundary

3B – RADIO WAVE PROPAGATION

between the sinking warm air and the cool air beneath. Over big cities, we call it smog. To the VHF and UHF DX enthusiast, we call it a for-sure key for VHF *tropospheric ducting* (key words) super range. Your author has communicated many times from his station in California to Hawaii on the 2-meter band during summer months when tropospheric ducting is at its best!

3B10 What happens to a radio wave as it travels in space and collides with other particles?
   A. Kinetic energy is given up by the radio wave.
   B. Kinetic energy is gained by the radio wave.
   C. Aurora is created.
   D. Nothing happens since radio waves have no physical substance

ANSWER A: It doesn't take much power for radio waves to travel over millions of miles in space. Look at the signals coming in from deep-space probes! But the further the radio wave travels, the more kinetic energy is given up.

3B11 When the earth's atmosphere is struck by a meteor, a cylindrical region of free electrons is formed at what layer of the ionosphere?
   A. The F1 layer
   B. The E layer
   C. The F2 layer
   D. The D layer

ANSWER B: High power meteor burst systems for remote data transmissions normally operate between 30 MHz and 50 MHz. These transmitting stations may be used to monitor climatic conditions in remote areas of the world, and it's surprising how predictable the incoming packets of digitized information come in from the meteor burst transmitter.

3B12 What is *transequatorial propagation*?
   A. Propagation between two points at approximately the same distance north and south of the magnetic equator
   B. Propagation between two points on the magnetic equator
   C. Propagation between two continents by way of ducts along the magnetic equator
   D. Propagation between any two stations at the same latitude

ANSWER A: A TE (transequatorial) signal is propagated between stations north and south of the equator. Some of the best contacts have been between Florida and Venezuela. If you look at a map, you will see that each station is about the same distance away from the equator—ideal conditions for a TE contact. Both CW and SSB are the normal modes for TE on VHF and UHF bands. Ship stations and aeronautical stations operating near the equator may be influenced by the ionosphere's strong effect in these regions. See figure at 3B13.

3B13 What is the maximum range for signals using *transequatorial propagation*?
   A. About 1,000 miles
   B. About 2,500 miles
   C. About 5,000 miles
   D. About 7,500 miles

ANSWER C: The total distance for a TE (transequatorial) contact is about 5,000 miles.

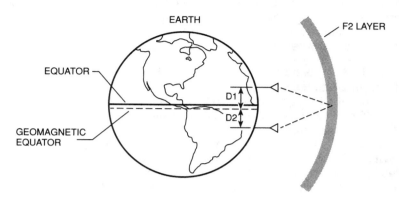

For transequatorial propagation to occur, stations at D1 and D2 must be equidistance from geomagnetic equator. Popular theory is that reflections occur in and off the F2 layer during peak sunspot activity.

Transequatorial Propagation

3B14 What is the best time of day for *transequatorial propagation?*
A. Morning
B. Noon
C. Afternoon or early evening
D. Transequatorial propagation only works at night.

ANSWER C: Similar to sporadic E, the best time of day for TE propagation is mid-afternoon or early evening. The time of day has allowed ultraviolet radiation from the sun to "warm up" the possible path.

3B15 What is knife-edge diffraction?
A. Allows normally line-of-sight signals to bend around sharp edges, mountain ridges, buildings and other obstructions
B. Arcing in sharp bends of conductors
C. Phase angle image rejection
D. Line-of-sight signals causing distortion to other signals

ANSWER A: Vertically polarized VHF and UHF signals will literally drag the bottom edge of the wave front over tall mountains that would normally block line-of-sight communications. Directional antennas can help fill in dead areas that you might think could never be reached by a VHF or UHF base station.

3B16 The average range for VHF communications is:
A. 5 miles
B. 15 miles
C. 30 miles
D. 100 miles

ANSWER C: 30 miles is a good figure to give if your customer asks how far their maritime or land mobile VHF system will operate over flat terrain from a roof-top high-gain antenna system.

3B17 The bending of radio waves passing over the top of a mountain range that disperses a weak portion of the signal behind the mountain is:
   A. Eddy-current phase effect
   B. Knife-edge diffraction
   C. Shadowing
   D. Mirror refraction effect

ANSWER B: Knife-edge diffraction and knife-edge refraction may fill in VHF and UHF signals into shadow areas of supposedly no coverage.

3B18 Knife-edge refraction:
   A. Is the bending of UHF frequency radio waves around a building, mountain or obstruction
   B. Causes the velocity of wave propagation to be different than original wave
   C. Both A and B above
   D. Attenuates UHF signals

ANSWER C: The radio wave can bend around a building, and bend over a mountain or obstruction because the velocity of the lower portion of the vertically-polarized wave front is slower than the upper part of the wave. Since the upper part of the wave is traveling faster, the lower part causes the signal to literally drag over the mountain top, and come back down quite sharply into an area of supposedly no coverage.

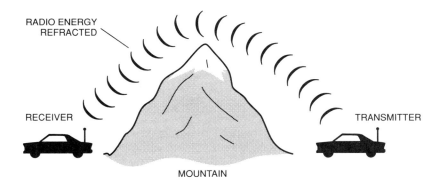

Knife-Edge Refraction

3B19 If the elapsed time for a radar echo is 62 microseconds, what is the distance in nautical miles to the object?
   A. 5
   B. 87
   C. 37
   D. 11.5

ANSWER A: It takes a radar wave 12.4 microseconds to travel one nautical mile. In 62 microseconds, the transmitted signal (or its echo) will travel 5 nautical miles (62 ÷ 12.4 = 5).

5 GENERAL RADIOTELEPHONE OPERATOR LICENSE

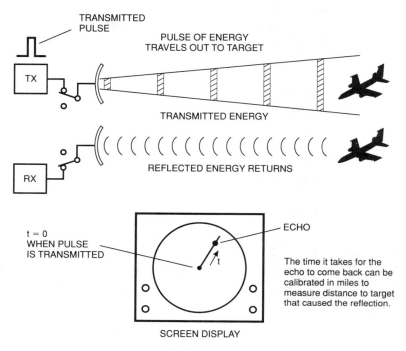

Radar Echo
Source: *Technology Dictionary,* ©1987, Master Publishing, Inc.

3B20 What is the wavelength of a signal at 500 MHz?
A. 0.062 cm
B. 6 meters
C. 60 cm
D. 60 meters

ANSWER C: To calculate the wavelength in meters of a signal if you know the frequency in MHz, or to calculate MHz if you know the wavelength in meters, simply remember the number 300. Then divide 300 by the frequency in MHz. In this problem, 300 ÷ 500 = 0.6 meter = 60 cm. You have to convert meters to centimeters (multiply by 100) to get the right answer.

3B21 The radar range in nautical miles to an object can be found by measuring the elapsed time during a radar pulse and dividing this quantity by:
A. 0.87 seconds
B. 1.15 microseconds
C. 12.36 microseconds
D. 1.73 microseconds

ANSWER C: We know that it takes exactly 12.36 microseconds for a radar wave to travel one nautical mile. In this question, they simply want to know the nautical mile range to an object, disregarding the return echo. If they had asked the range calculation to an object one mile away, you would need to multiply by 2 to calculate the time going to, and the return echo from, the object one mile away.

3B22 The band of frequencies least susceptible to atmospheric noise and interference is:
  A. 30 – 300 kHz
  B. 300 – 3000 kHz
  C. 3 – 30 MHz
  D. 300 – 3000 MHz

ANSWER D: Atmospheric noise and onboard electrical noise from ship and aircraft stations will dominate the low-frequency, medium-frequency, high-frequency, and even VHF-frequency bands up to 300 MHz. Signals between 300 MHz and 3 GHz are least susceptible to atmospheric noise.

Subelement 3C – Radio Practice (5 questions)

3C1 What is a *frequency standard*?
  A. A well-known (standard) frequency used for transmitting certain messages
  B. A device used to produce a highly accurate reference frequency
  C. A device for accurately measuring frequency to within 1 Hz
  D. A device used to generate wide-band random frequencies

ANSWER B: You may already have a frequency standard in your test equipment pool, but the world's most expensive frequency standards are stations WWV and WWVH. You can pick up time signals from them on 5 MHz, 10 MHz, 15 MHz, and 20 MHz. These signals are absolutely on frequency, 24 hours a day, for your equipment and test apparatus calibration.

3C2 What is a *frequency-marker generator*?
  A. A device used to produce a highly accurate reference frequency
  B. A sweep generator
  C. A broadband white noise generator
  D. A device used to generate wide-band random frequencies

ANSWER A: Older radio equipment might have a "CAL" band switch position. This is the calibrate position which energizes an internal marker generator. Every 100 kHz there would be a relatively accurate signal. But guess how you calibrate the internal marker generator? That's right, with the on-the-air WWV time signals referred to in question 3CI.

3C3 How is a *frequency-marker generator* used?
  A. In conjunction with a grid-dip meter
  B. To provide reference points on a receiver dial
  C. As the basic frequency element of a transmitter
  D. To directly measure wavelength

ANSWER B: The older marker generators came in handy when your digital readout went haywire on your older transceiver. You could generally find where you were on the dial by listening to the signals occurring every 100 kHz.

5 GENERAL RADIOTELEPHONE OPERATOR LICENSE

3C4 What is a *frequency counter?*
A. A frequency measuring device
B. A frequency marker generator
C. A device that determines whether or not a given frequency is in use before automatic transmissions are made
D. A broadband white noise generator

ANSWER A: Every radio shop should own a frequency counter. You can buy a good one for under $100. Portable LED or LCD frequency indicating devices may help you determine how accurately your client's transmitter is set on frequency. The more expensive the counter you buy, the more accurate the time base for more accurate frequency counts. Some frequency counters have a built-in crystal oven, and this is one of the best ways to give you a very stable reference to a frequency counter that may be calibrated directly to WWV and WWVH standards.

3C5 How is a *frequency counter* used?
A. To provide reference points on an analog receiver dial
B. To generate a frequency standard
C. To measure the deviation in an FM transmitter
D. To measure frequency

ANSWER D: You must never directly couple a frequency counter to your transmitter output. Most portable counters have a small rubber antenna that will pick up the signal from a nearby antenna quite nicely. Most counters can read out a signal within 50 feet. The more expensive the counter, the further away it can read a transmitting signal. If you plan to use the counter outside, make sure you purchase a counter with a LCD display. Stay away from LED displays for outside use because you can't see their readout in the bright sunlight!

Frequency Counter
Courtesy of OPTOELECTRONICS

3C6 What is the most the actual transmitter frequency could differ from a reading of 156,520,000 hertz on a frequency counter with a time base accuracy of ±1.0 ppm?
A. 165.2 Hz
B. 15.652 kHz
C. 156.52 Hz
D. 1.4652 MHz

ANSWER C: You will probably have 1 out of the following 6 questions on your test regarding the error readout on a frequency counter with a specific part per million time base. It's easy to solve these problems with a simple calculator.

3C – RADIO PRACTICE

Since they're asking parts per million, first convert the frequency given in Hz to MHz. You do this simply by moving the decimal point 6 places to the left. Now multiply the frequency in MHz times the parts per million time base accuracy (in this case 1 ppm). This will give you the error they are looking for in the answer.

Try this on your calculator: Clear 156.52 × 1 = 156.52 Hz.

3C7 What is the most the actual transmitter frequency could differ from a reading of 156,520,000 hertz on a frequency counter with a time base accuracy of ±0.1 ppm?
- A. 15.652 Hz
- B. 0.1 MHz
- C. 1.4652 Hz
- D. 1.5652 kHz

ANSWER A: Changing hertz to MHz will give you a frequency of 156.52. Multiply that by 0.1 parts per million, and see if your calculator reads 15.652 Hz. The keystrokes are: Clear 156.52 × 0.1 = 15.652.

3C8 What is the most the actual transmitter frequency could differ from a reading of 156,520,000 hertz on a frequency counter with a time base accuracy of ±10 ppm?
- A. 146.52 Hz
- B. 10 Hz
- C. 156.52 kHz
- D. 1565.20 Hz

ANSWER D: Keystroke the following on your calculator: Clear 156.52 × 10 = 1565.2 Hz. You are multiplying the frequency in MHz by the time base accuracy of 10 ppm.

3C9 What is the most the actual transmitter frequency could differ from a reading of 462,100,000 hertz on a frequency counter with a time base accuracy of ±1.0 ppm?
- A. 46.21 MHz
- B. 10 Hz
- C. 1.0 MHz
- D. 462.1 Hz

ANSWER D: How refreshing, a new frequency. From Hz to MHz, we end up with 462.1 MHz, which multiplied by 1.0 ppm gives you the answer. The following are the keystrokes: Clear 462.1 × 1 = 462.1 Hz.

3C10 What is the most the actual transmit frequency could differ from a reading of 462,100,000 hertz on a frequency counter with a time base accuracy of ±0.1 ppm?
- A. 46.21 Hz
- B. 0.1 MHz
- C. 462.1 Hz
- D. 0.2 MHz

ANSWER A: Remember, multiply the frequency in MHz times the time base accuracy. Try this on your calculator: Clear 462.1 × 0.1 = 46.21.

3C11 What is the most the actual transmit frequency could differ from a reading of 462,100,000 hertz on a frequency counter with a time base accuracy of ±10 ppm?
A. 10 MHz
B. 10 Hz
C. 4621 Hz
D. 462.1 Hz

ANSWER C: Here is the last problem. Again, frequency in MHz times the time base accuracy is performed on your calculator like this: Clear 462.1 × 10 = 4621 Hz.

When you are shopping for a frequency counter, remember that a counter with 0.1 ppm time base will give you a much more accurate reading than a counter with a 10 ppm time base.

3C12 What is a dip-*meter*?
A. A field strength meter
B. An SWR meter
C. A variable LC oscillator with metered feedback current
D. A marker generator

ANSWER C: When you tune your mobile antenna system, you watch for a dip on your SWR bridge. But you don't necessarily need a transmitter to generate a signal—you could do it with a portable variable LC oscillator that has a SWR meter that will measure the feedback current (reflective power). Your *variable* (key word) oscillator may be hooked up directly to a mobile antenna, and you can sweep through the frequencies to see where the antenna is resonant.

SWR Analyzer
Courtesy of MFJ Enterprises, Inc.

3C13 Why is a dip-meter used by many technicians?
A. It can measure signal strength accurately.
B. It can measure frequency accurately.
C. It can measure transmitter output power accurately.
D. It can give an indication of the resonant frequency of a circuit.

ANSWER D: If ever you need to perform service on older marine and aviation high-frequency transceivers, chances are you may need to use a dip meter to give an indication of the resonant frequency of specific circuits under test.

3C – RADIO PRACTICE

3C14 How does a dip-meter function?
A. Reflected waves at a specific frequency desensitize the detector coil.
B. Power coupled from an oscillator causes a decrease in metered current.
C. Power from a transmitter cancels feedback current.
D. Harmonics of the oscillator cause an increase in resonant circuit Q.

ANSWER B: Your little portable dip-meter, sometimes labeled as an SWR analyzer, couples power from its built-in oscillator to your antenna. At resonance, the antenna grabs all of the current, and this causes a decrease in metered current, seen as a dip on the indicating meter. The meter dips because all of the current is getting soaked up at resonance within the antenna system. If ever you are performing maintenance on a high-frequency antenna system aboard an aircraft, a boat, or an authorized base station, the dip meter is an invaluable tool to carry in your fix-it bag.

3C15 What two ways could a dip-meter be used in a radio station?
A. To measure resonant frequency of antenna traps and to measure percentage of modulation
B. To measure antenna resonance and to measure percentage of modulation
C. To measure antenna resonance and to measure antenna impedance
D. To measure resonant frequency of antenna traps and to measure a tuned circuit resonant frequency

ANSWER D: The dip-meter is a handy way to determine the resonant frequency of a box full of unmarked antenna traps. You can also measure a tuned circuit's resonant frequency, such as an antenna coupler. The dip-meter is a great one to measure a *tuned circuit* (key words).

3C16 What types of coupling occur between a dip-meter and a tuned circuit being checked?
A. Resistive and inductive
B. Inductive and capacitive
C. Resistive and capacitive
D. Strong field

ANSWER B: An older dip-meter may use plug-in coils for testing tuned circuits. The coils are coupled inductively, and capacitance is seen between the two coils that are close together. Even though you don't hold the coils right on top of each other, there is still some capacitive coupling along with the inductive coupling.

3C17 How tightly should the dip-meter be coupled with the tuned circuit being checked?
A. As loosely as possible, for best accuracy
B. As tightly as possible, for best accuracy
C. First loose, then tight, for best accuracy
D. With a soldered jumper wire between the meter and the circuit to be checked, for best accuracy

ANSWER A: If you get too close to the circuit you are testing, you could de-tune it with your dip-meter coil. For best accuracy, you want to keep the coupling as loose as possible.

3C18 What happens in a dip-meter when it is too tightly coupled with the tuned circuit being checked?
A. Harmonics are generated
B. A less accurate reading results
C. Cross modulation occurs
D. Intermodulation distortion occurs

ANSWER B: If you put your dip-meter right on top of a coil, you will have a *less accurate reading* (key words)—only get close enough to see a smooth dip on your readout. On the SWR analyzer type of dip-meters, everything is done for you automatically inside the box. All you need to do is to screw on the antenna cable, and sweep the frequency dial to see where the antenna is resonant.

3C19 What factors limit the accuracy, frequency response, and stability of an oscilloscope?
A. Sweep oscillator quality and deflection amplifier bandwidth
B. Tube face voltage increments and deflection amplifier voltage
C. Sweep oscillator quality and tube face voltage increments
D. Deflection amplifier output impedance and tube face frequency increments

ANSWER A: Every service shop should have a full-featured oscilloscope ("scope" for short). Many new marine and aeronautical transceivers may be employing digital signal processing (DSP). This demands a stable, *high-quality sweep oscillator* (key words) circuit inside the scope, and the scope must also have enough *bandwidth* (key word) to handle high-frequency applications, as well as VHF and UHF.

3C20 What factors limit the accuracy, frequency response, and stability of a D'Arsonval movement type meter?
A. Calibration, coil impedance and meter size
B. Calibration, series resistance and electromagnet current
C. Coil impedance, electromagnet voltage and movement mass
D. Calibration, mechanical tolerance and coil impedance

ANSWER D: Marine and aviation radios are now produced without the traditional needle movement meter. However, you may be working on older equipment that uses an electromechanical meter movement. Mechanical abuse and overload are the primary causes of inaccuracy and failure in these meters.

3C21 What factors limit the accuracy, frequency response, and stability of a frequency counter?
A. Number of digits in the readout, speed of the logic and time base stability
B. Time base accuracy, speed of the logic and time base stability
C. Time base accuracy, temperature coefficient of the logic and time base stability
D. Number of digits in the readout, external frequency reference and temperature coefficient of the logic

ANSWER B: Here we are back to counters again! Everyone should have one. Be sure to get one that has *accuracy, speed* and *stability* (key words).

3C – RADIO PRACTICE

3C22 How can the frequency response of an oscilloscope be improved?
A. By using a triggered sweep and a crystal oscillator as the time base
B. By using a crystal oscillator as the time base and increasing the vertical sweep rate
C. By increasing the vertical sweep rate and the horizontal amplifier frequency response
D. By increasing the horizontal sweep rate and the vertical amplifier frequency response

ANSWER D: When you buy an oscilloscope, try to get the widest bandwidth vertical amplifier you can afford. You will find that increasing the horizontal sweep rate will allow you to look at individual cycles of high frequencies.

3C23 How can the accuracy of a frequency counter be improved?
A. By using slower digital logic
B. By improving the accuracy of the frequency response
C. By increasing the accuracy of the time base
D. By using faster digital logic

ANSWER C: After reading several brochures for frequency counters, you will find that the *time base* (key words) is the key element for increased accuracy of the frequency readout. For shop use, a frequency counter with a built-in crystal oven will insure a stable and accurate readout, even though you might have just turned it on. As long as it is plugged into 110 VAC, the oven keeps the crystal warm even when the counter's power switch is turned off.

3C24 What is the name of the condition that occurs when the signals of two transmitters in close proximity mix together in one or both of their final amplifiers, and unwanted signals at the sum and difference frequencies of the original transmissions are generated?
A. Amplifier desensitization
B. Neutralization
C. Adjacent channel interference
D. Intermodulation interference

ANSWER D: Land mobile repeater stations and remote bases require regular maintenance. A problem called *intermodulation* may be caused by other close proximity transmitters. When you listen to the repeater's local speaker, you may hear two signals coming in at once. This is the effect commonly called intermodulation.

3C25 How does *intermodulation interference* between two transmitters usually occur?
A. When the signals from the transmitters are reflected out of phase from airplanes passing overhead
B. When they are in close proximity and the signals mix in one or both of their final amplifiers
C. When they are in close proximity and the signals cause feedback in one or both of their final amplifiers
D. When the signals from the transmitters are reflected in phase from airplanes passing overhead

ANSWER B: "Intermod," in its true sense, occurs within the repeater final amplifier. When signals *mix* in one or both repeater amplifiers, sum and difference frequencies on many different frequencies can be heard.

3C26 How can *intermodulation interference* between two transmitters in close proximity often be reduced or eliminated?
A. By using a Class C final amplifier with high driving power
B. By installing a terminated circulator or ferrite isolator in the feed line to the transmitter and duplexer
C. By installing a band-pass filter in the antenna feed line
D. By installing a low-pass filter in the antenna feed line

ANSWER B: At repeater and base station installations atop buildings and mountains, you may need to install *circulators* and *isolators* in the feed line to the transmitter and duplexer to minimize other nearby base stations getting into your system. This is an ongoing problem because most mountain tops will have different radios systems that will be coming and going, causing your repeater or remote base station to continually hang up on specific signals from a nearby antenna setup. A spectrum analyzer is a good piece of test equipment to use when you suspect the problem is intermodulation interference.

3C27 What can occur when a non-linear amplifier is used with a single-sideband phone transmitter?
A. Reduced amplifier efficiency
B. Increased intelligibility
C. Sideband inversion
D. Distortion

ANSWER D: Maritime mobile stations operating on marine SSB may legally employ single-sideband amplifiers for additional power output. The added power amplifier must be linear to prevent the signal from becoming *distorted* (key word). Older tube-type amplifiers may not be rated for good linearity on an SSB emission.

3C28 How can even-order harmonics be reduced or prevented in transmitter amplifier design?
A. By using a push-push amplifier
B. By using a push-pull amplifier
C. By operating class C
D. By operating class AB

ANSWER B: Most worldwide amplifiers run a pair of output tubes. The tubes are in push-pull, and this reduces even-order harmonics. Watch out for answer A—if you don't read it carefully, you could mistake it for "push-pull."

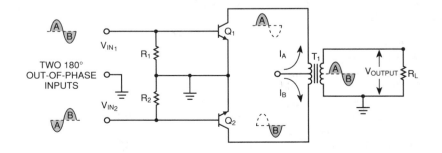

Push-Pull Amplifier

3C – RADIO PRACTICE

3C29 What is *receiver desensitizing*?
A. A burst of noise when the squelch is set too low
B. A burst of noise when the squelch is set too high
C. A reduction in receiver sensitivity because of a strong signal on a nearby frequency
D. A reduction in receiver sensitivity when the AF gain control is turned down

ANSWER C: You are up servicing a base station on a building top, and all of a sudden the incoming signal abruptly drops, and then pops up again at its normal level. Chances are the *receiver* of the equipment you are working on is losing *sensitivity* because of the presence of a *strong signal* (key words) on an adjacent channel. Time to consult your spectrum analyzer, or a sensitive portable frequency counter, and determine what steps you are going to take to minimize receiver desensitization.

3C30 What is the term used to refer to the reduction of receiver gain caused by the signals of a nearby station transmitting in the same frequency band?
A. Desensitizing
B. Quieting
C. Cross modulation interference
D. Squelch gain rollback

ANSWER A: Many times a land mobile installation on a roof top or mountain may require the use of cavities to notch out the offending incoming signal.

3C31 What is the term used to refer to a reduction in receiver sensitivity caused by unwanted high-level adjacent channel signals?
A. Intermodulation distortion
B. Quieting
C. Desensitizing
D. Overloading

ANSWER C: If signals mysteriously come in strong, then abruptly get weak, and then come back strong again, chances are there is someone nearby transmitting on an adjacent frequency, desensitizing your receiver. The best way to check for adjacent signals is through the use of a frequency counter.

3C32 How can *receiver desensitizing* be reduced?
A. Ensure good RF shielding between the transmitter and receiver
B. Increase the transmitter audio gain
C. Decrease the receiver squelch gain
D. Increase the receiver bandwidth

ANSWER A: Sometimes receiver desensitization may occur within the transceiver base station itself. This means you will need to go in there and add additional shielding between different stages of the transmitter and receiver.

3C33 What is *cross-modulation interference*?
A. Interference between two transmitters of different modulation type
B. Interference caused by audio rectification in the receiver preamp
C. Harmonic distortion of the transmitted signal
D. Modulation from an unwanted signal is heard in addition to the desired signal

ANSWER D: If you have driven your little handheld transceiver down to a big city, chances are you have heard the effects of cross-modulation interference. "Cross-mod" brings in modulation from an *unwanted signal* (key words) that rides on top of the signal you are presently listening to. This mix occurs inside your receiver with handheld transceivers. Do not confuse cross-modulation with intermodulation. "Intermod" occurs in the transmitters, and cross-modulation occurs in the receivers.

3C34 What is the term used to refer to the condition where the signals from a very strong station are superimposed on other signals being received?
 A. Intermodulation distortion
 B. Cross-modulation interference
 C. Receiver quieting
 D. Capture effect

ANSWER B: If in your receiver's signal you hear more than one transmitted frequency, chances are it is cross-modulation interference. Remember, if they say the word *received*, it means "cross-mod."

3C35 How can *cross-modulation* in a receiver be reduced?
 A. By installing a filter at the receiver
 B. By using a better antenna
 C. By increasing the receiver's RF gain while decreasing the AF gain
 D. By adjusting the pass-band tuning

ANSWER A: Cross-modulation at a land mobile radio installation may be reduced by installing cavity filters at the receiver antenna input. Inside the receiver section, you may wish to retune the helical resonators in order to minimize this off-channel interference.

3C36 What is the result of *cross-modulation*?
 A. A decrease in modulation level of transmitted signals
 B. Receiver quieting
 C. The modulation of an unwanted signal is heard on the desired signal
 D. Inverted sidebands in the final stage of the amplifier

ANSWER C: When you hear an *unwanted signal* (key words) on top of the desired signal, it is the result of cross-modulation.

3C37 What is the *capture effect*?
 A. All signals on a frequency are demodulated by an FM receiver
 B. All signals on a frequency are demodulated by an AM receiver
 C. The loudest signal received is the only demodulated signal
 D. The weakest signal received is the only demodulated signal

ANSWER C: A unique property of an FM receiver is the ability to hear only the loudest incoming FM signal. If a distant station is transmitting, and a strong local station comes on the air, you will only hear the local station. The stronger signal is "capturing" your receiver. We will consider local signals as being the "loudest" signal to correspond with answer C.

3C38 What is the term used to refer to the reception blockage of one FM-phone signal by another FM-phone signal?
 A. Desensitization
 B. Cross-modulation interference

C. Capture effect
D. Frequency discrimination

ANSWER C: When an incoming FM signal is blocked by a stronger FM signal, it is called "capture effect." Sometimes you hear the two signals coming in at once, sometimes you hear nothing more than a garbled message, and if there is a big difference between the two signal strengths, all you will hear is the stronger signal completely. In the AM aviation radio service, two signals coming in at once will usually lead to a tell-tale heterodyne tone. When the tower hears a heterodyne tone, the personnel know that two stations are attempting to transmit at once. In AM radio, you do not normally experience "capture effect."

3C39 With which emission type is the capture-effect most pronounced?
A. FM
B. SSB
C. AM
D. CW

ANSWER A: Capture effect is most common on VHF and UHF FM mobile, maritime mobile, and base station equipment.

3C40 How does a *spectrum analyzer* differ from a conventional time-domain oscilloscope?
A. The oscilloscope is used to display electrical signals while the spectrum analyzer is used to measure ionospheric reflection
B. The oscilloscope is used to display electrical signals in the frequency domain while the spectrum analyzer is used to display electrical signals in the time domain
C. The oscilloscope is used to display electrical signals in the time domain while the spectrum analyzer is used to display electrical signals in the frequency domain
D. The oscilloscope is used for displaying audio frequencies and the spectrum analyzer is used for displaying radio frequencies

ANSWER C: The spectrum analyzer displays the strength of signals at particular frequencies according to frequency (in a frequency domain) along a horizontal axis. A signal can be locked in and centered in the middle of the screen with 25 kHz to the left and 25 kHz to the right for a close examination of any off-frequency spurs it might have. A simple oscilloscope *cannot* display the signal in the frequency domain—it displays the signals on a time axis (in a time domain) going from left to right. With your output signal displayed on a spectrum analyzer, you may adjust your transmitter output network for minimum spurious signals. See figure at 3C41.

3C41 What does the horizontal axis of a *spectrum analyzer* display?
A. Amplitude
B. Voltage
C. Resonance
D. Frequency

ANSWER D: The spectrum analyzes the frequency of signals by displaying the amplitude of signals at particular frequencies on a frequency axis.

5 GENERAL RADIOTELEPHONE OPERATOR LICENSE

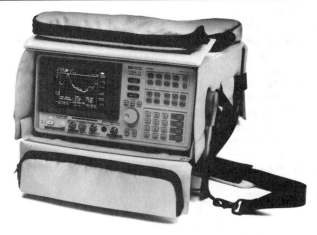

Spectrum Analyzer
Courtesy of Hewlett-Packard Company

3C42 What does the vertical axis of a *spectrum analyzer* display?
A. Amplitude
B. Duration
C. Frequency
D. Time
ANSWER A: The vertical axis of the spectrum analyzer displays the amplitude of the signal.

3C43 What test instrument can be used to display spurious signals in the output of a radio transmitter?
A. A spectrum analyzer
B. A wattmeter
C. A logic analyzer
D. A time-domain reflectometer
ANSWER A: You can check for spurious signals with a spectrum analyzer.

3C44 What test instrument is used to display intermodulation distortion products from an SSB transmitter?
A. A wattmeter
B. A spectrum analyzer
C. A logic analyzer
D. A time-domain reflectometer
ANSWER B: Only the spectrum analyzer allows you to see your output versus frequency by displaying it in the frequency domain. This allows you to check for harmonics that cause distortion in your SSB transmitter.

3C45 What advantage does a *logic probe* have over a voltmeter for monitoring logic states in a circuit?
A. A logic probe has fewer leads to connect to a circuit than a voltmeter
B. A logic probe can be used to test analog and digital circuits
C. A logic probe can be powered by commercial AC lines.
D. A logic probe is smaller and shows a simplified readout

ANSWER D: Now that most transceivers incorporate numerous digital circuits, the simple logic probe indicating that circuits are either *on* (low state) or *off* (high state) is a quick way to compare circuit operation to that listed in the technical manuals. Incidentally, technical manuals are normally *not* included with each new piece of equipment—you must buy them separately if you plan to service your own equipment.

Logic Probe

3C46 What piece of test equipment can be used to directly indicate high and low logic states?
A. A galvanometer
B. An electroscope
C. A logic probe
D. A Wheatstone bridge

ANSWER C: The logic probe, about the same size as a fountain pen, indicates high and low logic states. High is a 1, and low is a 0.

3C47 What is a logic probe used to indicate?
A. A short-circuit fault in a digital logic circuit
B. An open-circuit failure in a digital logic circuit
C. A high-impedance ground loop
D. High and low logic states in a digital logic circuit

ANSWER D: You can use a logic probe to help determine whether or not an integrated circuit (IC) chip is good or bad.

3C48 What piece of test equipment besides an oscilloscope can be used to indicate pulse conditions in a digital logic circuit?
A. A logic probe
B. A galvanometer
C. An electroscope
D. A Wheatstone bridge

ANSWER A: An oscilloscope is always your best piece of test equipment—but the logic probe is a close second because of all the electronic functions now being performed digitally.

3C49 What is one of the most significant problems you might encounter when you try to receive signals with a mobile station?
A. Ignition noise
B. Doppler shift
C. Radar interference
D. Mechanical vibrations

5 GENERAL RADIOTELEPHONE OPERATOR LICENSE

ANSWER A: Ignition noise is a significant problem with maritime mobile and mobile vehicle high-frequency and certain VHF installations. After you install your client's equipment, start the engine and check for noise. Noise filters are available to knock down severe electrical interference. Never consider an installation "complete" until you have specifically checked for engine interference.

3C50 What is the proper procedure for suppressing electrical noise in a mobile station?
  A. Apply shielding and filtering where necessary
  B. Insulate all plane sheet metal surfaces from each other
  C. Apply antistatic spray liberally to all non-metallic surfaces
  D. Install filter capacitors in series with all DC wiring

ANSWER A: Don't just turn on your noise blanker and put up with ignition interference; rather, consider shielding and filtering of the noise sources.

3C51 How can ferrite beads be used to suppress ignition noise?
  A. Install them in the resistive high voltage cable every 2 years
  B. Install them between the starter solenoid and the starter motor
  C. Install them in the primary and secondary ignition leads
  D. Install them in the antenna lead to the radio

ANSWER C: It is the primary and secondary ignition leads which radiate strong spark pulses that create ignition noise. Ferrite beads strung on the leads may help reduce the higher-frequency noises.

3C52 How can conducted and radiated noise caused by an alternator be suppressed?
  A. By installing filter capacitors in series with the DC power lead and by installing a blocking capacitor in the field lead
  B. By connecting the radio's power leads to the battery by the longest possible path and by installing a blocking capacitor in series with the positive lead
  C. By installing a high pass filter in series with the radio's power lead to the vehicle's electrical system and by installing a low-pass filter in parallel with the field lead
  D. By connecting the radio's power leads directly to the battery and by installing coaxial capacitors in the alternator leads

ANSWER D: Connecting a mobile transceiver directly to the vehicle's battery will minimize picking up alternator noise. The battery will act as a giant capacitor, hopefully minimizing noise coming up the line. However, severe cases of radiated alternator whine need to be cured by installing coaxial capacitors in the alternator leads. These need to go as close to the alternator as possible. If you can stop the noise at the exact spot it starts from, you can easily resolve most noise interference problems.

Noise Filter
Courtesy of Newmar®

3C – RADIO PRACTICE

3C53 What is a major cause of atmospheric static?
 A. Sunspots
 B. Thunderstorms
 C. Airplanes
 D. Meteor showers
ANSWER B: Thunderstorms will create lightning, and lightning may emit broadband atmospheric static that travels hundreds of miles from the source.

3C54 How can you determine if a line-noise interference problem is being generated within a building?
 A. Check the power-line voltage with a time-domain reflectometer
 B. Observe the AC wave form on an oscilloscope
 C. Turn off the main circuit breaker and listen on a battery-operated radio
 D. Observe the power-line voltage on a spectrum analyzer
ANSWER C: A small transistorized portable radio that tunes the AM broadcast band is a good noise "sniffer". Shut off the electricity to the building, and then listen to see whether or not your battery-operated radio is still picking up the interference. Use that same radio to "home in" on electrical interference within a building.

3C55 What causes *receiver desensitizing*?
 A. Audio gain adjusted too low
 B. Squelch gain adjusted too high
 C. The presence of a strong signal on a nearby frequency
 D. Squelch gain adjusted too low
ANSWER C: Receiver desensitizing is caused by the presence of a strong signal on an adjacent channel or nearby frequency.

3C56 How can *alternator whine* be minimized?
 A. By connecting the radio's power leads to the battery by the longest possible path
 B. By connecting the radio's power leads to the battery by the shortest possible path
 C. By installing a high pass filter in series with the radio's DC power lead to the vehicle's electrical system
 D. By installing filter capacitors in series with the DC power lead
ANSWER B: Many technicians have quickly resolved alternator whine coming in the line by running the red and black power cables directly to the battery posts. Always be sure to fuse both the red lead and the black lead directly at the posts to minimize damage in case of an electrical short.

3C57 An electrical relay:
 A. Is a current limiting device
 B. Is a device used for supplying 3 or more voltages to a circuit
 C. Is concerned mainly with HF audio amplifiers
 D. Provides either an open or closed circuit to some position of a device
ANSWER D: An electrical relay is usually used as a switch. Its open contacts provide a near infinite resistance open circuit, and its closed contacts provide a near zero resistance closed circuit.

3C58 A high standing-wave ratio on a transmission line can be caused by:
A. Excessive modulation
B. An increase in output power
C. Detuned antenna coupling
D. Poor B+ regulation

ANSWER C: Water inside an antenna mount will many times detune the antenna circuit. This causes high SWR, which can be detected with an SWR meter or analyzer.

3C59 The best insulation at UHF is:
A. Black rubber
B. Bakelite
C. Paper
D. Mica

ANSWER D: At UHF frequencies, mica is a good insulator—so is Teflon®.

3C60 The electrolyte in a lead storage battery is:
A. Potassium hydrate
B. Pure spongy lead
C. Sulphuric acid
D. Iron oxide

ANSWER C: Sulfuric acid is the electrolyte used in lead storage batteries. It is caustic, so keep it away from your skin and clothes. It corrodes tht terminals.

3C61 A dynamotor is used to:
A. Step up AC voltage
B. Step down AC voltage
C. Step down DC voltage
D. Step up DC voltage

ANSWER D: While you won't find many of these in traditional land mobile radio installations, the old dynamotor was used to step up DC voltages.

3C62 The SHF band is:
A. 3000 to 30,000 MHz
B. Above 300,000 MHz
C. 300 to 3000 MHz
D. 30,000 to 300,000 MHz

ANSWER A: SHF stands for "super high frequency." 3000 MHz to 30,000 MHz may also be stated as 3 GHz to 30 GHz. This is the band commonly used by point-to-point microwave links.

3C – RADIO PRACTICE

Frequency MHz	Band	ABRV	Wavelength Meters
0.003-0.03	Very Low Frequency	VLF	100,000-10,000
0.03-0.3	Low Frequency	LF	10,000-1,000
0.3-3.0	Medium Frequency	MF	1,000-100
3.0-30	High Frequency	HF	100-10
30-300	Very High Frequency	VHF	10-1
300-3,000	Ultra High Frequency	UHF	1-0.1
3,000-30,000	Super High Frequency	SHF	0.1-0.01
30,000-300,000	Extremely High Frequency	EHF	0.01-0.001

Radio Frequency Bands

3C63 When soldering electrical connections, care should be taken not to:
A. Heat both connections
B. Move the wire until solder has solidified
C. Have solder on the connections
D. Use rosin core solder

ANSWER B: When soldering, make sure that you do not move the soldered joint until the solder has solidified and the joint has cooled down for a few seconds.

3C64 If there are too many harmonics from a transmitter, check the:
A. Coupling
B. Tuning of circuits
C. Shielding
D. Any of the above

ANSWER D: A transceiver or maritime SSB installation emitting too many harmonics on transmit would require you to check the coupling between the stages, proper tuning of the circuits, and proper shielding inside the transmitter section. All of these are steps to help minimize harmonics.

3C65 One piece of equipment to indicate neutralization is:
A. A neon bulb
B. A tachometer
C. Wave trap
D. Filter

ANSWER A: A simple neon bulb held close to the metal chassis of a tube-type amplifier may begin to glow if the amplifier is not properly neutralized and is going into oscillation.

3C66 Interference to radio receivers in automobiles can be reduced by:
A. Connecting resistances in series with the spark plugs
B. Using heavy conductors between the starting battery and the starting motor
C. Connecting resistances in series with the starting battery
D. Grounding the negative side of the starting battery

ANSWER A: Resistor spark plugs will help decrease engine noise caused by the spark plugs in the engine. Answer B will help minimize voltage spikes, but won't help reduce interference from plug noise.

3C67 Magnetron oscillators are used for:
A. Generating SHF signals
B. Multiplexing
C. Generating rich harmonics
D. FM demodulation

ANSWER A: The magnetron is a high-power microwave tube. One of its uses is to generate radar pulses at super-high frequencies. X-band radar operates near 10 GHz. A magnetron is also used in a microwave oven.

3C68 Normally in a mobile radio installation the E output of a dynamotor is:
A. Adjusted by armature rheostat
B. Adjusted by field rheostat
C. Not adjustable
D. Adjusted by battery rheostat

ANSWER C: The output voltage of a dynamotor is not easily changed, except by changing the primary input voltage source. The output voltage is generally considered "not adjustable." This is why you won't see many dynamotors in two-way radio applications these days.

3C69 If a shunt motor, running with a load, has its shunt field opened, how would this affect the speed of the motor?
A. Slow down
B. Stop suddenly
C. Speed up then slow down
D. Unaffected

ANSWER C: A shunt motor running with a load creates a counter voltage which will help the motor run at a constant speed, even though the load may be slightly increasing or decreasing. If the shunt field is opened, the back EMF will instantly drop, causing the motor to quickly speed up. But then the motor will slow down as the load begins to reduce its RPM. With no load, the motor will literally run away, and ultimately destroy itself.

3C70 The state of charge of a Lead-Acid storage cell is determined by:
A. Hydrometer reading
B. Its open circuit voltage
C. Its short circuit discharge current
D. Ohmmeter

ANSWER A: We use a hydrometer reading to check the state of charge on a lead-acid storage cell. The specific gravity test is a comparison of the weight of the mixed electrolyte to the weight of an equal volume of distilled water. The specific gravity of distilled water is 1.000. The specific gravity of the electrolyte in a lead-acid cell is approximately 1.265 when the cell is fully charged. A cell with a specific gravity of less than 1.190 may need to be charged before attempting to operate radio equipment from that battery.

3C – RADIO PRACTICE

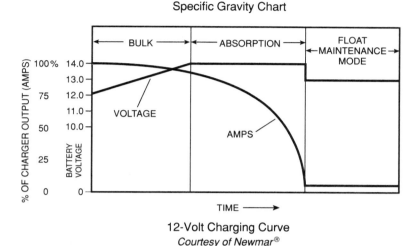

Charge Level	Specific Gravity*	Voltage (V)
100%	1.265	12.68
75%	1.225	12.45
50%	1.190	12.24
25%	1.155	12.06
Discharged	1.120	11.89

Specific Gravity Chart

*Specific gravity of distilled water is 1.0.

12-Volt Charging Curve
Courtesy of Newmar®

3C71 An absorption wave meter is useful in measuring:
- A. Field strength
- B. Output frequencies to conform with FCC tolerance
- C. Standing wave frequencies
- D. The resonant frequency of LC tank circuit

ANSWER D: A wave meter has a fixed coil and a variable capacitor with an indicating microammeter to measure a peak in current when the variable capacitor is adjusted to a LC tank circuit under test. Loose coupling between the wave meter probe and the circuit will insure a well-defined peak in the microammeter movement.

3C72 If a power supply fails, a new high-vacuum tube glows red-hot, smoke curls from around the filter choke:
- A. The transformer secondary has shorted
- B. The filter choke has an open turn
- C. One of the filter capacitors has shorted
- D. A bleeder resistor has opened

ANSWER C: Although we don't see many power supplies still using vacuum tubes, there was a lot that you could tell about what was happening within the power supply by looking at how the vacuum tube was glowing. If the plates got red-hot, something was about to happen. When smoke starts curling around the base of the tubes, chances are it's probably coming from too much current passing through the filter choke, and this usually means a dried out electrolytic capacitor. When an old electrolytic capacitor dries out, it develops a low resistance circuit which causes high current to pass through the power supply rectifier circuit. Immediately shut down the equipment, unsolder the leads from the filter capacitors, and check all for a short or very low dc resistance.

3C73 Velocity of radio wave propagation in free space does what?
A. Varies in speed
B. Is constant
C. Is same as speed of light
D. Both B and C above

ANSWER D: The velocity of radio waves *in free space* is constant, and is the same as the speed of light. This is the correct answer. But radio waves, entering the ionosphere and the lower atmosphere may be reflected and refracted creating subtle signal delays that must be measured down to a femtosecond (10^{-15}). This becomes a factor for our global positioning system time measurements.

3C74 In storing a fully-charged battery (lead-acid) for a long period of time (8 to 12 months) one should:
A. Drain and flush out the electrolyte then refill with distilled water
B. Keep it in a warm place
C. Jar it at least once a week to prevent sulfation
D. Maintain a constant trickle charge

ANSWER A: You can prolong the ultimate life of a lead-acid storage battery during periods of no use by draining into a suitable container the electrolyte, irrigating the plates with distilled water, and then refilling with distilled water. This should be done just after the battery has been fully charged.

3C75 A screen grid resistor burns out after smoking. First thing to check would be:
A. Short in screen grid by-pass capacitor
B. An open turn in plate coil
C. Short in B+ by-pass capacitor
D. Open grid leak resistor

ANSWER A: When a bypass capacitor shorts out, this will lead to high current in the screen resistor, and the resistor will go up in smoke in a matter of seconds. Whenever replacing burned-out components, always look for the original source of the obvious component failure.

3C76 Neglecting line losses, the RMS voltage along an RF transmission line having no standing waves:
A. Is equal to the impedance
B. Is one-half of the surge impedance
C. Is the product of the surge impedance times the line current
D. Varies sinusoidally along the line

ANSWER C: To calculate the average voltage along an RF coax cable or parallel conductor transmission line, multiply the characteristic impedance (surge impedance) by the amount of RF line current. Keep in mind that air is the perfect dielectric, and maximum wire diameter may minimize RF line losses. This is why some high-frequency radio stations feed distant antenna systems with open wire transmission line, held in place by expensive Teflon® spreaders, and use extremely large diameter wire to minimize I^2R losses.

3C77 Waveguides are:
A. A hollow tube that carries RF
B. Solid conductor of RF

C. Coaxial cables
D. Copper wire

ANSWER A: A waveguide is usually designed to carry RF energy at frequencies above 1 GHz. Most waveguides are rectangular, but the words "hollow tube" simply describe that waveguides do not rely on a center conductor to carry the RF.

3C78 When a vacuum tube operates at VHF or higher as compared to lower frequencies:
A. Transit time of electrons becomes important
B. It is necessary to make larger components
C. It is necessary to increase grid spacing
D. Only pentode is satisfactory

ANSWER A: Vacuum tubes operating in VHF or UHF power amplifiers are extremely expensive, and internal component size and make-up is small and critical. One important consideration is the transit time of electrons within the tube at VHF/UHF and microwave frequencies.

3C79 The output voltage of a separately-excited AC generator (running at a constant speed) could be controlled by:
A. Changing the dielectric constant of the armature
B. Changing the material of the armature
C. The field current
D. Operating the generator in a no-load condition

ANSWER C: When more current passes through the field coils, the increased current develops more magnetic lines of force, which creates a higher voltage output in the spinning rotor.

3C80 The FCC deleted this question.

3C81 A circulator:
A. Cools DC motors during heavy loads
B. Allows two or more antennas to feed one transmitter
C. Allows one antenna to feed two separate microwave transmitters and receivers at the same time
D. Insulates UHF frequencies on transmission lines

ANSWER C: The antenna circulator has several ports to feed separate microwave transmitters and receivers, all at the same time without significant interaction. The energy circulates in one direction with little significant loss, but is attenuated if circulated in the opposite direction.

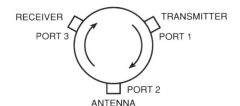

Energy transfers with little significant loss (0.5-1.0dB) from Port 1 to Port 2 and from Port 2 to Port 3, but has significant isolation (20-30dB) if the direction is reversed, i.e., Port 3 to Port 2, and Port 2 to Port 1.

Circulator

3C82 Static and interference from motors can be eliminated by:
A. Grounding the battery with a 2" copper strip
B. Installing faraday shields around connectors
C. Installing RF chokes across power line to ground
D. Installing bypass capacitors from the power line to grounded parts

ANSWER D: The motor brushes may generate noise pulses which travel along the DC input power line to the radio equipment. Since the noise pulses are AC, they can be shunted to ground with a bypass capacitor of the appropriate size.

3C83 Transmission line shielding is grounded:
A. At the input only
B. At both input and output
C. At the output only
D. If the antenna is a Marconi design

ANSWER B: A coaxial cable feed line in vehicles, boats, airplanes, and all land mobile installations, *must* have the shielding grounded at both ends of the circuit. Leaving the shield ungrounded at the antenna end of the circuit will cause the feed line to radiate on transmit, and pick up excessive noise on receive. Both ends must be grounded.

3C84 Motorboating (low frequency oscillations) in an amplifier can be stopped by:
A. Grounding the screen grid
B. Bypassing the screen grid resistor with a .1 mfd capacitor
C. Connect a capacitor between the B+ lead and ground
D. Grounding the plate

ANSWER C: Occasionally certain older RF amplifiers may go into a low frequency oscillation at specific frequencies. Connecting a capacitor between the B+ lead and chassis ground, keeping the leads extremely short, will help minimize this problem.

3C85 Auto interference to radio reception can be eliminated by:
A. Installing resistive spark plugs
B. Installing capacitive spark plugs
C. Installing capacitors in series with the spark plugs
D. installing two copper-braid ground strips

ANSWER A: High performance ignition systems will develop an extremely high voltage spark along the ignition primary plug wires. Resistive spark plugs may help shape the ignition interference to reduce the problem of spark plug noise pulses getting into the two-way radio receiver. On marine and aviation high-frequency systems, a noise blanker will help minimize this problem—but a good technician will always try to eliminate the noise at the source, not at the radio receiver.

3C86 DC series motor speed is affected by:
A. The load
B. The ripple frequency
C. The number of brushes
D. Stray RF fields

ANSWER A: As the load on a DC series motor increases, the speed decreases. Conversely, as the load decreases, the motor speed increases.

3C – RADIO PRACTICE **5**

3C87 An antenna radiates a primary signal of 500 watts output. If there is a 2nd harmonic output of 0.5 watt, what attenuation of the 2nd harmonic has occurred?
A. 3 dB
B. 10 dB
C. 20 dB
D. 30 dB

ANSWER D: It's relatively simple to calculate a dB power change. Remember this table:

DB	Power Change $\frac{P_1}{P_2}$	
3 dB	2X	Power change
6 dB	4X	Power change
9 dB	8X	Power change
10 dB	10X	Power change
20 dB	100X	Power change
30 dB	1000X	Power change
40 dB	10,000X	Power change

In this question, a signal of 500 watts output has a second harmonic at 0.5 watt. This is a 1000X power change, so the second harmonic is down 30 dB.

Derivation:

If $dB = 10 \log_{10} \frac{P_1}{P_2}$

then what power ratio is 20 dB?

$20 = 10 \log_{10} \frac{P_1}{P_2}$

$\frac{20}{10} = \log_{10} \frac{P_1}{P_2}$

$2 = \log_{10} \frac{P_1}{P_2}$

Remember: logarithm of a number is the exponent to which the base must be raised to get the number.

$\therefore 10^2 = \frac{P_1}{P_2}$

$100 = \frac{P_1}{P_2}$

Or $P_1 = 100\, P_2$

20 dB means P_1 is 100 times P_2

Decibels

3C88 On runway approach, an ILS Localizer shows:
A. Deviation left or right of runway center line
B. Deviation up or down from ground speed
C. Deviation percentage from authorized ground speed
D. wind speed along runway

ANSWER A: Here is that question again about an ILS localizer. Remember, local means next to, so deviation left or right, not *up and down*, is the correct answer. Up and down would be the glide slope. See question 3A25.

3C89 Where should a heat sink or heat shunt (such as a clamp or pliers) be placed when desoldering an IC chip?
A. On the tip of the soldering iron
B. Between the IC chip terminal and the soldering iron
C. Place clamp across all terminals
D. Between solder iron tip and the solder connection

ANSWER B: Integrated circuits are susceptible to failure caused by extreme heat at their chip terminals while being serviced with a soldering iron. Heat absorbing clamps should be placed at the chip terminal before going to the board with your soldering or desoldering iron.

3C90 What type of antenna system allows you to receive and transmit at the same time in both directions?
A. Simplex
B. Duplex
C. Multiplex
D. Digital duplex

ANSWER B: Duplex allows you to transmit and receive at the same time in both directions. We would use a "duplexer" on the antenna system to keep the transmit signal from coming back down and getting into the receiver that may be several hundred kilohertz or a couple of megahertz away. You normally use duplexers in VHF and UHF systems.

3C91 A microwave device which allows RF energy to pass through in one direction with very little loss, but absorbs RF power in the opposite direction:
A. Circulator
B. Wave trap
C. Multiplex
D. Isolator

ANSWER D: In microwave systems, the isolator keeps RF energy going in one direction, but opposes it from coming back in the opposite direction.

3C92 The output voltage of a separately-excited AC generator with constant frequency is dependent upon the adjustment of:
A. Iron or ferrite choke core
B. Field current
C. Brush position
D. Phase angle

ANSWER B: Increasing the current through the alternator field coils causes a higher output voltage.

3C93 What radio navigation aid determines the distance from a transponder beacon by measuring the length of time the radio signal took to travel to the receiver?
A. Radar
B. Loran C
C. Distance Marking (DM)
D. Distance Measuring Equipment (DME)

ANSWER D: "DME" stands for distance measuring equipment. The digital readout in the aircraft cockpit will indicate in nautical miles how far the aircraft is from the DME station.

3C94 Which of the following is a feature of an instrument landing system (ILS)?
A. Localizer: shows aircraft deviation horizontally from center of runway
B. Glide slope (or glide path): shows aircraft vertical altitude of an aircraft during landing
C. Provides communication to aircraft
D. Both a and b

ANSWER D: Both the glide slope and localizer signals will allow an aircraft an instrument landing approach. Remember, localizers show horizontal deviation, and glide slope shows too high—just right—too low indications while coming in for a landing. See question 3A25.

3C95 What is azimuth on a radar antenna?
A. Diameter
B. Degrees elevation
C. Degrees horizon
D. Impedance

ANSWER C: We have 360 degrees of azimuth in a radar. This allows you to calculate the bearing to a target as degrees on the horizon.

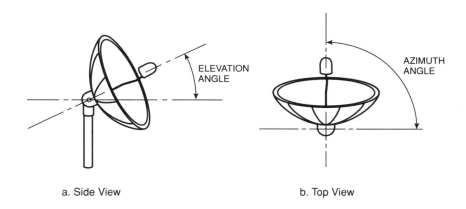

a. Side View   b. Top View

Antenna Adjustment Angles

3C96 Why do we tin component leads?
A. It helps to oxidize the wires
B. Prevents resting of circuit board
C. Decreases heating time and aids in connection
D. Provides ample wetting for good connection

ANSWER C: Tin is a component in solder which accelerates solder melting and aids in a good solder connection.

3C97 What is the purpose of using a small amount of solder on the tip of a soldering iron just prior to making a connection?
A. Removes oxidation
B. Burns up flux
C. Increases solder temperature
D. Aids in wetting the wires

ANSWER A: When soldering, develop a shiny wet solder tip before heating up the component leads you wish to solder. This will remove the oxidation that might prevent a quick and clean soldered connection.

Subelement 3D – Electrical Principles (16 questions)

3D1 What is *reactive power?*
A. Wattless, non-productive power
B. Power consumed in wire resistance in an inductor
C. Power lost because of capacitor leakage
D. Power consumed in circuit Q

ANSWER A: The letters AC are found in the word reactive (meaning out-of-phase). It indicates *non-productive power* (key words) in circuits containing inductors and capacitors. Since out-of-phase power in inductors will tend to cancel out-of-phase power in capacitors, reactive power is wattless. It is not converted to heat and dissipated. It is energy being stored temporarily in a field and then returned to the circuit. It circulates back and forth in a coil's magnetic field and/or a capacitor's electrostatic field.

3D2 What is the term for an *out-of-phase, non-productive power* associated with inductors and capacitors?
A. Effective power
B. True power
C. Peak envelope power
D. Reactive power

ANSWER D: If it is non-productive, it is reactive power in an AC circuit. It is energy stored temporarily in a coil's magnetic field or a capacitor's electrostatic field and then returned to the circuit. It is power not converted to heat and dissipated.

3D3 What is the term for energy that is stored in an electromagnetic or electrostatic field?
A. Potential energy
B. Amperes-joules
C. Joules-coulombs
D. Kinetic energy

ANSWER A: When you go into the examination room to take your Commercial General Radiotelephone Element 3 test, you will have a great deal of information energy stored as a result of the studying you have done. You will have *potential energy* (key words) all stored up, ready to react to your test.

3D4 What is responsible for the phenomenon when voltages across reactances in series can often be larger than the voltages applied to them?
A. Capacitance
B. Resonance
C. Conductance
D. Resistance

ANSWER B: Coils (having inductance) and capacitors (having capacitance) will be at resonance when the capacitive reactance equals the inductive reactance. At resonance, large voltages greater than the applied voltage can be present across these components in the circuit. You can see this with a mobile whip antenna. If something were to touch the whip in transmit, you may see quite a spark!

3D – ELECTRICAL PRINCIPLES

X_L = Inductive Reactance

$X_L = 2\pi f L$

X_C = Capacitive Reactance

$X_C = \dfrac{1}{2\pi f C}$

At Resonance $X_L = X_C$

$$2\pi f_r L = \dfrac{1}{2\pi f_r C}$$

Solving for f_r

$$f_r^2 = \dfrac{1}{(2\pi)^2 LC}$$

The Resonant Frequency, f_r, is:

$$f_r = \dfrac{1}{2\pi \sqrt{LC}}$$

Equation for Resonant Frequency

3D5 What is *resonance* in an electrical circuit?
A. The highest frequency that will pass current
B. The lowest frequency that will pass current
C. The frequency at which capacitive reactance equals inductive reactance
D. The frequency at which power factor is at a minimum

ANSWER C: At resonance, X_L (inductive reactance) equals X_C (capacitive reactance). In a mobile series resonant whip antenna, there will be maximum current through the loading coil when the whip is tuned to resonance. On your base station vertical trap antenna or trap beam antenna, parallel resonant circuits offer high impedance, trapping out a specific length of the antenna for a certain frequency. At resonance, the resonant parallel circuit looks like a high impedance to any of your power getting through to the longer length of the antenna system.

3D6 Under what conditions does resonance occur in an electrical circuit?
A. When the power factor is at a minimum
B. When inductive and capacitive reactances are equal
C. When the square root of the sum of the capacitive and inductive reactances is equal to the resonant frequency
D. When the square root of the product of the capacitive and inductive reactances is equal to the resonant frequency

ANSWER B: When a coil's inductive reactance equals a capacitor's capacitive reactance, resonance occurs. On antenna systems, you fine-tune your elements for resonance at a certain frequency.

3D7 What is the term for the phenomena which occurs in an electrical circuit when the inductive reactance equals the capacitive reactance?
A. Reactive quiescence
B. High Q
C. Reactive equilibrium
D. Resonance

ANSWER D: Resonance in a circuit occurs at the frequency where inductive reactance equals capacitive reactance.

5 GENERAL RADIOTELEPHONE OPERATOR LICENSE

3D8 What is the approximate magnitude of the impedance of a series R-L-C circuit at resonance?
A. High, as compared to the circuit resistance
B. Approximately equal to the circuit resistance
C. Approximately equal to XL
D. Approximately equal to XC

ANSWER B: On the 2-MHz marine band whips, you will see that manufacturers use an extremely large loading coil for this series resonant antenna. The bigger the loading coil, the less the coil resistance. At resonance, the magnitude of the impedance is equal to the *circuit resistance* (key words).

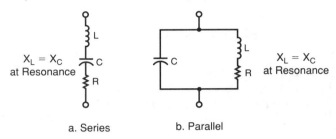

a. Series   b. Parallel

Series and Parallel Resonant Circuits

3D9 What is the approximate magnitude of the impedance of a parallel R-L-C circuit at resonance?
A. Approximately equal to the circuit resistance
B. Approximately equal to XL
C. Low, as compared to the circuit resistance
D. Approximately equal to XC

ANSWER A: Whether the system is parallel or series R-L-C resonant, the impedance will always be *equal to the circuit resistance* (key words).

3D10 What is the characteristic of the current flow in a series R-L-C circuit at resonance?
A. It is at a minimum
B. It is at a maximum
C. It is DC
D. It is zero

ANSWER B: A mobile whip antenna will have maximum current in a series R-L-C circuit at resonance. One way you can tell whether or not a mobile whip is working properly is to feel the coil after the transmitter has been shut down. If it's warm, there has been current through the coil and up to the whip tip stinger.

3D11 What is the characteristic of the current flow in a parallel R-L-C circuit at resonance?
A. The current circulating in the parallel elements is at a minimum
B. The current circulating in the parallel elements is at a maximum
C. The current circulating in the parallel elements is DC
D. The current circulating in the parallel elements is zero

ANSWER B: A parallel circuit at resonance is similar to a tuned trap in a multi-band trap antenna. At resonance, the trap keeps power from going any further to the longer antenna elements. But be assured there is maximum circulating current within the parallel R-L-C circuit. Now don't get confused with this question—a parallel circuit has maximum current within it, and minimum current going to the next stage. That's why they call those parallel resonant circuits "traps."

3D12 What is the *skin effect*?
A. The phenomenon where RF current flows in a thinner layer of the conductor, close to the surface, as frequency increases
B. The phenomenon where RF current flows in a thinner layer of the conductor, close to the surface, as frequency decreases
C. The phenomenon where thermal effects on the surface of the conductor increase the impedance
D. The phenomenon where thermal effects on the surface of the conductor decrease the impedance

ANSWER A: If you ever had a chance to inspect an old wireless station, you would have seen that the antenna "plumbing" leading out of the transmitter was usually constructed of hollow copper tubing. Radio-frequency current always travels along the thinner outside layer of a conductor, and as frequency increases, there is almost no current in the center of the conductor. The higher the frequency, the greater the skin effect. This is why we use wide copper ground foil to minimize the resistance to AC current that we need to pass to ground.

3D13 What is the term for the phenomenon where most of an RF current flows along the surface of the conductor?
A. Layer effect
B. Seeburg Effect
C. Skin effect
D. Resonance

ANSWER C: Marine single-sideband whip antennas may use hollow tubing as their radiators. This is because most of the RF current travels along the surface of the conductor. This phenomenon is called skin effect.

3D14 Where does practically all of the RF current flow in a conductor?
A. Along the surface
B. In the center of the conductor
C. In the magnetic field around the conductor
D. In the electromagnetic field in the conductor center

ANSWER A: As frequency increases, RF currents travel in the thin surface layers of the conductors, not within the body. On large high-power shipboard SSB transmitters, small copper tubing is an excellent way to pass the energy out of the ship's radio room, and up to the antenna system.

3D15 Why does practically all of an RF current flow within a few thousandths-of-an-inch of the conductor's surface?
A. Because of skin effect
B. Because the RF resistance of the conductor is much less than the DC resistance
C. Because of heating of the metal at the conductor's interior
D. Because of the AC resistance of the conductor's self inductance

ANSWER A: It is the skin effect that keeps the RF current traveling within a few thousandths of an inch of a conductor's surface.

5 GENERAL RADIOTELEPHONE OPERATOR LICENSE

3D16 Why is the resistance of a conductor different for RF current than for DC?
A. Because the insulation conducts current at radio frequencies
B. Because of the Heisenburg Effect
C. Because of skin effect
D. Because conductors are non-linear devices

ANSWER C: The higher the frequency of an alternating current (AC) signal, the greater the skin effect. For DC, the internal cross-sectional area of a conductor is very important, but for radio frequency signals, it's not important.

3D17 What is a *magnetic field*?
A. Current flow through space around a permanent magnet
B. A force set up when current flows through a conductor
C. The force between the plates of a charged capacitor
D. The force that drives current through a resistor

ANSWER B: In marine and aeronautical installations, you would always route your power cable well away from the magnetic or fluxgate compass assembly. As current passes through a conductor, the current produces a magnetic field as a force around the conductor. Stronger current will produce a stronger magnetic field, which could affect the reading of a magnetic or fluxgate compass.

3D18 In what direction is the magnetic field about a conductor when current is flowing?
A. In the same direction as the current
B. In a direction opposite to the current flow
C. In all directions; omnidirectional
D. In a direction determined by the left hand rule

ANSWER D: If you grab a big straight cable with your left hand, with your thumb pointing in the direction the electrons are flowing in the conductor (opposite from conventional current), your fingers will naturally point in the direction of the magnetic lines of force. This is known as the left-hand rule. Have you ever been to a junk yard and watched the electromagnet coil magically lift a car off the ground and drop it into a nearby boxcar? When the coil is energized, it becomes a magnet. As quickly as the energy is cut off, magnetism stops, and the car drops off.

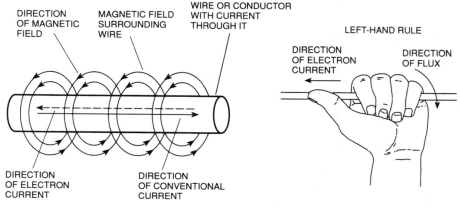

Left-Hand Rule

3D – ELECTRICAL PRINCIPLES

3D19 What device is used to store electrical energy in an electrostatic field?
A. A battery
B. A transformer
C. A capacitor
D. An inductor

ANSWER C: A capacitor is made up of parallel plates separated by a dielectric (non-conductor). When a voltage is placed across a capacitor, energy is stored in the electrostatic field developed between the capacitor plates. As a reminder, notice the letters "AC" in the word capacitor, and the letters "ATIC" in the word electrostatic. It should lead you to the correct answer.

3D20 What is the term used to express the amount of electrical energy stored in an electrostatic field?
A. Coulombs
B. Joules
C. Watts
D. Volts

ANSWER B: The joule is a measure of energy, or the capacity to do work. The amount of electrical energy stored in an electrostatic field associated with capacitors is expressed in joules. If a capacitor is charged with one coulomb of electrons in a second by a voltage of one volt, one joule of energy is stored in the capacitor's electrostatic field. Using energy at a joule per second is a watt of electrical power.

3D21 What factors determine the capacitance of a capacitor?
A. Area of the plates, voltage on the plates and distance between the plates
B. Area of the plates, distance between the plates and the dielectric constant of the material between the plates
C. Area of the plates, voltage on the plates and the dielectric constant of the material between the plates
D. Area of the plates, amount of charge on the plates and the dielectric constant of the material between the plates

ANSWER B: Some of the SSB equipment you may work on will still use variable capacitors. The amount of capacitance is determined by the area of the plates as you turn the variable capacitor, the distance between the plates, and the dielectric constant of the material between the plates, usually air in variable capacitors. As you engage more of the plates next to each other, capacitance increases.

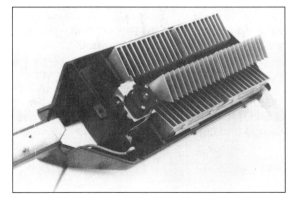

Variable Capacitor
Courtesy of AEA, Inc.

5 GENERAL RADIOTELEPHONE OPERATOR LICENSE

3D22 What is the dielectric constant for air?
   A. Approximately 1
   B. Approximately 2
   C. Approximately 4
   D. Approximately 0

ANSWER A: The dielectric constant for air is approximately 1. For atmospheric propagation studies at VHF frequencies, the dielectric constant of air at 1 is normally abbreviated "n", for normal. Glass has a dielectric constant of 8; Bakelite®, 5; and Teflon®, 2. Teflon is used extensively at VHF and UHF frequencies because it can firmly hold components in place, and has approximately the dielectric constant of air.

3D23 What determines the strength of the magnetic field around a conductor?
   A. The resistance divided by the current
   B. The ratio of the current to the resistance
   C. The diameter of the conductor
   D. The amount of current

ANSWER D: Back to the electromagnetic field that develops around conductors—the more current passing through a conductor, the greater the magnetic field strength around the conductor.

3D24 Why would the rate at which electrical energy is used in a circuit be less than the product of the magnitudes of the AC voltage and current?
   A. Because there is a phase angle that is greater than zero between the current and voltage
   B. Because there are only resistances in the circuit
   C. Because there are no reactances in the circuit
   D. Because there is a phase angle that is equal to zero between the current and voltage

ANSWER A: When the *phase angle is greater than zero* (key words) between voltage and current, the rate at which electrical energy is used in a circuit is going to be less than the product of the magnitudes of the AC voltage and current. Since the voltage and current aren't in phase, the true power will be less because there is some reactive power present. See question 3D25.

3D25 In a circuit where the AC voltage and current are out of phase, how can the true power be determined?
   A. By multiplying the apparent power times the power factor
   B. By subtracting the apparent power from the power factor
   C. By dividing the apparent power by the power factor
   D. By multiplying the RMS voltage times the RMS current

ANSWER A: If we multiply apparent power times the power factor of an AC voltage and current out of phase, we can determine true power.

3D – ELECTRICAL PRINCIPLES

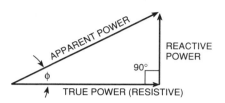

True power is power dissipated as heat, while reactive power of a circuit is stored in inductances or capacitances and then returned to the circuit.

Apparent power is voltage times current without taking into account the phase angle between them. According to right angle trigonometry,

$$\text{cosine } \phi = \frac{\text{side adjacent}}{\text{hypotenuse}}$$

therefore,

$$\text{cosine } \phi = \frac{\text{True Power}}{\text{Apparent Power}}$$

therefore,

True Power = Apparent Power × Cosine ϕ

Cosine ϕ is called the *power factor* (PF) of a circuit.

Since apparent power is E × I,

True Power = E × I × Cos ϕ

or

True Power = E × I × PF

Apparent and True Power

3D26 What does the power factor equal in an R-L circuit having a 60 degree phase angle between the voltage and the current?
 A. 1.414
 B. 0.866
 C. 0.5
 D. 1.73

ANSWER C: If you are using a scientific calculator, you will find the cosine of 60 degrees, given in the problem, comes out 0.5, answer C. The power factor is cos ϕ. If you don't have a scientific calculator, you might memorize this table:

ϕ	cos ϕ
30°	0.866
45°	0.707
60°	0.5

3D27 What does the power factor equal in an R-L circuit having a 45 degree phase angle between the voltage and the current?
 A. 0.866
 B. 1.0
 C. 0.5
 D. 0.707

ANSWER D: Using your scientific calculator, or from memory, remember that the cosine of 45 degrees is 0.707. Think of an airplane, the very popular 707 jet aircraft, taking off at an angle of 45 degrees. See table at 3D26.

5 GENERAL RADIOTELEPHONE OPERATOR LICENSE

3D28 What does the power factor equal in an R-L circuit having a 30 degree phase angle between the voltage and the current?
- A. 1.73
- B. 0.5
- C. 0.866
- D. 0.577

ANSWER C: At 30 degrees, the power factor is 0.866. If you don't pass your Commercial exam, you could be 86'd out of the room! See table at 3D26.

3D29 How many watts are being consumed in a circuit having a power factor of 0.2 when the input is 100 VAC and 4 amperes is being drawn?
- A. 400 watts
- B. 80 watts
- C. 2000 watts
- D. 50 watts

ANSWER B: Here come two simple problems that are based on the formula:

    True power = E (volts) × I (amps) × PF (power factor).

True power works out to be 100 volts × 4 amps × 0.2, giving us 80 watts. Easy enough, right?

3D30 How many watts are being consumed in a circuit having a power factor of 0.6 when the input is 200 VAC and 5 amperes is being drawn?
- A. 200 watts
- B. 1000 watts
- C. 1600 watts
- D. 600 watts

ANSWER D: True power = E × I × PF; therefore, it is 200 volts × 5 amps × 0.6 = 600 watts. Easy!

3D31 What is the effective radiated power of a repeater with 50 watts transmitter power output, 4 dB feedline loss, 3 dB duplexer and circulator loss, and 6 dB antenna gain?
- A. 158 watts, assuming the antenna gain is referenced to a half-wave dipole
- B. 39.7 watts, assuming the antenna gain is referenced to a half-wave dipole
- C. 251 watts, assuming the antenna gain is referenced to a half-wave dipole
- D. 69.9 watts, assuming the antenna gain is referenced to a half-wave dipole

ANSWER B: You will have one question out of the following 10 questions based on simple dB calculations at a repeater site. Don't worry about logarithms, you can do these in your head, and all you must remember is: 3 dB = a 2X increase; to twice the value; and −3 dB = a 2X loss to one half the value.

To solve the problems, just add and subtract all the dB's, and if it ends up about +3 dB, then double your transmitter power output to end up with effective radiated power. If your answer ends up around −3 dB gain, cut your transmitter power in half for effective radiated power. In this question, we have −7 dB loss, offset by 6 dB gain for 1 dB net loss. If we started out with 50 watts, our effective radiated power (ERP) will be slightly less, so 39.7 is the only answer in the ballpark.

3D – ELECTRICAL PRINCIPLES

3D32 What is the effective radiated power of a repeater with 50 watts transmitter power output, 5 dB feedline loss, 4 dB duplexer and circulator loss, and 7 dB antenna gain?
- A. 300 watts, assuming the antenna gain is referenced to a half-wave dipole
- B. 315 watts, assuming the antenna gain is referenced to a half-wave dipole
- C. 31.5 watts, assuming the antenna gain is referenced to a half-wave dipole
- D. 69.9 watts, assuming the antenna gain is referenced to a half-wave dipole

ANSWER C: Here we have 9 dB loss, and 7 dB gain. That's just about a half-power loss (−3 dB), and the only answer that is reasonably close is answer C, 31.5 watts.

3D33 What is the effective radiated power of a repeater with 75 watts transmitter power output, 4 dB feedline loss, 3 dB duplexer and circulator loss, and 10 dB antenna gain?
- A. 600 watts, assuming the antenna gain is referenced to a half-wave dipole
- B. 75 watts, assuming the antenna gain is referenced to a half-wave dipole
- C. 18.75 watts, assuming the antenna gain is referenced to a half-wave dipole
- D. 150 watts, assuming the antenna gain is referenced to a half-wave dipole

ANSWER D: We have 7 dB loss, and a great antenna system giving us 10 dB gain. This gives us a +3 dB, which is 2 times our power output. There it is, only one answer at 150 watts!

3D34 What is the effective radiated power of a repeater with 75 watts transmitter power output, 5 dB feedline loss, 4 dB duplexer and circulator loss, and 6 dB antenna gain?
- A. 37.6 watts, assuming the antenna gain is referenced to a half-wave dipole
- B. 237 watts, assuming the antenna gain is referenced to a half-wave dipole
- C. 150 watts, assuming the antenna gain is referenced to a half-wave dipole
- D. 23.7 watts, assuming the antenna gain is referenced to a half-wave dipole

ANSWER A: 9 dB loss, offset by 6 dB antenna gain gives a 3 dB net loss, or half power. Half of 75 watts is answer A, 37.6 watts.

3D35 What is the effective radiated power of a repeater with 100 watts transmitter power output, 4 dB feedline loss, 3 dB duplexer and circulator loss, and 7 dB antenna gain?
- A. 631 watts, assuming the antenna gain is referenced to a half-wave dipole
- B. 400 watts, assuming the antenna gain is referenced to a half-wave dipole

5 GENERAL RADIOTELEPHONE OPERATOR LICENSE

    C. 25 watts, assuming the antenna gain is referenced to a half-wave dipole

    D. 100 watts, assuming the antenna gain is referenced to a half-wave dipole

ANSWER D: 7 dB down the tubes, offset by 7 dB gain. No change on power output, giving us still 100 watts ERP.

3D36 What is the effective radiated power of a repeater with 100 watts transmitter power output, 5 dB feedline loss, 4 dB duplexer and circulator loss, and 10 dB antenna gain?

    A. 800 watts, assuming the antenna gain is referenced to a half-wave dipole

    B. 126 watts, assuming the antenna gain is referenced to a half-wave dipole

    C. 12.5 watts, assuming the antenna gain is referenced to a half-wave dipole

    D. 1260 watts, assuming the antenna gain is referenced to a half-wave dipole

ANSWER B: A total of 9-dB loss, offset by 10-dB gain, net +1 dB. Since we had 100 watts to start with and the net gain is 1 dB, the only close answer is answer B, 126 watts.

3D37 What is the effective radiated power of a repeater with 120 watts transmitter power output, 5 dB feedline loss, 4 dB duplexer and circulator loss, and 6 dB antenna gain?

    A. 601 watts, assuming the antenna gain is referenced to a half-wave dipole

    B. 240 watts, assuming the antenna gain is referenced to a half-wave dipole

    C. 60 watts, assuming the antenna gain is referenced to a half-wave dipole

    D. 379 watts, assuming the antenna gain is referenced to a half-wave dipole

ANSWER C: Here we have 9 dB loss with 6 dB gain. That's a net loss of 3 dB. It cuts our 120-watt transmitter to an effective radiated power of only 60 watts. Quit buying junk coax!

3D38 What is the effective radiated power of a repeater with 150 watts transmitter power output, 4 dB feedline loss, 3 dB duplexer and circulator loss, and 7 dB antenna gain?

    A. 946 watts, assuming the antenna gain is referenced to a half-wave dipole

    B. 37.5 watts, assuming the antenna gain is referenced to a half-wave dipole

    C. 600 watts, assuming the antenna gain is referenced to a half-wave dipole

    D. 150 watts, assuming the antenna gain is referenced to a half-wave dipole

ANSWER D: This one is a wash. You have −7 dB offset by +7 dB. This would make that 150-watt transmitter have an effective radiated power of 150 watts.

3D – ELECTRICAL PRINCIPLES

3D39 What is the effective radiated power of a repeater with 200 watts transmitter power output, 4 dB feedline loss, 4 dB duplexer and circulator loss, and 10 dB antenna gain?
- A. 317 watts, assuming the antenna gain is referenced to a half-wave dipole
- B. 2000 watts, assuming the antenna gain is referenced to a half-wave dipole
- C. 126 watts, assuming the antenna gain is referenced to a half-wave dipole
- D. 260 watts, assuming the antenna gain is referenced to a half-wave dipole

ANSWER A: An 8 dB loss, offset by 10 dB gain for a net of +2 dB is not quite a double in ERP. Your only close answer is a 200-watt transmitter with an effective radiated power of 317 watts. It's almost 3 dB, which would give you an answer of 400 watts. 317 watts is closer than an incorrect answer of 126 watts (actually a loss), and a slight gain of only 260 watts.

3D40 What is the effective radiated power of a repeater with 200 watts transmitter power output, 4 dB feedline loss, 3 dB duplexer and circulator loss, and 6 dB antenna gain?
- A. 252 watts, assuming the antenna gain is referenced to a half-wave dipole
- B. 63.2 watts, assuming the antenna gain is referenced to a half-wave dipole
- C. 632 watts, assuming the antenna gain is referenced to a half-wave dipole
- D. 159 watts, assuming the antenna gain is referenced to a half-wave dipole

ANSWER D: You have a total loss of 7 dB, and a puny antenna gain of only +6 dB. This means that you are down 1 dB, causing your 200-watt transmitter to radiate more like a 159-watt transmitter, the only logical answer below 200 watts.

3D41 What is the *photoconductive effect*?
- A. The conversion of photon energy to electromotive energy
- B. The increased conductivity of an illuminated semiconductor junction
- C. The conversion of electromotive energy to photon energy
- D. The decreased conductivity of an illuminated semiconductor junction

ANSWER B: The photo cell has high resistance when no light shines on it, and a varying resistance when light shines on it. It can be used in an ON or OFF state as a simple beam alarm across a doorway, or may be found in almost all modern SLR cameras as a sensitive light meter to judge the amount of light present.

3D42 What happens to photoconductive material when light shines on it?
- A. The conductivity of the material increases
- B. The conductivity of the material decreases
- C. The conductivity of the material stays the same
- D. The conductivity of the material becomes temperature dependent

ANSWER A: In the dark, a photo cell has high resistance. When you shine light on it, the conductivity increases as the resistance decreases.

3D43 What happens to the resistance of a photoconductive material when light shines on it?
A. It increases
B. It becomes temperature dependent
C. It stays the same
D. It decreases

ANSWER D: More light, less resistance. The resistance varies inversely with the light.

Light Incidence	Current I	Resistance	Test Conditions
Dark	5nA	2000MΩ	$V_R = 10V$ $*E_e = 0$
Light	15µA	666.7kΩ	$V_R = 10V$ $*E_e = 250µW/cm^2$ at 940nm

*Irradiance (E_e) is the radiant power per unit area incident on surface

Resistance Variation of a Photo Diode

3D44 What happens to the conductivity of a semiconductor junction when it is illuminated?
A. It stays the same
B. It becomes temperature dependent
C. It increases
D. It decreases

ANSWER C: Conductivity increases when light shines on a semiconductor junction—just like the photo cell. More light; less resistance.

3D45 What is an *optocoupler*?
A. A resistor and a capacitor
B. A frequency modulated helium-neon laser
C. An amplitude modulated helium-neon laser
D. An LED and a phototransistor

ANSWER D: Optocouplers are found in many modern marine and aviation transceivers. When you tune the dial for a new channel, you are no longer tuning a big variable capacitor. Rather, you are interrupting a light beam on a phototransistor.

3D46 What is an *optoisolator*?
A. An LED and a phototransistor
B. A P-N junction that develops an excess positive charge when exposed to light
C. An LED and a capacitor
D. An LED and a solar cell

ANSWER A: Another name for an optocoupler is an optoisolator. A LED shines through a window at a phototransistor in an optoisolator.

3D47 What is an *optical shaft encoder*?
A. An array of optocouplers chopped by a stationary wheel
B. An array of optocouplers whose light transmission path is controlled by a rotating wheel

3D – ELECTRICAL PRINCIPLES

C. An array of optocouplers whose propagation velocity is controlled by a stationary wheel
D. An array of optocouplers whose propagation velocity is controlled by a rotating wheel

ANSWER B: When you spin the big knob of your new high-frequency transceiver it is connected to an optical shaft encoder that determines the tuned frequency.

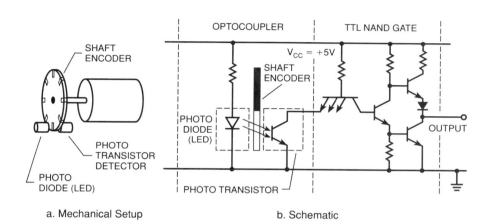

a. Mechanical Setup   b. Schematic

Optocoupler Used for Shaft Encoder

3D48 What does the *photoconductive effect* in crystalline solids produce a noticeable change in?
A. The capacitance of the solid
B. The inductance of the solid
C. The specific gravity of the solid
D. The resistance of the solid

ANSWER D: The resistance of the crystalline solid varies when light shines on it because of the photoconductive effect.

3D49 What is the meaning of the term *time constant* of an RC circuit?
A. The time required to charge the capacitor in the circuit to 36.8% of the supply voltage
B. The time required to charge the capacitor in the circuit to 36.8% of the supply current
C. The time required to charge the capacitor in the circuit to 63.2% of the supply current
D. The time required to charge the capacitor in the circuit to 63.2% of the supply voltage

ANSWER D: If you have a separate power supply feeding your equipment, when you turn off your power supply with the equipment still turned on, you will notice that the power supply output voltage decreases slowly. The slow decay is because of the voltage still across the large electrolytic filter capacitors on the output of the power supply. This is a visual example that capacitors charge up, and discharge, on a curve. The time required to charge (or discharge in this case) the capacitor in an RC (Resistance-Capacitance) circuit to 63.2 percent of the supply voltage is called a time constant, τ, where:

▶ τ = RC τ = Greek letter tau, the time constant in **seconds**
R = total resistance in **ohms**
C = capacitance in **farads**

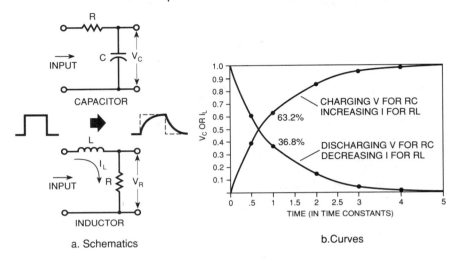

RC (R×C) and RL (L/R) Time Constant

3D50 What is the meaning of the term *time constant* of an RL circuit?
A. The time required for the current in the circuit to build up to 36.8% of the maximum value
B. The time required for the voltage in the circuit to build up to 63.2% of the maximum value
C. The time required for the current in the circuit to build up to 63.2% of the maximum value
D. The time required for the voltage in the circuit to build up to 36.8% of the maximum value

ANSWER C: In an RL circuit, the coil will develop a back EMF which will oppose the flow of current. In the RL circuit, the time required for the current to build up to 63.2 percent of the maximum value is also called a time constant, τ, where:

▶ τ = $\dfrac{L}{R}$ τ = Greek letter tau, the time constant in **seconds**
L = total inductance in **henries**
R = total resistance in **ohms**

3D51 What is the term for the time required for the capacitor in an RC circuit to be charged to 63.2% of the supply voltage?
A. An exponential rate of one
B. One time constant
C. One exponential period
D. A time factor of one

ANSWER B: In an RC circuit, assuming there is no initial charge on the capacitor, it takes one time constant to charge a capacitor to 63.2 percent of its final value.

3D – ELECTRICAL PRINCIPLES

3D52 What is the term for the time required for the current in an RL circuit to build up to 63.2% of the maximum value?
A. One time constant
B. An exponential period of one
C. A time factor of one
D. One exponential rate

ANSWER A: In an RL circuit, assuming there is no initial current through the inductance, it takes one time constant for the current in the circuit to build up to 63.2 percent of its final value. Measuring V_R tracks I_L.

3D53 What is the term for the time it takes for a charged capacitor in an RC circuit to discharge to 36.8% of its initial value of stored charge?
A. One discharge period
B. An exponential discharge rate of one
C. A discharge factor of one
D. One time constant

ANSWER D: Remember the example of the filter capacitor in the power supply, where the capacitors discharged when the input power was removed? The time it takes a capacitor in an RC circuit to discharge to 36.8 percent of its initial value of stored charge is one time constant.

3D54 What is meant by *back EMF?*
A. A current equal to the applied EMF
B. An opposing EMF equal to R times C (RC) percent of the applied EMF
C. A current that opposes the applied EMF
D. A voltage that opposes the applied EMF

ANSWER D: A back EMF which opposes the applied EMF is developed in an RL circuit by the inductor. This reduces the current flow in the inductor and the circuit.

3D55 After two time constants, the capacitor in an RC circuit is charged to what percentage of the supply voltage?
A. 36.8%
B. 63.2%
C. 86.5%
D. 95%

ANSWER C: To calculate the percentage of the supply voltage after two time constants in an RC circuit, write down the percent after a single time constant as 63.2 percent. The remaining percent of charge that the capacitor must charge is 36.8 percent. It is found by deducting 63.2 percent from 100 percent. Since 36.8 percent is the final value to which the capacitor will charge after it has charged to 63.2 percent, in the next time constant the capacitor will charge to 63.2 percent of the 36.8 percent. In other words, another percent charge is added to the capacitor in the second time constant equal to:

$$36.8\% \times 63.2\% = 23.26\% \text{ (rounded to 23.3\%)}$$

This really needs to be calculated in decimal format as:

$$0.368 \times 0.632 = 0.2326$$

Percent values are converted to decimals by moving the decimal point two places to the left. Decimal values are converted to percent values by multiplying by 100 (moving decimal point two places to the right.)

Since the final percent of charge in two time constants is desired, it is only necessary to add the two percent charges together:

	Percent	Decimal
Charge in first time constant =	63.2%	0.632
Charge in second time constant =	23.3%	0.233
Total charge after two time constants	86.5%	0.865

The process can be continued for the third time constant. Since the capacitor has charged to 86.5 percent in two time constants, if you start from that point, the final value of charge would be another 13.5 percent. Therefore, in the next time constant, the third, the capacitor would charge another:

13.5% × 63.2% = 8.53% (rounded to 8.5%)

After the third time constant, the total charge is:

86.5% + 8.5% = 95%

If the process is continued through five time constants, the capacitor charge is 99.3 percent. Therefore, in electronic calculations, the capacitor is considered to be fully charged (or discharged) after five time constants.

3D56 After two time constants, the capacitor in an RC circuit is discharged to what percentage of the starting voltage?
A. 86.5%
B. 63.2%
C. 36.8%
D. 13.5%

ANSWER D: This question is actually the reverse of the previous question. It is asking to what percent the capacitor has discharged. The capacitor discharges the same percent in a time constant as it charges (for the same RC circuit values, of course.) However, the question is asking for the amount of charge remaining. At one time constant, the capacitor discharges 63.2 percent (in the previous question, it charged 63.2 percent in one time constant); therefore, the percent charge remaining on the capacitor is 36.8 percent (100%-63.2%). From the previous question, the capacitor charges another 23.3 percent in the second time constant. Thus, the capacitor on discharge, discharges another 23.3 percent in the second time constant. After two time constants the capacitor has discharged by 86.5 percent; therefore, the remaining charge is 13.5 percent (100%-86.5%). In five time constants the capacitor would have discharged 99.3 percent of its charge, so 0.7 percent of charge remains—it is considered to be fully discharged.

3D57 What is the time constant of a circuit having a 100-microfarad capacitor in series with a 470-kilohm resistor?
A. 4700 seconds
B. 470 seconds
C. 47 seconds
D. 0.47 seconds

ANSWER C: First of all, this is an RC circuit, so remember the formula:

$\tau = RC$

Time constant (**seconds**) = Resistance (**ohms**) × Capacitance (**farads**)

Remember that "micro" means × 10^{-6} (the multiplication is accomplished by moving the decimal point 6 places to the left), and "kilo" means × 10^{+3} (the multiplication is accomplished by moving the decimal point 3 places to the right). The problem solution is as follows:

$\tau = 470 \times 10^{+3} \times 100 \times 10^{-6}$
$\tau = 4.7 \times 10^{+2} \times 10^{+3} \times 1 \times 10^{+2} \times 10^{-6}$
$\tau = 4.7 \times 10^{+1} = 47$ seconds

Although this is fairly easy, you can make it easier by using a calculator. Turn on your calculator, press the clear button (several times for good luck), and then execute the following keystrokes:

470000 × .0001 = and presto, 47 appears on the display.

This is the correct answer. Be sure you understand that 470 kilohms equals 470000 and that 100 microfarads equals .0001.

3D58 What is the time constant of a circuit having a 220-microfarad capacitor in parallel with a 1-megohm resistor?
   A. 220 seconds
   B. 22 seconds
   C. 2.2 seconds
   D. 0.22 seconds

ANSWER A: Since it is an RC circuit, use $\tau = RC$.

Using powers of 10:

$\tau = 1 \times 10^{+6} \times 220 \times 10^{-6}$ $\tau = 220$ seconds

Using only decimal values:

$\tau = 1000000 \times .000220$ $\tau = 220$ seconds

Remember that when using powers of ten: to multiply, you add the exponents; to divide, you subtract the exponent of the denominator (lower one) from the exponent of the numerator (upper one). Since "meg" = 10^{+6} and "micro" = 10^{-6}, multiplying "megs" times "micros" means you add +6 and −6, and the result is 0 or 10^0, which is 1. In this case, 220 × 1 = 220. Actually, this stuff is pretty simple once you get the hang of it.

To calculate using decimal values on your calculator, the calculator keystrokes are: Clear, 1000000 × .000220 = 220.

3D59 What is the time constant of a circuit having two 100-microfarad capacitors and two 470-kilohm resistors all in series?
   A. 470 seconds
   B. 47 seconds
   C. 4.7 seconds
   D. 0.47 seconds

5 GENERAL RADIOTELEPHONE OPERATOR LICENSE

ANSWER B: For an RC circuit $\tau = RC$, but first we must combine resistor and capacitor values to come up with a single value for each. Since this is a series circuit, the resistor values add ($R_T = R_1 + R_2$) to arrive at a total resistance, but for the capacitance total you have to solve the formula $C_T = (C_1 \times C_2) \div (C_1 + C_2)$. Therefore,

Powers of 10:

$$C_T = \frac{(1 \times 10^{+2} \times 10^{-6} \times 1 \times 10^{+2} \times 10^{-6})}{(100 \times 10^{-6} + 100 \times 10^{-6})}$$

$$C_T = \frac{(1 \times 10^{-8})}{(2 \times 10^{+2} \times 10^{-6})}$$

$$C_T = \frac{(1 \times 10^{-8})}{2 \times 10^{-4}} = 0.5 \times 10^{-4} = 50 \times 10^{-6}$$

$C_T = 0.00005$ farads or 50 microfarads

For the resistance:

$R_T = 470{,}000 + 470{,}000 = 940{,}000$
$R_T = 4.7 \times 10^{+5} + 4.7 \times 10^{+5} = 9.4 \times 10^{+5}$

Now that we have calculated the total series capacitance and the total resistance, we simply multiply $0.00005 \times 940{,}000 = 47$ seconds. In powers of ten this is $\tau = 0.5 \times 10^{-4} \times 9.4 \times 10^{+5} = 4.7 \times 10^{+1} = 47$.

Of course, the old timers that recognize that two equal capacitors in series equals a single capacitor at half the value would immediately jump to $\tau = 50$ microfarads $\times$ 940 kilohms, pull out the calculator and press the following: Clear, $.00005 \times 940000 = 47$. The time constant is in seconds. Notice if you miss a single decimal point, they have an incorrect answer for you. Watch out!

Remember:

Capacitors in Series:

$$C_T = \frac{C_1 \times C_2}{C_1 + C_2}$$

Capacitors in Parallel:

$$C_T = C_1 + C_2$$

C = Capacitance in **farads**

Resistors in Series:

$$R_T = R_1 + R_2$$

Resistors in Parallel:

$$R_T = \frac{R_1 \times R_2}{R_1 + R_2}$$

R = Resistance in **ohms**

Capacitors, Resistors in Series and Parallel

3D60 What is the time constant of a circuit having two 100-microfarad capacitors and two 470-kilohm resistors all in parallel?
    A. 470 seconds
    B. 47 seconds
    C. 4.7 seconds
    D. 0.47 seconds

3D – ELECTRICAL PRINCIPLES

ANSWER B: Again a RC circuit, only this time everything is in parallel. Since the capacitors are in parallel, the capacitance values add ($C_T = C_1 + C_2$), but for the total resistance you have to solve the formula $R_T = (R_1 \times R_2) \div (R_1 + R_2)$. Therefore,

Powers of 10:

$$R_T = \frac{(4.7 \times 10^{+5} \times 4.7 \times 10^{+5})}{(4.7 \times 10^{+5} + 4.7 \times 10^{+5})}$$

$$R_T = \frac{(22.09 \times 10^{+10})}{9.4 \times 10^{+5}} = 2.35 \times 10^{+5}$$

$$R_T = 235{,}000 \text{ ohms or 235 kilohms}$$

for the capacitance:

$$C_T = 0.000100 + 0.000100 = 0.000200$$
$$C_T = 100 \times 10^{-6} + 100 \times 10^{-6} = 200 \times 10^{-6}$$
$$C_T = 200 \text{ microfarads or } 0.0002 \; (2 \times 10^{-4}) \text{ farads}$$

Since the total resistance and the total capacitance are known, just multiply 235,000 × .000200 to end up with 47 seconds. Now that wasn't too hard, was it?

This is really easy for old timers. They know that the capacitor values add for parallel capacitors—so they have 200 microfarads. For the resistance, they know that two equal value resistors in parallel result in a single resistor at half the value or 235 kilohms. They would go immediately to τ = 200 microfarads × 235 kilohms. On their calculator they would press the following: Clear, .0002 × 235000 = 47. The time constant is in seconds, and keep exact track of the decimal points.

3D61 What is the time constant of a circuit having two 220-microfarad capacitors and two 1-megohm resistors all in series?
   A. 55 seconds
   B. 110 seconds
   C. 220 seconds
   D. 440 seconds

ANSWER C: This is another series RC circuit, however, now there are two capacitors in series and two resistors in series. Review the calculations in question 3D59 The procedure is the same. Just remember that two equal capacitors in series have a total capacitance of half of one of the capacitors, and two equal resistors in series have a total resistance that is twice the value of one of the resistors. Now you can do this one in your head. Two 220 microfarad capacitors in series is really one big 110 microfarad capacitor. Write down 0.000110 as you recall that 110 microfarads is 110×10^{-6}— "micro" means move the decimal point six places to the left. Two one-megohm resistors in series would be two megohms. Write down 2,000,000 as you recall that 2 megohms is $2 \times 10^{+6}$— "meg" means move the decimal point six places to the right. Since $\tau = 2 \times 10^{+6} \times 110 \times 10^{-6}$, you can see by the powers of ten that "megs" times "micros" cancel each other and you end up with 2 × 110 = 220 seconds. Anytime you have both "megs" and micros" multiplied together, all those zeroes to the left or right of the stated numerical value cancel.

5 GENERAL RADIOTELEPHONE OPERATOR LICENSE

3D62 What is the time constant of a circuit having two 220-microfarad capacitors and two 1-megohm resistors all in parallel?
   A. 22 seconds
   B. 44 seconds
   C. 220 seconds
   D. 440 seconds

ANSWER C: Another RC circuit with all components in parallel. Refer to question 3D60. It will be easy if you remember that two equal capacitors in parallel have a total capacitance of twice the value of one of the capacitors, and that two equal value resistors in series have a total resistance equal to one half the value of one of the resistors. And more good news—"megs" are being multiplied by "micros" (refer to question 3D61) so the powers of ten after the numerical value cancel. Two 220 microfarad capacitors in parallel is a larger 440 microfarad capacitor. Two one-megohm resistors in parallel is a resistance of 0.5 megohms. Solve for the time constant by multiplying .5 × 440 ("megs" times "micros") and you end up with 220 seconds as the correct answer. The calculator keystrokes are: Clear, .5 × 440 = 220.

3D63 What is the time constant of a circuit having one 100-microfarad capacitor, one 220-microfarad capacitor, one 470-kilohm resistor and one 1-megohm resistor all in series?
   A. 68.8 seconds
   B. 101.1 seconds
   C. 220.0 seconds
   D. 470.0 seconds

ANSWER B: This time another RC circuit with all components in series, but now the series capacitors are not equal and neither are the two resistors. You can't do this one in your head. You need to calculate the total capacitance and the total resistance. First resolve the two capacitors in series:

$$C_T = \frac{(100 \times 10^{-6} \times 220 \times 10^{-6})}{(100 \times 10^{-6} + 220 \times 10^{-6})}$$

$$C_T = \frac{(22{,}000 \times 10^{-12})}{320 \times 10^{-6}} = 68.75 \times 10^{-6}$$

$$C_T = 68.75 \text{ microfarads}$$

Now find the total resistance. Since the resistors are in series, the resistors add (470 k is 0.47 meg), therefore

$$R_T = (0.47 + 1) \text{ megohms}$$

Multiplying R times C to calculate the time constant (remembering "meg"s times "micros" cancel), gives τ = 1.47 × 68.75 = 101.1 seconds, the correct answer.

3D64 What is the time constant of a circuit having a 470-microfarad capacitor and a 1-megohm resistor in parallel?
   A. 0.47 seconds
   B. 47 seconds
   C. 220 seconds
   D. 470 seconds

ANSWER D: Easy one here—do it in your head. It's an RC circuit—one capacitor and one resistor in parallel—so τ = RC. The time constant calculations work

3D – ELECTRICAL PRINCIPLES

the same for series or parallel circuits. It's made easy by using "megs" times "micros"—the powers of ten cancel. Only one simple step: multiply 1 "megs" times 470 "micros" to get 470 seconds. There's your answer.

3D65 What is the time constant of a circuit having a 470-microfarad capacitor in series with a 470-kilohm resistor?
 A. 221 seconds
 B. 221000 seconds
 C. 470 seconds
 D. 470000 seconds

ANSWER A: Same circuit as previous question except the resistor and capacitor values are changed. Use "megs" times "micros" to make it easy, but you have one additional step—you must convert the 470 kilohm resistor value to megohms. 470,000 ohms is equal to 0.47 megohm ($0.47 \times 10^{+6}$). Therefore, $\tau = 0.47 \times 470 = 220.9$. Round this off to 221 seconds, which is answer A.

3D66 What is the time constant of a circuit having a 220-microfarad capacitor in series with a 470-kilohm resistor?
 A. 103 seconds
 B. 220 seconds
 C. 470 seconds
 D. 470000 seconds

ANSWER A: Same circuit as in previous question, so use "megs" and "micros" and convert the resistance value to megohms. 470,000 ohms is 0.47 megohm. The time constant $\tau = 0.47 \times 220 = 103.4$ seconds, rounded off to 103 seconds.

> Do you have RC time constants down pat now? You'll probably have one examination question that will require you to calculate an RC time constant.

3D67 How long does it take for an initial charge of 20 V DC to decrease to 7.36 V DC in a 0.01-microfarad capacitor when a 2-megohm resistor is connected across it?
 A. 12.64 seconds
 B. 0.02 seconds
 C. 1 second
 D. 7.98 seconds

ANSWER B: The questions now change emphasis a bit. The next five questions are still on RC circuits, but now a 0.01 microfarad capacitor is charged to 20 volts DC and then discharged through a 2-megohm resistor across it. The three values stay the same for the five problems, making it easy to calculate the progression of correct answers.

Note on the chart below that a capacitor discharges (or charges) 63.2 percent in one time constant. Thus, after discharging for one time constant, 36.8 percent of the charge remains on the capacitor. Therefore, in one time constant, the 20 volts will drop down to 36.8 percent or 7.36 volts ($.368 \times 20 = 7.36$). Write down one time constant = 7.36 volts for later reference. Since 7.36 volts matches the voltage referred to in the question, the time asked for is one time constant. Use the "megs" times "micros" again to calculate the time constant as $2 \times .01 = .02$ second. That's the correct answer.

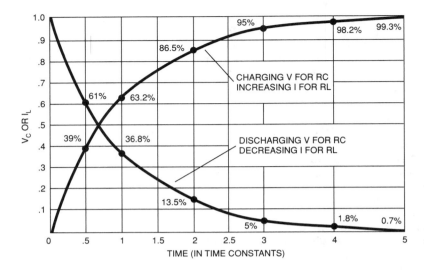

Time Constants Showing V and I Percentages

3D68 How long does it take for an initial charge of 20 V DC to decrease to 2.71 V DC in a 0.01-microfarad capacitor when a 2-megohm resistor is connected across it?
- A. 0.04 seconds
- B. 0.02 seconds
- C. 7.36 seconds
- D. 12.64 seconds

ANSWER A: Refer to the chart at question 3D67. From the previous question, the voltage across the capacitor after discharging for one time constant is 7.36 volts. If the capacitor continues to discharge for another time constant, it again will discharge 63.2 percent of its charge, and have 36.8 percent of the 7.36 volts left. Thus, after two time constants of discharge, the voltage on the capacitor will be 2.71 volts (7.36 × .368 = 2.71). Write down 2.71 = two time constants. It is the time the question is asking for. If one time constant is 0.02 second (as you calculated in question 3D67), two time constants will be 0.04 second, the correct answer.

3D69 How long does it take for an initial charge of 20 V DC to decrease to 1 V DC in a 0.01-microfarad capacitor when a 2-megohm resistor is connected across it?
- A. 0.01 seconds
- B. 0.02 seconds
- C. 0.04 seconds
- D. 0.06 seconds

ANSWER D: The discharge of the capacitor continues from the previous question. Refer to chart in question 3D67. The voltage across the capacitor is down to 2.71 volts at the start of this question. Repeating the same process as the last question, in one more time constant—the third—the capacitor will discharge another 63.2 percent. The voltage after three time constants will be 36.8 percent of 2.71 volts or 1.0 volt (2.71 × .368 = 1). Write down 1.0 V = three

3D – ELECTRICAL PRINCIPLES

time constants. 1.0 V DC matches the voltage asked for, so the time is three time constants. Three time constants = 3 × 0.02 = 0.06 second, your correct answer.

3D70 How long does it take for an initial charge of 20 V DC to decrease to 0.37 V DC in a 0.01-microfarad capacitor when a 2-megohm resistor is connected across it?
- A. 0.08 seconds
- B. 0.6 seconds
- C. 0.4 seconds
- D. 0.2 seconds

ANSWER A: The discharge continues. Refer to chart in question 3D67. Now the voltage asked for is down to 0.37 volt. Continuing the same thought process as the last three questions, a fourth time constant discharge reduces the voltage across the capacitor to 1 V × 0.368 = 0.368, rounded to 0.37 volt. Write down 0.37 volt = four time constants. Four time constants is 4 × 0.02 = 0.08 second, the correct answer. Remember the time constant for this circuit is 2 × 0.01 = 0.02 second.

3D71 How long does it take for an initial charge of 20 V DC to decrease to 0.13 V DC in a 0.01-microfarad capacitor when a 2-megohm resistor is connected across it?
- A. 0.06 seconds
- B. 0.08 seconds
- C. 0.1 seconds
- D. 1.2 seconds

ANSWER C: Now the discharge voltage asked for is all the way down to 0.13 volt DC. As in previous questions, this voltage may be calculated as the discharge voltage after one additional time constant from the previous question by multiplying 0.37 × 0.368 volt = 0.1362 volt. The question calls it 0.13 V DC. Write down 0.13 volt = five time constants. Five time constants = 5 × .02 = 0.1 second. If you plot the voltages against the time constants, you will have the same curve as the chart in question 3D67, except the chart's vertical scale is normalized, so the curve comes out in percentage.

3D72 How long does it take for an initial charge of 800 V DC to decrease to 294 V DC in a 450-microfarad capacitor when a 1-megohm resistor is connected across it?
- A. 80 seconds
- B. 294 seconds
- C. 368 seconds
- D. 450 seconds

ANSWER D: Since we're having so much fun, let's go for some new numbers, such as 800 volts DC and a 450 microfarad capacitor with a 1 megohm resistor connected across it. These values apply to the next five questions.

Let's continue the same steps as in the previous questions. Since the voltage across a capacitor drops to 36.8 percent of its initial value, what will the voltage be across the capacitor in one time constant? 800 × .368 = 294.4 volts. Write down 294 volts = one time constant. That's the voltage for this question, so we know the time wanted is one time constant. How long is the time constant? Using "megs" × "micros", the time constant = 1 × 450 = 450 seconds.

5 GENERAL RADIOTELEPHONE OPERATOR LICENSE

3D73 How long does it take for an initial charge of 800 V DC to decrease to 108 V DC in a 450-microfarad capacitor when a 1-megohm resistor is connected across it?
A. 225 seconds
B. 294 seconds
C. 450 seconds
D. 900 seconds

ANSWER D: The capacitor voltage is down to 294 volts from the previous question. What will the voltage be after discharging for another time constant—the second one? 294 volts × 0.368 = 108.3 volts, the voltage asked for. Write down 108 volts = two time constants. Two time constants = 2 × 450 = 900 seconds, the correct answer.

3D74 How long does it take for an initial charge of 800 V DC to decrease to 39.9 VDC in a 450-microfarad capacitor when a 1-megohm resistor is connected across it?
A. 1,350 seconds
B. 900 seconds
C. 450 seconds
D. 225 seconds

ANSWER A: You guessed it, the discharge continues for another time constant—the third. Let's check to see if this is the voltage asked for. The voltage from the last question is 108 volts times 36.8 percent is 39.9 volts. Sure enough, the voltage checks. Write down 39.9 volts = three time constants. 3 × "megs" × 450 × "micros" = 3 × 450 = 1350 seconds.

3D75 How long does it take for an initial charge of 800 V DC to decrease to 40.2 VDC in a 450-microfarad capacitor when a 1-megohm resistor is connected across it?
A. Approximately 225 seconds
B. Approximately 450 seconds
C. Approximately 900 seconds
D. Approximately 1,350 seconds

ANSWER D: This voltage is almost the same as 39.9 volts in the problem above. We'll still consider this the third time constant, so 3(1 × 450) = 1350 seconds is close enough, since "approximately" is used in the answer.

3D76 How long does it take for an initial charge of 800 V DC to decrease to 14.8 VDC in a 450-microfarad capacitor when a 1-megohm resistor is connected across it?
A. Approximately 900 seconds
B. Approximately 1,350 seconds
C. Approximately 1,804 seconds
D. Approximately 2,000 seconds

ANSWER C: The discharge of the capacitor is now into the fourth time constant. The initial voltage is 39.9 volts at the end of the third time constant. At the end of the fourth time constant the voltage is 39.9 volts × 0.368 = 14.67 volts, very close to the voltage required. Write down 14.7 volts = four time constants. 4 × 450 = 1800 seconds. Answer C with an approximate time of 1804 is the correct answer. Take a well-deserved break. You have finished the discussion of time constants!

3D – ELECTRICAL PRINCIPLES

3D77 What is the impedance of a network comprised of a 0.1-microhenry inductor in series with a 20-ohm resistor, at 30 MHz? (Specify your answer in rectangular coordinates.)
  A. 20 + j19
  B. 20 − j19
  C. 19 + j20
  D. 19 − j20

ANSWER A: Don't panic, we are not going to send you back to high school for algebra, trigonometry, and calculus to solve these problems. Here we have a circuit containing both reactance and resistance, giving us *impedance*, represented by the letter Z. Impedance has both a resistive (R) and a reactive part (X_C or X_L). The parts are represented by vectors that are at right angles (perpendicular) to each other. As a result, $Z = \sqrt{R^2 + X^2}$. Each part forms a leg of a right triangle and contributes to the magnitude of the impedance, which is the hypotenuse of the triangle. You determine the impedance's magnitude by calculating each part, R and X, and then Z. Let's first calculate the inductive reactance X_L of this question's circuit with the formula for inductive reactance:

▶ $X_L = 2\pi f L$ Where: L is inductance in **henries**
                                          f is frequency in **hertz**
                                          $\pi = 3.14$

At 30 MHz,
  $X_L = 2 \times 3.14 \times 30 \times 10^{+6} \times 0.1 \times 10^{-6}$
  $X_L = 18.84$ rounded to 19 ohms

The resistance, R is given as 20 ohms

Therefore,
  $Z = \sqrt{20^2 + 19^2} = \sqrt{400 + 361} = \sqrt{761}$
  $Z = 27.59$ rounded to 27.6 ohms

But the examination question reads, "Specify your answer in rectangular coordinates." This is good news because it allows us to simplify the problem solving—even to the point of doing them in our heads! Rectangular coordinates consist of the two parts of the impedance discussed above—the first part is the resistance, and the second part is the reactance. The reactance is preceded by a (+j) or a (−j) which indicates whether the circuit's reactance is inductive (+j) or capacitive (−j). The resistance is placed first followed by the reactance; therefore, the impedance of an inductive circuit is $Z = R + jX_L$, and for a capacitive circuit it is $Z = R - jX_C$.

This first problem has an inductor in series with a 20-ohm resistor. Since it's an inductor, its reactance is inductive and a +j precedes it. Therefore, for this problem, the impedance in rectangular coordinates is 20 + j19. That's not so hard, right?

3D78 What is the impedance of a network comprised of a 0.1-microhenry inductor in series with a 30-ohm resistor, at 5 MHz? (Specify your answer in rectangular coordinates.)
  A. 30 − j3
  B. 30 + j3
  C. 3 + j30
  D. 3 − j30

5 GENERAL RADIOTELEPHONE OPERATOR LICENSE

ANSWER B: Here is another inductive circuit; therefore, in the rectangular coordinates required, the reactance is inductive and will have a +j in front of it. Thus, two of the possible answers are already eliminated (A and D because they have a −j reactance). Normally, you would calculate the inductance from $X_L = 2\pi fL$ and place a +j in front of it to determine the correct answer, but in this case, it's much easier. The question gives the resistance as 30 ohms, so you know the answer is 30 + jsomething. Answer B is the only one that has "30 +j". It is the correct answer.

Rectangular Coordinates

A quantity can be represented by points that are measured on an X-Y plane with an X and Y axis perpendicular to each other. The point where the axes cross is the origin, the points on the X and Y axis are called coordinates, and the magnitude is represented by the length of a vector from the origin to the unique point defined by the *rectangular coordinates*.

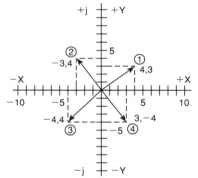

To be able to identify ac quantities and their phase angle, the X axis is called the real axis and the Y axis is called the imaginary axis. There are positive and negative real axis coordinates and positive and negative imaginary coordinates. The imaginary axis coordinates have a **j** operator to identify they are imaginary coordinates.

In rectangular coordinates:
Vector 1 is 4 + j3
Vector 2 is −3 + j4
Vector 3 is −4 − j4
Vector 4 is 3 − j4

Polar Coordinates

A quantity can be represented by a vector starting at an origin, the length of which is the magnitude of the quantity, and rotated from a zero axis through 360°.

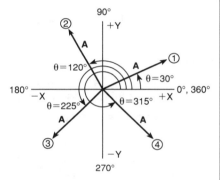

The position of the Vector A can be defined by the angle through which it is rotated. When we do so we are defining the ac quantity in *polar coordinates*.

In polar coordinates:
Vector 1 is A /30°
Vector 2 is A /120°
Vector 3 is A /225°
Vector 4 is A /315°

Representing an AC Quantity

3D – ELECTRICAL PRINCIPLES

3D79 What is the impedance of a network comprised of a 10-microhenry inductor in series with a 40-ohm resistor, at 500 MHz? (Specify your answer in rectangular coordinates.)
  A. 40 + j31400
  B. 40 − j31400
  C. 31400 + j40
  D. 31400 − j40

ANSWER A: This will also be a (+j) answer because the network is inductive. The resistance is given as 40 ohms, so you already have the resistor value for the rectangular coordinates. The answer will be of the form "40 +j", so look for the answer that begins "40 +j". Only correct answer A has that form. Look, you've picked the correct answer without calculating a thing! You could doublecheck your work by computing the inductive reactance in ohms, $X_L = 2\pi fL$, to agree with the j operator. Remember, f = 500 MHz. $X_L = 2 \times 3.14 \times 500 \times 10^{+6} \times 10 \times 10^{-6}$.

3D80 What is the impedance of a network comprised of a 100-picofarad capacitor in parallel with a 4000-ohm resistor, at 500 kHz? (Specify your answer in polar coordinates.)
  A. 2490 ohms, /51.5 degrees
  B. 4000 ohms, /38.5 degrees
  C. 5112 ohms, /−38.5 degrees
  D. 2490 ohms, /−51.5 degrees

ANSWER D: Shift gears now and be alert because this test question requires your answer in polar coordinates. Refer to the figure at question 3D77. Disconnect yourself from the thought process used for rectangular coordinates in the previous question. Because the resistance is given as 4,000, *DO NOT* look at the 4,000-ohm resistor as your best choice for the correct answer. Polar coordinate values are calculated differently. In polar coordinates, the ac quantity—let's say impedance Z in this case—is described as a vector at a particular phase angle. You solve for the vector magnitude Z separately and for the phase angle separately. The vector Z (which is the hypotenuse of a right triangle) is defined by a resistance vector (R) and a reactance vector (X) at right angles to each other. If the reactance vector is inductive, it is a $+X_L$; if it is capacitive the reactance is $-X_C$. Positive inductive reactances produce positive angles and negative capacitive reactances produce negative angles. Let's begin the necessary calculations by finding the value of the capacitive reactance with the formula:

▶ $X_C = \dfrac{1}{2\pi fC}$ Where: C is capacitance in **farads**
                                        f is frequency in **hertz**
                                        $\pi = 3.14$

$X_C = \dfrac{1}{2 \times 3.14 \times 0.5 \times 10^{+6} \times 100 \times 10^{-12}}$

Clearing the powers of ten:

▶ $X_C = \dfrac{10^{+6}}{6.28 \times 0.5 \times 100}$ This equation can be used directly when:
                                                        f is frequency in **MHz**
                                                        C is in **picofarads**

5 GENERAL RADIOTELEPHONE OPERATOR LICENSE

$$X_c = \frac{1,000,000}{314}$$

$$X_c = 3184.7 \text{ ohms}$$

The resistance is already given at 4,000 ohms, so R = 4000.

You now need to find the magnitude of the impedance, Z in ohms, of the *parallel* circuit. This must be done with vector (phasor) algebra, as follows:

$$Z = \frac{Z_1 \times Z_2}{Z_1 + Z_2} \qquad \text{Where } Z_1 = R \,\underline{/0°}$$
$$\text{and } Z_2 = X_c \,\underline{/-90°}$$

$$Z = \frac{R\,\underline{/0°} \times X_c\,\underline{/-90°}}{(R+j0) + (0-jX_c)}$$

$$Z = \frac{4000\,\underline{/0°} \times 3185\,\underline{/-90°}}{(4000 - j3185)} = \frac{4000 \times 3185\,\underline{/-90°}}{\sqrt{4000^2 + 3185^2}\,\underline{/\arctan -3185/4000}}$$

$$Z = \frac{12.74 \times 10^{+6}\,\underline{/-90°}}{\sqrt{16 \times 10^{+6} + 10.144 \times 10^{+6}}\,\underline{/\arctan -0.7963}}$$

$$Z = \frac{12.74 \times 10^{+6}\,\underline{/-90°}}{5.113 \times 10^3\,\underline{/-38.5°}}$$

$$Z = 2.493 \times 10^3\,\underline{/-90° + 38.5°} = 2493\,\underline{/-51.5°}$$

Answer D is the correct answer. You will need a scientific calculator to look up the tangent of the phase angles. The solution is not easy. You must keep your wits about you because you cannot add and subtract vectors directly. You must deal with the magnitude and the angle in polar coordinates and the real (resistance) and imaginary (reactance) parts in rectangular coordinates. See the discussions at questions 3D87 and 3D91. You can further doublecheck your answer by recognizing that there must be a minus sign in front of the phase angle because the circuit has capacitive reactance.

3D81 What is the impedance of a network comprised of a 0.001-microfarad capacitor in series with a 400-ohm resistor, at 500 kHz? (Specify your answer in rectangular coordinates.)
  A. 400 − j318
  B. 318 − j400
  C. 400 + j318
  D. 318 + j400

ANSWER A: Now you are back to rectangular coordinates again. Easy one here—do it in your head. It's a 400 ohm resistor in series with a capacitor. You know that gives you a −j, thus, answer A because none of others have "400 −j". If you want to check X_c use the formula in question 3D80.

3D82 What is the impedance of a network comprised of a 100-ohm-reactance inductor in series with a 100-ohm resistor? (Specify your answer in polar coordinates.)
  A. 121 ohms, $\underline{/35 \text{ degrees}}$
  B. 141 ohms, $\underline{/45 \text{ degrees}}$
  C. 161 ohms, $\underline{/55 \text{ degrees}}$
  D. 181 ohms, $\underline{/65 \text{ degrees}}$

ANSWER B: This one is relatively easy—impedance equals the square root of the sum of the resistance squared and the inductive reactance squared. Z in polar coordinates is:

$$Z = \sqrt{100^2 + 100^2} \; / \arctan 100/100$$
$$Z = \sqrt{20000} \; / \arctan 1$$
$$Z = 141.4 \; / +45°$$

You might recognize the square root easier if you work in powers of 10. The magnitude of $Z = \sqrt{(1 \times 10^2)^2 + (1 \times 10^2)^2} = \sqrt{2 \times 10^4} = \sqrt{2} \times 10^2 = 1.414 \times 10^2$. You may recognize the $\sqrt{2}$ as 1.414, so the answer is 141.4. From just the magnitude, you could have selected the correct answer, or you might have used the phase angle. A right triangle with equal R and X has a phase angle of 45° because, as you probably recall, the tangent of 45° is 1.

3D83 What is the impedance of a network comprised of a 100-ohm-reactance inductor, a 100-ohm-reactance capacitor, and a 100-ohm resistor all connected in series? (Specify your answer in polar coordinates.)
  A. 100 ohms, / 90 degrees
  B. 10 ohms, / 0 degrees
  C. 100 ohms, / 0 degrees
  D. 10 ohms, / 100 degrees

ANSWER C: Easy one again—since inductive reactance cancels capacitive reactance, in this circuit with both reactances equal, the final reactance is zero and all we have left is 100 ohms resistance. Since there is no reactance vector, there is no phase difference, so the phase angle is zero degrees. The resistance vector lies on the 0° axis.

One of the easiest ways to check this is to put the impedance in rectangular coordinates: Z = 100 +j100 -j100 = 100 +j0, just a resistance of 100 ohms lying on the zero axis.

3D84 What is the impedance of a network comprised of a 400-ohm-reactance capacitor in series with a 300-ohm resistor? (Specify your answer in polar coordinates.)
  A. 240 ohms, / 36.9 degrees
  B. 240 ohms, / −36.9 degrees
  C. 500 ohms, / 53.1 degrees
  D. 500 ohms, / −53.1 degrees

ANSWER D: Here is a series problem which is easily solved by first calculating the magnitude of the impedance which equals the square root of resistance squared plus reactance squared. The square root of 300 squared plus 400 squared is the square root of 250,000 or 500 ohms. Since the circuit is capacitive, the phase angle will have a minus sign in front of it. If you remember any trigonometry, you know that a right triangle with a 3-4-5 relationship of the sides is an easy one to determine the side values. With sides of 300 and 400, the hypotenuse will be 500, with a phase angle whose tangent is 3/4 or 4/3 depending on the relationship of the sides. In this case it's 4/3 and the angle is −53.1 degrees.

5 GENERAL RADIOTELEPHONE OPERATOR LICENSE

3D85 What is the impedance of a network comprised of a 300-ohm-reactance capacitor, a 600-ohm-reactance inductor, and a 400-ohm resistor, all connected in series? (Specify your answer in polar coordinates.)
 A. 500 ohms, /37 degrees
 B. 400 ohms, /27 degrees
 C. 300 ohms, /17 degrees
 D. 200 ohms, /10 degrees

ANSWER A: Here's one that even though the answer is requested in polar coordinates, it is best to start out using rectangular coordinates. Z = (400 +j600 −j300) = (400 +j300). Now you can use the 3-4-5 ratio we talked about in the previous question. With resistance side of a right triangle at 400 ohms and the reactance side at 300 ohms, you know that the hypotenuse magnitude of the impedance is Z = 500. The angle is a positive one with an arctan of (an angle whose tangent is) 300/400, or 0.75. Grab a trig table or scientific calculator for an angle of 36.8° rounded to 37°.

3D86 What is the impedance of a network comprised of a 400-ohm-reactance inductor in parallel with a 300-ohm resistor? (Specify your answer in polar coordinates.)
 A. 240 ohms, /36.9 degrees
 B. 240 ohms, /−36.9 degrees
 C. 500 ohms, /53.1 degrees
 D. 500 ohms, /−53.1 degrees

ANSWER A: We know right off the bat that the answer is going to be either A or C because an inductive reactance will have a positive polar coordinate angle. To come up with 240 ohms at an angle of +36.9° as the correct impedance, let's start out by defining the two impedances that are in parallel:

Z_1 = 300 /0° since it is a resistance in **ohms**
Z_2 = 400 /+90° since it is an inductive reactance in **ohms**

▶ $Z = \dfrac{Z_1 \times Z_2}{Z_1 + Z_2}$ (Resultant Z in **ohms** from two parallel Zs)

Additions and subtractions are easier in rectangular coordinates, and multiplication and division are easier in polar coordinates.

$Z = \dfrac{300 \times 400 \,/\, 0° + 90°}{(300 + j0) + (0 + j400)} = \dfrac{120000 \,/\, +90°}{500 \,/\, \text{arctan } 400/300}$

Z = 240 /+90° − 53.1°
Z = 240 /36.9°

To solve for answer A, we used the 3-4-5 ratio, and both rectangular and polar coordinates.

3D87 What is the impedance of a network comprised of a 1.0-millihenry inductor in series with a 200-ohm resistor, at 30 kHz? (Specify your answer in rectangular coordinates.)
 A. 200 − j188
 B. 200 + j188
 C. 188 + j200
 D. 188 − j200

3D – ELECTRICAL PRINCIPLES 5

ANSWER B: At last, another easy one. This is an inductive circuit with a resistance of 200 ohms; therefore, in rectangular coordinates the answer is 200 +j. Answer B is the only one 200 +j answer. Calculate inductive reactance to check your work. $X_L = 2\pi f L$ (where f = 30 kHz) = $2 \times 3.14 \times 30 \times 10^3 \times 1 \times 10^{-3}$ = 188.4.

Recall from algebra that $+2 \times +2 = +4$ and $-2 \times -2 = +4$ $\therefore \sqrt{4}$ has two roots $+2$ or -2. In ac circuit analysis there is a need to take the square roof of a negative number. For example, $\sqrt{-4}$ has a root $\sqrt{-1} \times \sqrt{4} = j2$ and $-j2$. The root of a negative number is called an *imaginary number* and this is indicated by writing the operator j in front of the root. j is equal to $\sqrt{-1}$ and is the basic imaginary quantity. Since $j = \sqrt{-1}$, it is interesting to note the following because they will be encountered in ac circuit analysis:

$$j^2 = j \times j = \sqrt{-1} \times \sqrt{-1} = -1$$

$$j^3 = j \times j \times j = j^2 \times j = -1 \times j = -j$$

$$j^4 = j \times j \times j \times j = j^2 \times j^2 = -1 \times -1 = +1$$

$$\frac{1}{j} = \frac{1}{j} \times \frac{j}{j} = \frac{j}{j^2} = \frac{j}{-1} = -j$$

Summary

$j = \sqrt{-1}$

$j^2 = -1$

$j^3 = -j$

$j^4 = +1$

$\frac{1}{j} = -j$

j Operator as Vector Rotator

Note that an imaginary quantity can be considered as a vector being rotated by the operator j.

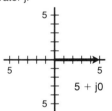

Vector starts at 0°.

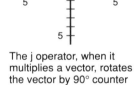

The j operator, when it multiplies a vector, rotates the vector by 90° counter clockwise.

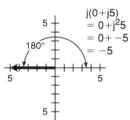

A second 90° rotation puts the vector at 180°.

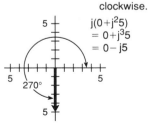

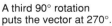

A third 90° rotation puts the vector at 270°.

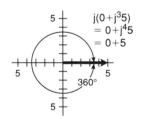

A fourth 90° rotation puts the vector back at 0°.

Complex Numbers (Real and Imaginary) and Operator j

5 GENERAL RADIOTELEPHONE OPERATOR LICENSE

3D88 What is the impedance of a network comprised of a 10-millihenry inductor in series with a 600-ohm resistor, at 10 kHz? (Specify your answer in rectangular coordinates.)
 A. 628 + j600
 B. 628 − j600
 C. 600 + j628
 D. 600 − j628

ANSWER C: Another easy one here. The circuit is inductive, so look for the (+j) right after the 600 ohm resistor. No calculations required, but if you'd like to check, $X_L = 2 \times 3.14 \times 10 \times 10^{+3} \times 10 \times 10^{-3} = 628$.

3D89 What is the impedance of a network comprised of a 0.01-microfarad capacitor in parallel with a 300-ohm resistor, at 50 kHz? (Specify your answer in rectangular coordinates.)
 A. 150 − j159
 B. 150 + j159
 C. 159 + j150
 D. 159 − j150

ANSWER D: Here is another one you will need to drag out the calculator on. Come on, isn't this fun? First find capacitive reactance. Refer back to question 3D80, if the capacitance is in **picofarads** and the frequency is in **megahertz**, then

$$X_C = \frac{10^{+6}}{2\pi f C}$$

50 kHz is 0.05 MHz and 0.01 microfarads is 10,000 picofarads, so you will divide 6.28 × .05 × 10,000 into 1 million, and end up with 318.47 ohms.

Now solve for impedance:

$$Z = \frac{Z_1 \times Z_2}{Z_1 + Z_2}$$

$$Z = \frac{300 \,\underline{/0°} \times 318.5 \,\underline{/-90°}}{(300 + j0) + (0 - j318.5)}$$

$$Z = \frac{95550 \,\underline{/-90°}}{\sqrt{300^2 + 318.5^2} \,\underline{/\arctan 318.5/300}} = \frac{95550 \,\underline{/-90°}}{438 \,\underline{/-46.5°}}$$

$$Z = 218.3 \,\underline{/-90° + 46.5°} = 218.3 \,\underline{/-43.5°}$$

Your answer comes out 218 ohms. The answer needs to be in rectangular coordinates. To convert to rectangular coordinates, you will be finding the resistance side and the reactance side of a right triangle with a hypotenuse of 218.3 and a contained angle $\Theta = -43.5°$. The parts are:

 R = 218.3 Cos Θ = 218.3 × 0.7253 = 158.3 (answer calls it 159)
 X_C = 218.3 Sin Θ = 218.3 × 0.6883 = 150.2 (answer calls it 150)

Therefore, in rectangular coordinates:

 Z = 158.3 −j150.2

Match this to the correct answer D of Z = 159 −j150.

3D – ELECTRICAL PRINCIPLES

3D90 What is the impedance of a network comprised of a 0.1-microfarad capacitor in series with a 40-ohm resistor, at 50 kHz? (Specify your answer in rectangular coordinates.)
   A. 40 + j32
   B. 40 − j32
   C. 32 − j40
   D. 32 + j40

ANSWER B: Take a rest on this one. This circuit is capacitive, so look for a (40−j) as the only answer that will agree with a 40-ohm resistor in series with a capacitor. Answer B is your match.

3D91 What is the impedance of a network comprised of a 1.0-microfarad capacitor in parallel with a 30-ohm resistor, at 5 MHz? (Specify your answer in rectangular coordinates.)
   A. 0.000034 + j.032
   B. 0.032 + j.000034
   C. 0.000034 − j.032
   D. 0.032 − j.000034

ANSWER C: Here comes another parallel circuit, so pull out that calculator. Use the following formula from question 3D80 to calculate the capacitive reactance. Remember frequency must be in **megahertz** and capacitance in **picofarads**:

$$X_C = \frac{10^{+6}}{2 \times 3.14 \times 5 \times 1 \times 10^{+6}} = \frac{1}{31.4} = 0.032$$

Therefore,

$$X_C = Z_2 = 0.032 \,\underline{/-90°}$$

$Z_1 = 30 \,\underline{/0°}$, the resistance given in the question.

Therefore,

$$Z = \frac{30 \,\underline{/0°} \times 0.032 \,\underline{/-90°}}{(30 + j0) + (0 - j0.032)}$$

$$Z = \frac{0.96 \,\underline{/-90°}}{\sqrt{30^2 + 0.032^2} \,\underline{/\arctan 0.032/30}}$$

$$Z = \frac{0.96 \,\underline{/-90°}}{\sqrt{900 + 0.001} \,\underline{/-0.1°}} = \frac{0.96 \,\underline{/-90°}}{30 \,\underline{/-0.1°}}$$

$$Z = 0.032 \,\underline{/-90° + 0.1°}$$
$$Z = 0.032 \,\underline{/-89.9°}$$

The impedance Z = 0.032 ohms at a phase angle of −89.9°.

The question requires rectangular coordinates. They are:

   R = 0.032 Cos (−89.9°) = 0.000034
   X_C = 0.032 Sin (−89.9°) = 0.032

Therefore,

   Z = 0.000034 − j0.032, answer C.

5 GENERAL RADIOTELEPHONE OPERATOR LICENSE

Converting Rectangular Coordinates to Polar Coordinates:

Convert $Z = R + jX$ to $Z\underline{/\theta}$ $\qquad Z = \sqrt{R^2 + X^2}\underline{/\arctan \frac{X}{R}}$

Example:

$Z = 4 + j3 = \sqrt{4^2 + 3^2}\underline{/\arctan \frac{3}{4}}$ $\qquad$ arctan = an angle whose tangent is

$\qquad = 5\underline{/37°}$ $\qquad\qquad\qquad\qquad = \dfrac{\text{side opposite}}{\text{side adjacent}} = \dfrac{X}{R}$

Converting Polar Coordinates to Rectangular Coordinates:

Convert $Z\underline{/\theta}$ to $Z = R + jX$

Example:

$Z = 5\underline{/37°} \qquad X = 5 \sin 37°$ $\qquad\qquad\qquad\qquad X = Z \sin \theta$
$\qquad\qquad\qquad = 5 \times 0.6 = 3$ $\qquad\qquad\qquad\qquad R = Z \cos \theta$
$\qquad\qquad R = 5 \cos 37°$ $\qquad\qquad\qquad\qquad\qquad Z = 4 + j3$
$\qquad\qquad\qquad = 5 \times 0.8 = 4$

Addition and subtraction are performed best in rectangular coordinates. The real terms and imaginary terms must be combined separately:

Add

$(R_1 + jX_1) + (R_2 + jX_2) = (R_1 + R_2) + (jX_1 + jX_2)$

Example: $(5 + j10) + (6 + j6) = (5 + 6) + (j10 + j6) = (11 + j16)$

Subtract

$(R_1 + jX_1) - (R_2 - jX_2)$

Set up as: $\qquad\qquad\qquad\qquad\qquad\qquad$ Example:

$\begin{array}{r} R_1 + jX_1 \\ -(R_2 - jX_2) \end{array} = \begin{array}{r} R_1 + jX_1 \\ +(-R_2 + jX_2) \\ \hline (R_1-R_2) + j(X_1+X_2) \end{array} \qquad \begin{array}{r} 5 + j10 \\ -(3 - j3) \\ \hline (5-3) + j(10+3) \end{array} = \begin{array}{r} 5 + j10 \\ -3 + j3 \\ \hline 2 + j13 \end{array}$

Multiplication and division are performed best in polar coordinates. The magnitude and angular portions are combined separately:

Multipy

$V_1\underline{/\theta_1} \times V_2\underline{/\theta_2} = V_1 \times V_2 \underline{/\theta_1 + \theta_2}$ $\qquad$ Example:

$V_3\underline{/-\theta_3} \times V_4\underline{/\theta_4} = V_3 \times V_4 \underline{/-\theta_3 + \theta_4}$ $\qquad 5\underline{/20°} \times 10\underline{/20°} = 50\underline{/40°}$

$\qquad\qquad\qquad\qquad\qquad\qquad\qquad\qquad\qquad 4\underline{/-40°} \times 2\underline{/60°} = 8\underline{/20°}$

Multiply the magnitude, add the angles.

Divide

$\dfrac{V_1\underline{/\theta_1}}{V_2\underline{/\theta_2}} = \dfrac{V_1}{V_2}\underline{/\theta_1 - \theta_2} \qquad\qquad\qquad\qquad \dfrac{10\underline{/40°}}{5\underline{/20°}} = 2\underline{/20°}$

$\dfrac{V_3\underline{/-\theta_3}}{V_4\underline{/-\theta_4}} = \dfrac{V_3}{V_4}\underline{/-\theta_3-(-\theta_4)} = \dfrac{V_3}{V_4}\underline{/-\theta_3+\theta_4} \qquad \dfrac{100\underline{/-60°}}{20\underline{/-20°}} = 5\underline{/-60° + 20°}$

$\qquad\qquad\qquad\qquad\qquad\qquad\qquad\qquad\qquad\qquad\qquad = 5\underline{/-40°}$

Divide the magnitude, subtract the angles.

Complex Number Arithmetic

3D – ELECTRICAL PRINCIPLES

3D92 What is the impedance of a network comprised of a 100-ohm-reactance capacitor in series with a 100-ohm resistor? (Specify your answer in polar coordinates.)
   A. 121 ohms, $\angle -25$ degrees
   B. 141 ohms, $\angle -45$ degrees
   C. 161 ohms, $\angle -65$ degrees
   D. 191 ohms, $\angle -85$ degrees
ANSWER B: R = 100 ohms and $-X_C$ = 100 ohms. Calculate $Z = \sqrt{100^2 + 100^2} = \sqrt{2 \times 10^4}$. This works out to be the square root of 20,000, which is 141 ohms. The phase angle will be negative so the correct answer B can be selected without going further. However, the arctan is 100/100 or 1, and from this we know that the angle is 45°.

3D93 What is the impedance of a network comprised of a 100-ohm-reactance capacitor in parallel with a 100-ohm resistor? (Specify your answer in polar coordinates.)
   A. 31 ohms, $\angle -15$ degrees
   B. 51 ohms, $\angle -25$ degrees
   C. 71 ohms, $\angle -45$ degrees
   D. 91 ohms, $\angle -65$ degrees
ANSWER C: Polar coordinates again, but you are given X_C so you don't have to calculate it. First calculate impedance:

$$Z = \frac{100 \angle 0° \times 100 \angle -90°}{\sqrt{100^2 + 100^2} \angle \arctan -1} = \frac{10000 \angle -90°}{141.4 \angle -45°}$$

$$Z = 70.7 \angle -45°$$

You can determine the correct answer as C just from the impedance magnitude, but since the arctan is 1, you know the phase angle is 45°.

3D94 What is the impedance of a network comprised of a 300-ohm-reactance inductor in series with a 400-ohm resistor? (Specify your answer in polar coordinates.)
   A. 400 ohms, $\angle 27$ degrees
   B. 500 ohms, $\angle 37$ degrees
   C. 600 ohms, $\angle 47$ degrees
   D. 700 ohms, $\angle 57$ degrees
ANSWER B: Easy one here—let's find the magnitude of the impedance and see if the correct answer can be determined just from it. The phase angle will be positive because the circuit is inductive. You don't really have to calculate if you recognize that the right triangle sides have a ratio of 3-4-5. Since resistance is 400 ohms, inductive reactance 300 ohms, Z will have a magnitude of 500 ohms. You don't need to know more than that—the correct answer is B.

3D95 What is the impedance of a network comprised of a 100-ohm-reactance inductor in parallel with a 100-ohm resistor? (Specify your answer in polar coordinates.)
   A. 71 ohms, $\angle 45$ degrees
   B. 81 ohms, $\angle 55$ degrees
   C. 91 ohms, $\angle 65$ degrees
   D. 100 ohms, $\angle 75$ degrees

ANSWER A: Another one like the previous question except for a parallel circuit. It looks like if you determine the magnitude of the impedance you can select the correct answer. Just calculate the magnitude, don't worry about the angle. The impedance magnitude is:

$$Z = \frac{100 \times 100}{\sqrt{100^2 + 100^2}} = \frac{10,000}{\sqrt{2 \times 10^2}}$$

$$Z = \frac{10,000}{141.4} = 70.71 \text{ ohms}$$

Match 70.71 rounded to the 71 in answer A. And this is further confirmed because the angle is positive to support that it is an inductive circuit.

3D96 What is the impedance of a network comprised of a 300-ohm-reactance capacitor in series with a 400-ohm resistor? (Specify your answer in polar coordinates.)
  A. 200 ohms, /−10 degrees
  B. 300 ohms, /−17 degrees
  C. 400 ohms, /−27 degrees
  D. 500 ohms, /−37 degrees

ANSWER D: Hey, here's another of those 3-4-5 ratios, and another one where the magnitude will probably lead to the correct answer. Let's try it. The circuit is a series circuit with the rectangular coordinates as R = 400 ohms an $-jX_C$ = 300 ohms. Therefore, Z = 500 ohms. You don't even have to know the angle because answer D is the only one with 500 ohms magnitude. You know the angle has a tangent = 0.75, so the angle is −37° because the circuit is capacitive.

Whew! Quite a mathematical experience!
Here are some hints to remember:

1. Use rectangular coordinates with series circuits and eliminate the wrong answers by determining whether or not the correct answer should be (+j) inductive, or (−j) capacitive. Usually the resistance is given so half of the impedance is known.
2. When working polar coordinates, place the values given on a right triangle for a visual picture. If the vertical line reactance and the base line resistance are equal in value, the phase angle will be 45 degrees. Also, in a right triangle, if either leg has a value of 3 and the other leg has a value of 4, the impedance will have a value of 5. Finally, see whether or not inductive reactances are greater than capacitive reactances, or vice versa. This will help you determine whether the impedance has a minus j for capacitive reactance, or a plus j for inductive reactance as part of your answer. If inductive reactance and capacitive reactance cancel, then the phase angle will be zero.
3. If you just can't seem to calculate an answer that agrees with the four choices on your test, cancel out the incorrect answers by qualifying the circuit as capacitive or inductive. If it's inductive, look for a +j or a positive angle in the answer; and if it's capacitive, look for a −j or a negative angle in the answer. Above all, never leave an answer blank! Your test will probably have only two of these problems on it.

3D – ELECTRICAL PRINCIPLES

3D97 The speed of a series DC motor will be affected by the:
   A. line voltage frequency
   B. number of windings
   C. load
   D. rotor coil windings

ANSWER C: As you apply additional loading to a series DC motor, the speed will be affected by the increased load and the motor will run slower.

3D98 The expression 'voltage regulation' as it applies to a shunt-wound DC generator operating at a constant frequency refers to:
   A. voltage output efficiency
   B. voltage in the secondary compared to the primary
   C. voltage fluctuations from load to no-load
   D. rotor winding voltage ratio

ANSWER C: The shunt motor incorporates the field coil in parallel with both the DC source and the armature. Changes in the load demand will be balanced out by the shunt motor back EMF that will vary with the load. Less load, greater back EMF to keep the motor from running away. Larger loads will be compensated with less back EMF.

3D99 When an emergency transmitter uses 325 watts and a receiver uses 50 watts, how many hours can a 12.6-volt, 55-ampere-hour battery supply full power to both units?
   A. 6 hours
   B. 3 hours
   C. 1.8 hours
   D. 1.2 hours

ANSWER C: For this question, we will assume for our calculations that the transmitter and receiver are running constantly because no duty cycle was stated, nor was there an indication that the receiver cycles out on TX. First compute the emergency transmitter current consumption at 12.6 volts DC. Step 1 is to add the receiver power requirements and transmitter power requirements together; 325 + 50 = 375 watts. Now convert this power requirement to current demand.

$$P = E \times I \quad \text{or} \quad I = \frac{P}{E} \qquad I = \frac{375}{12.6} = 29.76 \text{ amps}$$

Now divide the 55 ampere-hour capacity of the battery by 29.76 amps:

$$\frac{55}{29.76} = 1.8 \text{ hours}$$

3D100 The expression "voltage regulation" as it applies to a generator operating at a constant frequency refers to:
   A. full load to no load
   B. limited load to peak load
   C. source input supply frequency
   D. field frequency

ANSWER A: The term "voltage regulation" in this question indicates the capability of a generator operating at a constant frequency to handle load variations from its rated maximum load to no load with little change in output voltage.

5 GENERAL RADIOTELEPHONE OPERATOR LICENSE

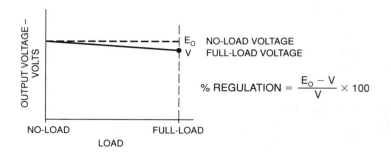

AC Generator Regulation Operating at Rated Frequency

3D101 The output of a separately-excited AC generator running at a constant speed can be controlled by:
A. armature
B. brushes
C. field current
D. exciter

ANSWER C: We can control the output voltage of a separately-excited AC generator running at a set speed by varying the field current with a large rheostat. As we increase field current, the magnetic field increases and the armature coil will couple a greater voltage.

3D102 What occurs if the load is removed from an operating series DC motor?
A. it will stop running
B. speed will increase slightly
C. no change occurs
D. it will accelerate until it flies apart

ANSWER D: The simple series DC motor's speed is controlled by the load. If the load is removed from the motor, it will go faster and faster until it burns up or self destructs into little pieces.

3D103 The speed of a DC motor varies with the:
A. load
B. number of brushes
C. ripple frequency
D. RF field voltage

ANSWER A: A changing load varies the speed of a series DC motor.

3D104 A 12.6-volt, 8-ampere-hour battery is supplying power to a receiver which uses 50 watts and a radar system that uses 300 watts. How long will the battery last?
A. 100.8 hours
B. 27.7 hours
C. 1 hour
D. 17 minutes or 0.3 hours

ANSWER D: Use Ohm's law to compute current demand. The receiver's 50 watts plus the radar's 300 watts = 350 watts. 350 watts divided by 12.6 volts = 27.77 amps. If we are pulling 27 amps from a small 8 ampere-hour battery, we know this system won't run more than just a few minutes—certainly a lot less time than the other incorrect answers. 8 ÷ 27 = 0.29 hours which rounds to 0.3 hour.

3D105 What is the total voltage when 12 Nickel-Cadmium batteries are connected in series?
  A. 12 volts
  B. 12.6 volts
  C. 15 volts
  D. 72 volts

ANSWER C: Nickel-cadmium batteries are found in many portable two-way radio and GPS receivers. Unlike alkaline cells which operate at 1.5 volts DC, nickel-cadmium batteries operate at 1.25 volts DC. 12 × 1.25 = 15 total volts if they are connected in series.

3D106 A ship radar unit uses 315 watts and a radio uses 50 watts. If the equipment is connected to a 50-ampere-hour battery rated at 12.6 volts, how long will the battery last?
  A. 28.97 hours
  B. 29 minutes
  C. 1 hour 43 minutes
  D. 10 hours 50 minutes

ANSWER C: Total power consumption is 315 + 50 = 365 watts. To calculate current, the formula is I = P ÷ E. I = 365 ÷ 12.6 = 28.9 amps. If we are running this load off of a 50 ampere-hour battery, 50 ÷ 28.9 = 1.7 hours. Convert to minutes: 1.7 × 60 = 103 minutes = 60 + 43 minutes = 1 hour 43 minutes.

3D107 The output voltage of a separately-excited AC generator (running at a constant speed) is controlled by:
  A. input voltage frequency
  B. the load
  C. primary voltage
  D. field current

ANSWER D: We control the output voltage of a separately-excited AC generator by varying the field current.

3D108 There is an improper impedance match between a 30-watt transmitter and the antenna and 5 watts is reflected. How much power is actually radiated?
  A. 35 watts
  B. 30 watts
  C. 25 watts
  D. 20 watts

ANSWER C: If the load is reflecting back 5 watts from a received 30 watts from the transmitter, 25 watts will be radiated.

5 GENERAL RADIOTELEPHONE OPERATOR LICENSE

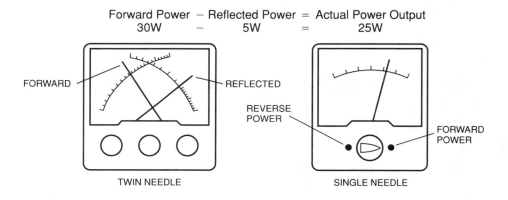

Directional Watt Meter

3D109 How long will a 12.6-volt, 50 ampere-hour battery last if it supplies power to an emergency transmitter rated at 531 watts of plate input power and other emergency equipment with a combined power rating of 530 watts?
- A. 6 hours
- B. 4 hours
- C. 1 hour
- D. 35 minutes

ANSWER D: Let's first calculate total power consumption: 531 + 530 = 1061 watts. Now calculate current: I = 1061 ÷ 12.6 = 84.2 amps at 12.6 VDC. Immediately we know we can't pull 84 amps for a full hour from a 50 ampere-hour battery, so the quick answer is D. To prove it, calculate: 50 ÷ 84 = 0.59 × 60 minutes = 35 minutes.

3D110 A 12.6-volt, 55-ampere-hour battery is connected to a radar unit rated at 325 watts and a receiver that uses 20 watts. How long will radar unit and receiver be able to draw full power from the battery?
- A. 6 hours
- B. 4 hours
- C. 2.3 hours
- D. 2 hours

ANSWER D: Total power consumption = 345 watts. Calculate current: I = 345 ÷ 12.6 = 27.38 amps at 12.6 VDC. 55 ampere-hour divided by 27.38 amps = 2 hours.

3D111 The speed of a series wound DC motor varies with:
- A. the number of commutator bars
- B. the number of slip rings
- C. the load applied to the motor
- D. the direction of rotation

ANSWER C: The motor speed varies with the load applied to the motor.

3D – ELECTRICAL PRINCIPLES

3D112 A 6-volt battery with 1.2 ohms internal resistance is connected across two 3-watt bulbs. What is the current flow?
- A. .57 amps
- B. .83 amps
- C. 1.0 amps
- D. 6.0 amps

ANSWER B: The circuit schematic is shown in the figure below. The internal resistance is identified as R_{INT}. The light bulbs are identified as resistors R_{L1} and R_{L2}. It is assumed that the bulb resistance remains constant from cold to hot. To solve for the circuit current, you must know the total resistance; therefore, you must first find the equivalent resistance of the two light bulbs which are in parallel. The wattage of each light bulb is given as 3 watts, so you can find the resistance of each bulb by using $R = E^2 \div P = 36 \div 3 = 12$ ohms for each bulb. Two 12-ohm resistors in parallel = 6 ohms. Thus, the total resistance across the battery is 1.2 ohms + 6 ohms = 7.2 ohms. To find the current, use I = E ÷ R = 6 ÷ 7.2 = 0.83 amp.

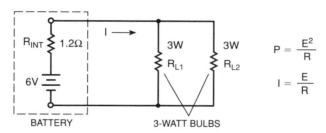

Finding Circuit Current

3D113 A dynamotor is approximately:
- A. 100% efficient
- B. 85% efficient
- C. 65% efficient
- D. 40% efficient

ANSWER C: The older dynamotor found in early shipboard AM transmitters and radar units was approximately 65 percent efficient. Dynamotors have now been replaced by solid-state circuits whose efficiency and reliability are much greater.

3D114 The power input to a 52-ohm transmission line is 1,872 watts. The current flowing through the line is:
- A. 6 amps
- B. 144 amps
- C. 0.06 amps
- D. 28.7 amps

ANSWER A: Recall the formula for calculating current when power and resistance are known:

$$I^2 = \frac{P}{R} = \frac{1872}{52} = 36$$

$I = \sqrt{36} = 6$ amps, which is answer A

3D115 If a marine radiotelephone receiver uses 75 watts of power and a transmitter uses 325 watts, how long can they both operate before discharging a 50-ampere-hour 12-volt battery?
A. 40 minutes
B. 1 hour
C. 1 1/2 hours
D. 6 hours

ANSWER C: First calculate total power consumption: 75 + 325 = 400 total watts. The watt-hour capability of the battery is 12V × 50A = 600 watt-hours. Divide total watts used, 400 watts, into 600 watt-hours and you get 1.5 hours (answer C). Or you could use I = P ÷ E = 400 ÷ 12 = 33.3 amps. Divide 33.3 amps into 50 ampere-hours and you also end up with 1.5 hours.

Subelement 3E – Circuit Components (13 questions)

3E1 Structurally, what are the two main categories of semiconductor diodes?
A. Junction and point contact
B. Electrolytic and junction
C. Electrolytic and point contact
D. Vacuum and point contact

ANSWER A: The semiconductor diode is a common component found in mobile and base station radio equipment. Depending on how the diode is biased, it may look like an open circuit or a closed circuit. The junction and point contact are two main categories of semiconductor diodes.

3E2 What are the two primary classifications of Zener diodes?
A. Hot carrier and tunnel
B. Varactor and rectifying
C. Voltage regulator and voltage reference
D. Forward and reversed biased

ANSWER C: The Zener diode is found in both handheld transceivers as well as mobile radio equipment. The Zener may be used for voltage regulation, or as a voltage reference.

3E3 What is the principal characteristic of a Zener diode?
A. A constant current under conditions of varying voltage
B. A constant voltage under conditions of varying current
C. A negative resistance region
D. An internal capacitance that varies with the applied voltage

ANSWER B: Remember that a Zener diode is for voltage regulation, for constant voltage even though current changes.

3E4 What is the range of voltage ratings available in Zener diodes?
A. 2.4 volts to 200 volts
B. 1.2 volts to 7 volts
C. 3 volts to 2000 volts
D. 1.2 volts to 5.6 volts

ANSWER A: There is only one answer with two 2's in it, so 2.4 volts to 200 volts is the range of ratings of Zener diodes.

3E – CIRCUIT COMPONENTS

ANODE — CATHODE

Here is the schematic symbol of a diode. Current will only flow ONE WAY in a diode. You can remember this diode diagram as a one-way arrow (key words).

Semiconductor Diode

ANODE — CATHODE

Here is the schematic symbol of a Zener diode. Since a diode only passes energy in one direction, look for that one-way arrow, plus a "Z" indicating it is a Zener diode. Doesn't that vertical line look like a tiny "Z".

Zener Diode

3E5 What is the principal characteristic of a tunnel diode?
A. A high forward resistance
B. A very high PIV
C. A negative resistance region
D. A high forward current rating

ANSWER C: As a tunnel diode is conducting current, there is a spot where current increases as the voltage drop across the diode decreases. They call this a negative resistance region.

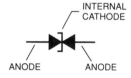

INTERNAL CATHODE
ANODE   ANODE

Here is the schematic symbol of a tunnel diode. Memorize the schematic symbol for a tunnel diode. It has arrow-heads tip to tip, as if they are stuck in a tunnel.

Tunnel Diode

ANODE — CATHODE

Here is the schematic symbol for a varactor diode. The internal capacitance of a varactor diode varies as the voltage applied to its terminals changes. Your author builds microwave equipment with varactor diodes, so be sure to identify the proper symbol. The symbol is essentially a capacitor and a diode combined.

Varactor Diode

3E6 What special type of diode is capable of both amplification and oscillation?
A. Point contact diodes
B. Zener diodes
C. Tunnel diodes
D. Junction diodes

ANSWER C: The tunnel diode is commonly used in both amplifier and oscillator circuits. The negative resistance region is particularly useful in oscillators.

5 GENERAL RADIOTELEPHONE OPERATOR LICENSE

3E7 What type of semiconductor diode varies its internal capacitance as the voltage applied to its terminals varies?
A. A varactor diode
B. A tunnel diode
C. A silicon-controlled rectifier
D. A Zener diode

ANSWER A: We use the varactor diode to tune VHF and UHF circuits by varying the voltage applied to the varactor diode.

3E8 What is the principal characteristic of a varactor diode?
A. It has a constant voltage under conditions of varying current
B. Its internal capacitance varies with the applied voltage
C. It has a negative resistance region
D. It has a very high PIV

ANSWER B: Change the voltage, and the diode's internal capacitance will vary.

3E9 What is a common use of a varactor diode?
A. As a constant current source
B. As a constant voltage source
C. As a voltage-controlled inductance
D. As a voltage-controlled capacitance

ANSWER D: Change the voltage, and the internal capacitance will change—this is very useful in varying the resonant frequency of tuned circuits.

3E10 What is a common use of a hot-carrier diode?
A. As balanced mixers in SSB generation
B. As a variable capacitance in an automatic frequency control circuit
C. As a constant voltage reference in a power supply
D. As VHF and UHF mixers and detectors

ANSWER D: If your job requires servicing VHF and UHF equipment, you may find a circuit with a hot carrier diode because of its low noise-figure characteristics.

3E11 What limits the maximum forward current in a junction diode?
A. The peak inverse voltage
B. The junction temperature
C. The forward voltage
D. The back EMF

ANSWER B: Guess what will kill most electronic components? High temperature! The limit of the maximum forward current in a junction diode is the junction temperature. This is why you will find these diodes mounted to the chassis of your equipment. The chassis acts as a heat sink to keep the junction temperature from exceeding its maximum limit.

3E12 How are junction diodes rated?
A. Maximum forward current and capacitance
B. Maximum reverse current and PIV
C. Maximum reverse current and capacitance
D. Maximum forward current and PIV

ANSWER D: When choosing a junction diode, you will need to know how much forward current will be passing through the diode, and the amount of peak inverse voltage (PIV) that the diode must stand in its non-conducting direction.

3E – CIRCUIT COMPONENTS

A ROW OF TINY GLASS SIGNAL DIODES

Diodes Mounted on a Circuit Board

3E13 What is a common use for point contact diodes?
A. As a constant current source
B. As a constant voltage source
C. As an RF detector
D. As a high voltage rectifier
ANSWER C: The little point contact diode, since it only conducts in one direction, may be used as an RF detector in some VHF and UHF equipment.

3E14 What type of diode is made of a metal whisker touching a very small semiconductor die?
A. Zener diode
B. Varactor diode
C. Junction diode
D. Point-contact diode
ANSWER D: Your author remembers the early crystal radios. They used a point-contact diode made of a metal whisker touching sensitive areas of a very small semiconductor surface.

3E15 What is one common use for PIN diodes?
A. As a constant current source
B. As a constant voltage source
C. As an RF switch
D. As a high voltage rectifier
ANSWER C: PIN diodes may be used in small handheld transceivers for RF switching. This gets away from mechanical relays for switching.

3E16 What special type of diode is often used in RF switches, attenuators, and various types of phase shifting devices?
A. Tunnel diodes
B. Varactor diodes
C. PIN diodes
D. Junction diodes
ANSWER C: The PIN diode may also be used in attenuators, along with RF switching.

3E17 What are the three terminals of a bipolar transistor?
A. Cathode, plate and grid
B. Base, collector and emitter
C. Gate, source and sink
D. Input, output and ground

ANSWER B: Almost like the alphabet, B for base, C for collector, (skip D), and E for emitter.

Transistors Mounted on a Circuit Board

3E18 What is the meaning of the term alpha with regard to bipolar transistors?
A. The change of collector current with respect to base current
B. The change of base current with respect to collector current
C. The change of collector current with respect to emitter current
D. The change of collector current with respect to gate current

ANSWER C: In bipolar transistors, the term "alpha" is the variation of COLLEC-TOR *current* with respect to *EMITTER current* (CCEC, key words). This is an important consideration when designing a circuit that will utilize a bipolar transistor.

3E19 What is the term used to express the ratio of change in DC collector current to a change in emitter current in a bipolar transistor?
A. Gamma
B. Epsilon
C. Alpha
D. Beta

ANSWER C: A for alpha, (no B), C for collector current, (no D), E for emitter current. This is as simple as ACE. You will ace your examination!

3E20 What is the meaning of the term beta with regard to bipolar transistors?
A. The change of collector current with respect to base current
B. The change of base current with respect to emitter current
C. The change of collector current with respect to emitter current
D. The change in base current with respect to gate current

ANSWER A: When we talk about the term "beta" for a bipolar transistor, think of beta with a "B" for base current. This is the change of *COLLECTOR* current with respect to *BASE* current. B for base; B for beta.

3E – CIRCUIT COMPONENTS

3E21 What is the term used to express the ratio of change in the DC collector current to a change in base current in a bipolar transistor?
- A. Alpha
- B. Beta
- C. Gamma
- D. Delta

ANSWER B: Bipolar, base current, beta.

3E22 What is the meaning of the term alpha cutoff frequency with regard to bipolar transistors?
- A. The practical lower frequency limit of a transistor in common emitter configuration
- B. The practical upper frequency limit of a transistor in common base configuration
- C. The practical lower frequency limit of a transistor in common base configuration
- D. The practical upper frequency limit of a transistor in common emitter configuration

ANSWER B: When we speak of a cutoff frequency, we speak of the upper frequency limit of a transistor. *Upper frequency limit* (key words)—in this case, for alpha which is the relation of collector current to emitter current in a common base circuit.

3E23 What is the term used to express that frequency at which the grounded base current gain has decreased to 0.7 of the gain obtainable at 1 kHz in a transistor?
- A. Corner frequency
- B. Alpha cutoff frequency
- C. Beta cutoff frequency
- D. Alpha rejection frequency

ANSWER B: *Grounded base* is the same as a common base configuration. Common base means *alpha*. Since 1 kHz is mentioned in this question, they are referring to the alpha cutoff frequency in a transistor.

3E24 What is the meaning of the term beta cutoff frequency with regard to a bipolar transistor?
- A. That frequency at which the grounded base current gain has decreased to 0.7 of that obtainable at 1 kHz in a transistor
- B. That frequency at which the grounded emitter current gain has decreased to 0.7 of that obtainable at 1 kHz in a transistor
- C. That frequency at which the grounded collector current gain has decreased to 0.7 of that obtainable at 1 kHz in a transistor
- D. That frequency at which the grounded gate current gain has decreased to 0.7 of that obtainable at 1 kHz in a transistor

ANSWER B: In a bipolar transistor, the *beta cutoff frequency* is that frequency at which the *grounded emitter current gain* (key words) has decreased to a certain value at a certain frequency.

5 GENERAL RADIOTELEPHONE OPERATOR LICENSE

3E25 What is the meaning of the term transition region with regard to a transistor?
A. An area of low charge density around the P-N junction
B. The area of maximum P-type charge
C. The area of maximum N-type charge
D. The point where wire leads are connected to the P- or N-type material

ANSWER A: When a transistor goes into saturation, transition region is the term for an area of low charge density around the P-N junction. Just like when you go into saturation reading over all of these questions and answers, you may need to put the book down, and get "recharged" from a *"low charge"* (key words) to get out of saturation!

3E26 What does it mean for a transistor to be fully saturated?
A. The collector current is at its maximum value
B. The collector current is at its minimum value
C. The transistor's Alpha is at its maximum value
D. The transistor's Beta is at its maximum value

ANSWER A: When a transistor is fully saturated, the collector current is maximum.

3E27 What does it mean for a transistor to be cut off?
A. There is no base current
B. The transistor is at its operating point
C. No current flows from emitter to collector
D. Maximum current flows from emitter to collector

ANSWER C: At cutoff, there is no current from emitter to the collector.

3E28 What are the elements of a unijunction transistor?
A. Base 1, base 2 and emitter
B. Gate, cathode and anode
C. Gate, base 1 and base 2
D. Gate, source and sink

ANSWER A: The two horizontal lines coming out of the unijunction transistor represent *base 1 and base 2* (key words). Watch out for answer C. The angled arrowhead line is an *emitter* (key word).

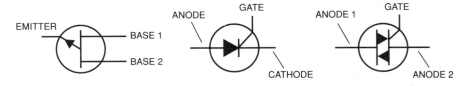

Notice there is an upper and lower line within the unijunction transistor diagram. Think of a uniform—upper and lower—part. There also is only one angled arrowhead line on the opposite side.

The SCR diagram looks a little bit like a switch, doesn't it? The SCR's gate element allows it to control other parts of a circuit. It is a diode with another gate control that turns on or off the SCR.

The triac is best viualized as two diodes connected in parallel, one in one direction, the other in the opposite direction, in a single package. Look at the diagram — this is the way it looks, doesn't it?

Unijunction Transistor   SCR   TRIAC

3E – CIRCUIT COMPONENTS

3E29 For best efficiency and stability, where on the load-line should a solid-state power amplifier be operated?
 A. Just below the saturation point
 B. Just above the saturation point
 C. At the saturation point
 D. At 1.414 times the saturation point

ANSWER A: Solid-state amplifiers are now common in commercial boat installations where high power SSB equipment is desired. Most sideband sets run only 100 watts output, so a solid-state power amplifier, operated off your SSB set just below the saturation point, will put out a hefty signal. You must carefully monitor your drive power, as well as the voltage feeding the power amplifier circuit. Overdriving the SSB solid-state power amplifier could cause distortion, and possibly burn out one of the expensive output transistors.

3E30 What two elements widely used in semiconductor devices exhibit both metallic and non-metallic characteristics?
 A. Silicon and gold
 B. Silicon and germanium
 C. Galena and germanium
 D. Galena and bismuth

ANSWER B: Both silicon and germanium exhibit a combination of metallic and non-metallic properties. Think of *silicon* rectifiers and *germanium* diodes.

3E31 What are the three terminals of an SCR?
 A. Anode, cathode and gate
 B. Gate, source and sink
 C. Base, collector and emitter
 D. Gate, base 1 and base 2

ANSWER A: Like any diode, it has an anode and a cathode, but it also has a special gate element. A control signal on the gate allows current from anode to cathode if the diode is forward biased. See figure at question 3E28.

3E32 What are the two stable operating conditions of an SCR?
 A. Conducting and nonconducting
 B. Oscillating and quiescent
 C. Forward conducting and reverse conducting
 D. NPN conduction and PNP conduction

ANSWER A: The SCR may be considered either conducting or not conducting. It's either on or off. Black or white. One or zero.

3E33 When an SCR is in the triggered or on condition, its electrical characteristics are similar to what other solid-state device (as measured between its cathode and anode)?
 A. The junction diode
 B. The tunnel diode
 C. The hot-carrier diode
 D. The varactor diode

ANSWER A: When the SCR is on, it's similar to a junction diode that will pass current in one direction.

3E34 Under what operating condition does an SCR exhibit electrical characteristics similar to a forward-biased silicon rectifier?
  A. During a switching transition
  B. When it is used as a detector
  C. When it is gated "off"
  D. When it is gated "on"
ANSWER D: When it is in the "on" position, it looks like a forward-biased silicon rectifier.

3E35 What is the transistor called which is fabricated as two complementary SCRs in parallel with a common gate terminal?
  A. TRIAC
  B. Bilateral SCR
  C. Unijunction transistor
  D. Field effect transistor
ANSWER A: Since the TRIAC looks like two diodes, it could be compared to two complementary SCR's in parallel with a common gate terminal.

3E36 What are the three terminals of a TRIAC?
  A. Emitter, base 1 and base 2
  B. Gate, anode 1 and anode 2
  C. Base, emitter and collector
  D. Gate, source and sink
ANSWER B: The triac has one gate and two anodes. The angled line is a gate. See figure at question 3E28.

3E37 What is the normal operating voltage and current for a light-emitting diode?
  A. 60 volts and 20 mA
  B. 5 volts and 50 mA
  C. 1.7 volts and 20 mA
  D. 0.7 volts and 60 mA
ANSWER C: The light-emitting diode (LED) is found on many types of marine and aviation two-way radio equipment. The LED draws approximately 20 milliamps of current, and requires only 1.7 volts to illuminate. Many manufacturers are staying away from LED readouts which may be used in an open cockpit of a boat. These are hard to see in the direct sunlight. There is nothing that can be done to make them more brilliant.

3E38 What type of bias is required for an LED to produce luminescence?
  A. Reverse bias
  B. Forward bias
  C. Zero bias
  D. Inductive bias
ANSWER B: The LED requires forward bias in order to illuminate. The LED is a diode, and it must be forward biased to make it give off light. See the figure at question 3E39.

3E – CIRCUIT COMPONENTS 5

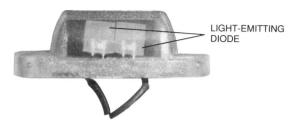

Two LEDs in a Single Package

3E39 What are the advantages of using an LED?
A. Low power consumption and long life
B. High lumens per cm per cm and low power consumption
C. High lumens per cm per cm and low voltage requirement
D. A current flows when the device is exposed to a light source

ANSWER A: The light-emitting diode draws very little current and has long life. Unfortunately, it does NOT have high lumens and is almost impossible to see in the direct sunlight.

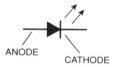

ANODE   CATHODE

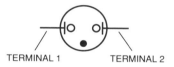

TERMINAL 1   TERMINAL 2

The light-emitting diode (LED) is a giveaway on the schematic diagram! Look at the two arrows indicating an illuminating device.

You can spot a neon lamp schematic symbol by the black dot at the bottom of the circle, and the two little "bulbs" on the end of the horizontial lines going into the circle.

LED   Neon Lamp

3E40 What colors are available in LEDs?
A. Yellow, blue, red and brown
B. Red, violet, yellow and peach
C. Violet, blue, orange and red
D. Red, green, orange and yellow

ANSWER D: Four of the popular colors for indicators are available in LEDs. Maybe a little reverse psychology is required here. Blue and peach are not available. Peach—who would want peach?

3E41 How can a neon lamp be used to check for the presence of RF?
A. A neon lamp will go out in the presence of RF
B. A neon lamp will change color in the presence of RF
C. A neon lamp will light only in the presence of very low frequency RF
D. A neon lamp will light in the presence of RF

ANSWER D: A small neon lamp is handy when checking out marine radio, land mobile, or aeronautical radio installations. It should glow in the presence of strong RF. It may also be used to see whether or not a power amplifier has gone into oscillation. If the lamp continues to glow after you have cut off the power amplifier exciter, chances are you have an amplifier in oscillation. Shut it down—quick!

5 GENERAL RADIOTELEPHONE OPERATOR LICENSE

3E42 What would be the bandwidth of a good crystal lattice band-pass filter for a single-sideband phone emission?
  A. 6 kHz at −6 dB
  B. 2.1 kHz at −6 dB
  C. 500 Hz at −6 dB
  D. 15 kHz at −6 dB
ANSWER B: A single-sideband emission is normally about 3 kHz wide. Without making the voice sound pinched, a 2.1-kHz filter is what comes with most worldwide sets.

3E43 What would be the bandwidth of a good crystal lattice band-pass filter for a double-sideband phone emission?
  A. 1 kHz at −6 dB
  B. 500 Hz at −6 dB
  C. 6 kHz at −6 dB
  D. 15 kHz at −6 dB
ANSWER C: Both aeronautical VHF stations operate full carrier, double sideband, and so do many shortwave broadcast stations. Double sideband requires twice the bandwidth of single sideband. To receive the entire double-sideband signal, you would need a 6 kHz band-pass filter with −6 dB skirts to pass the signal.

3E44 What is a crystal lattice filter?
  A. A power supply filter made with crisscrossed quartz crystals
  B. An audio filter made with 4 quartz crystals at 1-kHz intervals
  C. A filter with infinitely wide and shallow skirts made using quartz crystals
  D. A filter with narrow bandwidth and steep skirts made using quartz crystals
ANSWER D: Marine and aircraft SSB transceivers usually have excellent SSB and CW filters built in. The filters have a nice narrow bandwidth and steep skirts. The filters are made up of quartz crystals. Some equipment may also offer accessory filters for a tighter response, such as for the reception of data or weather facsimile signals. Extremely sharp filters may sometimes be selected from the front panel of certain SSB receivers and transceivers.

3E45 What technique can be used to construct low cost, high performance crystal lattice filters?
  A. Splitting and tumbling
  B. Tumbling and grinding
  C. Etching and splitting
  D. Etching and grinding
ANSWER D: The process of etching and grinding crystal lattice filters allows you to custom-tailor their response. Your author remembers the old days of putting a penciled × on the face of the quartz crystal to lower its frequency, and using toothpaste to grind it down for a higher frequency. Anyone remember those days of changing frequency when "rock bound"?

3E – CIRCUIT COMPONENTS

3E46 What determines the bandwidth and response shape in a crystal lattice filter?
  A. The relative frequencies of the individual crystals
  B. The center frequency chosen for the filter
  C. The amplitude of the RF stage preceding the filter
  D. The amplitude of the signals passing through the filter

ANSWER A: The bandwidth and response of a crystal filter will relate to the frequency of the individual crystals within that filter unit.

3E47 What is an *enhancement-mode* FET?
  A. An FET with a channel that blocks voltage through the gate
  B. An FET with a channel that allows a current when the gate voltage is zero
  C. An FET without a channel to hinder current through the gate
  D. An FET without a channel; no current occurs with zero gate voltage

ANSWER D: Handheld manufacturers have *enhanced* their marine transceivers by using enhancement-mode field-effect transistors that won't consume current when there is no voltage applied to the gate. As soon as a voltage of the right polarity is applied to the gate, a channel is formed, and there is current between source and drain. This type of "on demand" device is one more way that handheld transceivers try to conserve battery voltage.

3E48 What is a *depletion-mode* FET?
  A. An FET that has a channel with no gate voltage applied; a current flows with zero gate voltage
  B. An FET that has a channel that blocks current when the gate voltage is zero
  C. An FET without a channel; no current flows with zero gate voltage
  D. An FET without a channel to hinder current through the gate

ANSWER A: Your handheld battery may become *depleted* if depletion-mode FETs are used in many of the circuits. Even though there is no gate voltage applied, current still continues to flow.

3E49 Why do many MOSFET devices have built-in gate-protective Zener diodes?
  A. The gate-protective Zener diode provides a voltage reference to provide the correct amount of reverse-bias gate voltage
  B. The gate-protective Zener diode protects the substrate from excessive voltages
  C. The gate-protective Zener diode keeps the gate voltage within specifications to prevent the device from overheating
  D. The gate-protective Zener diode prevents the gate insulation from being punctured by small static charges or excessive voltages

ANSWER D: The "front-end" transistors of the modern transceiver must sometimes sustain major amounts of incoming signals from nearby transmitters, or maybe even a nearby lightning strike. Lightning *static* could destroy a MOSFET if it weren't for the built-in, gate-protective zener diode.

3E50 What do the initials CMOS stand for?
  A. Common mode oscillating system
  B. Complementary mica-oxide silicon

C. Complementary metal-oxide semiconductor
D. Complementary metal-oxide substrate

ANSWER C: When we talk about transistorized devices, they are in fact semiconductors. A CMOS is made from layers of metal-oxide semiconductor material.

3E51 Why are special precautions necessary in handling FET and CMOS devices?
A. They are susceptible to damage from static charges
B. They have fragile leads that may break off
C. They have micro-welded semiconductor junctions that are susceptible to breakage
D. They are light sensitive

ANSWER A: Any equipment using FET and CMOS transistors can become damaged easily from a nearby static discharge.

3E52 How does the input impedance of a *field-effect transistor* compare with that of a bipolar transistor?
A. One cannot compare input impedance without first knowing the supply voltage
B. An FET has low input impedance; a bipolar transistor has high input impedance
C. The input impedance of FETs and bipolar transistors is the same
D. An FET has high input impedance; a bipolar transistor has low input impedance

ANSWER D: An FET is much easier to work with in circuits than the simple bipolar transistor. Its higher input impedance is less likely to load down a circuit that it is attached to, and the FET can also handle big swings in signal levels.

3E53 What are the three terminals of a field-effect transistor?
A. Gate 1, gate 2, drain
B. Emitter, base, collector
C. Emitter, base 1, base 2
D. Gate, drain, source

ANSWER D: GaDS, better remember the three terminals of a field-effect transistor—gate, drain and source.

3E54 What are the two basic types of junction *field-effect transistors*?
A. N-channel and P-channel
B. High power and low power
C. MOSFET and GaAsFET
D. Silicon FET and germanium FET

ANSWER A: Be sure to note carefully the direction of the gate arrow to determine whether it is N-channel or P-channel.

3E55 What is an *operational amplifier*?
A. A high-gain, direct-coupled differential amplifier whose characteristics are determined by components external to the amplifier unit
B. A high-gain, direct-coupled audio amplifier whose characteristics are determined by components external to the amplifier unit
C. An amplifier used to increase the average output of frequency modulated signals
D. A program subroutine that calculates the gain of an RF amplifier

3E – CIRCUIT COMPONENTS

ANSWER A: The operational amplifier (op amp) is a direct-coupled *differential* amplifier that offers high gain and high input impedance. The "differential" wording refers to the op-amp input design where the output is determined by the difference voltage between the two inputs. Different op-amp circuits and characteristics are determined by the *external* components connected to the amplifier.

3E56 What would be the characteristics of the ideal op-amp?
A. Zero input impedance, infinite output impedance, infinite gain, flat frequency response
B. Infinite input impedance, zero output impedance, infinite gain, flat frequency response
C. Zero input impedance, zero output impedance, infinite gain, flat frequency response
D. Infinite input impedance, infinite output impedance, infinite gain, flat frequency response

ANSWER B: For an ideal amplifier, the output impedance of the op amp is just opposite of the input impedance—zero output impedance versus infinite input impedance. Zero output impedance contributes to a flat frequency response in audio amplifier circuits.

3E57 What determines the gain of a closed-loop op-amp circuit?
A. The external feedback network
B. The collector-to-base capacitance of the PNP stage
C. The power supply voltage
D. The PNP collector load

ANSWER A: The amplifier gain without feedback is very high. Feedback, applied to set the gain to particular amounts, is determined by an external feedback network. It may be one or more fixed resistors, or a type of variable circuit.

3E58 What is meant by the term *op-amp offset voltage?*
A. The output voltage of the op-amp minus its input voltage
B. The difference between the output voltage of the op-amp and the input voltage required in the following stage
C. The potential between the amplifier-input terminals of the op-amp in a closed-loop condition
D. The potential between the amplifier-input terminals of the op-amp in an open-loop condition

ANSWER C: If the op amp has been constructed properly, you should see almost no voltage between the amplifier input terminals when the feedback loop is *closed*.

3E59 What is the input impedance of a theoretically ideal op-amp?
A. 100 ohms
B. 1000 ohms
C. Very low
D. Very high

ANSWER D: Remember that the input impedance to an op amp is always very high.

3E60 What is the output impedance of a theoretically ideal op-amp?
A. Very low
B. Very high
C. 100 ohms
D. 1000 ohms

ANSWER A: Remember that the output impedance of an op amp is always very low.

3E61 What is a *phase-locked loop* circuit?
A. An electronic servo loop consisting of a ratio detector, reactance modulator, and voltage-controlled oscillator
B. An electronic circuit also known as a monostable multivibrator
C. An electronic circuit consisting of a precision push-pull amplifier with a differential input
D. An electronic servo loop consisting of a phase detector, a low-pass filter and voltage-controlled oscillator

ANSWER D: There are few circuits that have made such a big difference in communications as the phase-locked loop circuit. In fact, frequency selection from a transceiver with PLL synthesis came to a point where the operator actually had *too much control* of where the set might transmit and receive. Most land mobile equipment now locks out the consumer from getting into and making changes with the PLL. Same thing for marine radios—while mariners may tune in authorized ITU bands, the PLL circuitry has been locked out from the front panel so it can't be changed to go in any other region.

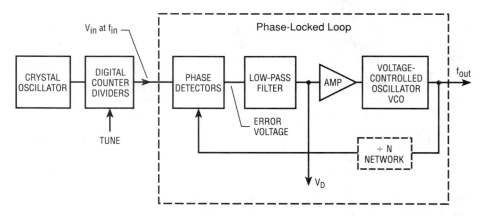

In a phase-locked loop the output frequency of the VCO is locked by phase and frequency to the frequency of the input signal V_{in}.

By using a very accurate crystal controlled source and dividing it down (or multiplying it up), the output frequency, f_{out}, of the PLL will be controlled very accurately and with rigid stability to the input frequency, f_{in}. The phase detector develops an error voltage that is determined by the frequencies and phase of the two inputs. The feedback system wants to reduce the error voltage to zero.

V_D is a voltage proportional to the frequency changes occuring in V_{in}; therefore, the PLL can be used as an FM demodulator.

With divide by N network, f_{out} will be f_{in} multiplied by N.

Phase-Locked Loop (PLL)

3E – CIRCUIT COMPONENTS

3E62 What functions are performed by a *phase-locked loop*?
A. Wideband AF and RF power amplification
B. Comparison of two digital input signals, digital pulse counter
C. Photovoltaic conversion, optical coupling
D. Frequency synthesis, FM demodulation

ANSWER D: In building his first external frequency synthesizer, your author was able to "synthesize" hundreds of channels with just a couple of crystals. How many of you remember being "rock bound" with an old rig? See figure at 3E61 for FM demodulation.

3E63 A circuit compares the output from a voltage-controlled oscillator and a frequency standard. The difference between the two frequencies produces an error voltage that changes the voltage-controlled oscillator frequency. What is the name of the circuit?
A. A doubly balanced mixer
B. A phase-locked loop
C. A differential voltage amplifier
D. A variable frequency oscillator

ANSWER B: In a phase-locked loop, the *"difference between"* two frequencies will produce an error voltage to the voltage-controlled oscillator. *Difference* is a key word.

3E64 What do the initials *TTL* stand for?
A. Resistor-transistor logic
B. Transistor-transistor logic
C. Diode-transistor logic
D. Emitter-coupled logic

ANSWER B: If you regularly work with digital modes, transistor-transistor logic is commonplace within this type of transceiver. You will need a logic probe to troubleshoot TTL circuits.

3E65 What is the recommended power supply voltage for *TTL* series integrated circuits?
A. 12.00 volts
B. 50.00 volts
C. 5.00 volts
D. 13.60 volts

ANSWER C: If you ever worked on TTL circuits, you know the importance of those 5-volt regulators that always seem to burn up at the wrong time. TTL circuits run on +5.0 volts.

3E66 What logic state do the inputs of a *TTL* device assume if they are left open?
A. A high logic state
B. A low logic state
C. The device becomes randomized and will not provide consistent high or low logic states
D. Open inputs on a TTL device are ignored

ANSWER A: Here are several key words—inputs *open* and *high* state. Remember four letters make up the word *open*, and four letters make up the word *high*.

3E67 What level of input voltage is *high* in a *TTL* device operating with a 5-volt power supply?
 A. 2.0 to 5.5 volts
 B. 1.5 to 3.0 volts
 C. 1.0 to 1.5 volts
 D. –5.0 to –2.0 volts
ANSWER A: If the input voltage is at the *high* level, depending on the number of circuits connected at the point, by specification the level will be between 2.0 volts and 5.5 volts.

3E68 What level of input voltage is *low* in a *TTL* device operating with a 5-volt power supply?
 A. –2.0 to –5.5 volts
 B. 2.0 to 5.5 volts
 C. –0.6 to 0.8 volts
 D. –0.8 to 0.4 volts
ANSWER C: If the input voltage to a TTL IC is at the *low* level, it will usually be around +0.2 volt. It can be as high as +0.8 volt, but the noise margin is down to zero.

3E69 Why do circuits containing *TTL* devices have several bypass capacitors per printed circuit board?
 A. To prevent RFI to receivers
 B. To keep the switching noise within the circuit, thus eliminating RFI
 C. To filter out switching harmonics
 D. To prevent switching transients from appearing on the supply line
ANSWER D: Since TTL devices are switching circuits with a rise and fall time so sharp they create troublesome transients, bypass capacitors will knock down the "spikes" that can create tremendous RFI. Remember, capacitors used as bypass want to keep the voltage across them constant.

3E70 What is a *CMOS IC*?
 A. A chip with only P-channel transistors
 B. A chip with P-channel and N-channel transistors
 C. A chip with only N-channel transistors
 D. A chip with only bipolar transistors
ANSWER B: The *C* in CMOS means *complementary*, which identifies an IC circuit that uses both P-channel and N-channel MOS transistors. MOS means Metal-Oxide-Silicon; therefore, this is MOS with both N and P transistors. Also, remember the problems associated with CMOS devices—you must make sure to minimize any static charges that could wipe them out in an instant.

3E71 What is one major advantage of *CMOS* over other devices?
 A. Small size
 B. Low current consumption
 C. Low cost
 D. Ease of circuit design
ANSWER B: That sensitive CMOS IC features low current consumption.

3E – CIRCUIT COMPONENTS

3E72 Why do *CMOS* digital integrated circuits have high immunity to noise on the input signal or power supply?
A. Larger bypass capacitors are used in CMOS circuit design
B. The input switching threshold is about two times the power supply voltage
C. The input switching threshold is about one-half the power supply voltage
D. Input signals are stronger

ANSWER C: The CMOS device is relatively immune to noise because its switching threshold is one-half the power supply voltage. As a result, power supply noise transients or input transients much larger than other logic circuit types will not cause a transition in the input state.

3E73 Signal energy is coupled into a traveling-wave tube at:
A. collector end of helix
B. anode end of the helix
C. cathode end of the helix
D. focusing coils

ANSWER C: The traveling-wave tube (TWT) is found in microwave telemetry equipment, and a low-power exciter signal is coupled to the TWT's cathode end of the helix. Care must be taken never to get near the waveguide output of a TWT system because there are dangerous microwaves when the circuit is switched on.

3E74 Permanent magnetic field that surrounds a traveling-wave tube (TWT) is intended to:
A. provide a means of coupling
B. prevent the electron beam from spreading
C. prevent oscillations
D. prevent spurious oscillations

ANSWER B: The permanent magnetic field that surrounds a TWT helps focus the electron beam.

3E75 Electromagnetic coils encase a traveling wave tube to:
A. provide a means of coupling energy
B. prevent the electron beam from spreading
C. prevent oscillation
D. prevent spurious oscillation

ANSWER B: The electromagnetic coils surrounding the traveling wave tube develop an axial magnetic focusing field to keep the electron beam diameter small and to keep the beam in the center of the helical coil inside the traveling wave tube.

Subelement 3F – Practical Circuits (22 questions)

3F1 What is a linear electronic voltage regulator?
A. A regulator that has a ramp voltage as its output
B. A regulator in which the pass transistor switches from the "off" state to the "on" state
C. A regulator in which the control device is switched on or off, with the duty cycle proportional to the line or load conditions
D. A regulator in which the conduction of a control element is varied in direct proportion to the line voltage or load current

ANSWER D: A linear electronic voltage regulator varies the conduction of a circuit in direct proportion to variations in the *line voltage* (key words) to, or the *load current* (key words) from, the device. You will find in your base station power supplies, which utilize a big heavy transformer, a sophisticated voltage regulation circuit.

3F2 What is a switching electronic voltage regulator?
A. A regulator in which the conduction of a control element is varied in direct proportion to the line voltage or load current
B. A regulator that provides more than one output voltage
C. A regulator in which the control device is switched on or off, with the duty cycle proportional to the line or load conditions
D. A regulator that gives a ramp voltage at its output

ANSWER C: The switching voltage regulator actually switches the control device completely *on or off* (key words).

3F3 What device is usually used as a stable reference voltage in a linear voltage regulator?
A. A Zener diode
B. A tunnel diode
C. An SCR
D. A varactor diode

ANSWER A: For voltage regulation and a stable reference voltage, Zener diodes are found in aeronautical handheld transceivers, aircraft radio, marine radio, and business band equipment. The most common caused of voltage regulator failure is a voltage transient or over-voltage. Whenever replacing a defective voltage regulator, monitor other voltage sources to see why the regulator failed.

3F4 What type of linear regulator is used in applications requiring efficient utilization of the primary power source?
A. A constant current source
B. A series regulator
C. A shunt regulator
D. A shunt current source

ANSWER B: A series regulator normally runs cool. It's more efficient than a shunt regulator. Everything passes through the series regulator.

3F – PRACTICAL CIRCUITS

3F5 What type of linear voltage regulator is used in applications where the load on the unregulated voltage source must be kept constant?
 A. A constant current source
 B. A series regulator
 C. A shunt current source
 D. A shunt regulator

ANSWER D: In circuits where the load on the unregulated input source must be kept constant, we use the slightly less-efficient shunt regulator. These can sometimes get quite warm, and are usually mounted to the chassis of the equipment. The chassis acts as a heat sink.

3F6 To obtain the best temperature stability, what should be the operating voltage of the reference diode in a linear voltage regulator?
 A. Approximately 2.0 volts
 B. Approximately 3.0 volts
 C. Approximately 6.0 volts
 D. Approximately 10.0 volts

ANSWER C: We normally run the operating voltage of the reference diode at around 6 volts. This gives the circuit good temperature stability.

3F7 What is the meaning of the term remote sensing with regard to a linear voltage regulator?
 A. The feedback connection to the error amplifier is made directly to the load
 B. Sensing is accomplished by wireless inductive loops
 C. The load connection is made outside the feedback loop
 D. The error amplifier compares the input voltage to the reference voltage

ANSWER A: Marine SSB transceivers, running 100 watts PEP output, must be run off of a stable power supply. Remote sensing of a portion of the output voltage directly at the load feeds back to the error amplifier for accurate conditions at the load. This keeps the output voltage from having wide variations due to an additional voltage drop between the regulator and the load.

3F8 What is a three-terminal regulator?
 A. A regulator that supplies three voltages with variable current
 B. A regulator that supplies three voltages at a constant current
 C. A regulator containing three error amplifiers and sensing transistors
 D. A regulator containing a voltage reference, error amplifier, sensing resistors and transistors, and a pass element

ANSWER D: The common voltage regulator found in mobile radio equipment is the 3-terminal regulator. When working on equipment, you can spot these regulators up alongside the metal chassis, using the chassis as a heat sink. Inside that small plastic IC is a voltage regulator containing a voltage reference, an error amplifier, sensing resistors and transistors, and a pass element. These regulators normally run relatively hot, so don't consider them necessarily bad if they seem hot to the touch. It's when you find these regulators stone cold that you may need to troubleshoot the circuit.

5 GENERAL RADIOTELEPHONE OPERATOR LICENSE

Voltage Regulator IC Mounted in Circuit Board

3F9 What are the important characteristics of a three-terminal regulator?
    A. Maximum and minimum input voltage, minimum output current and voltage
    B. Maximum and minimum input voltage, maximum output current and voltage
    C. Maximum and minimum input voltage, minimum output current and maximum output voltage
    D. Maximum and minimum input voltage, minimum output voltage and maximum output current

ANSWER B: Two important characteristics of a 3-terminal regulator are maximum and minimum input voltage, which are given in all answers. The other three are minimum output current and minimum and maximum output voltage. Only one answer indicates *maximum output current and voltage* (key words), and these are important considerations when choosing a 3-terminal regulator for a particular circuit.

3F10 What is the distinguishing feature of a Class A amplifier?
    A. Output for less than 180 degrees of the signal cycle
    B. Output for the entire 360 degrees of the signal cycle
    C. Output for more than 180 degrees and less than 360 degrees of the signal cycle
    D. Output for exactly 180 degrees of the input signal cycle

ANSWER B: The Class A amplifier works during the entire 360-degree signal cycle. It offers good linearity, but with all its constant work, may have low efficiency. See figure at question 3F11.

3F11 What class of amplifier is distinguished by the presence of output throughout the entire signal cycle and the input never goes into the cutoff region?
    A. Class A
    B. Class B
    C. Class C
    D. Class D

ANSWER A: If it works off an entire signal cycle (key words), it's a Class A amplifier.

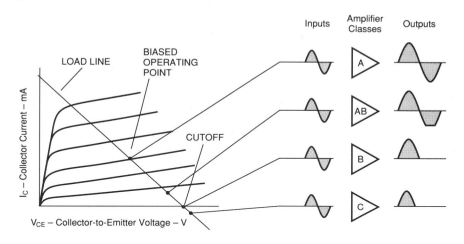

Various Classes of Transistorized Amplifiers

3F12 What is the distinguishing characteristic of a Class B amplifier?
  A. Output for the entire input signal cycle
  B. Output for greater than 180 degrees and less than 360 degrees of the input signal cycle
  C. Output for less than 180 degrees of the input signal cycle
  D. Output for 180 degrees of the input signal cycle
ANSWER D: The Class B amplifier only works during half of the input cycle, or for a total of 180 degrees. We usually run Class B amplifiers in push/pull pairs. Their efficiency is higher than Class A amplifiers because each tube or transistor gets to rest for half of the cycle. See figure at question 3F11.

3F13 What class of amplifier is distinguished by the flow of current in the output essentially in 180 degree pulses?
  A. Class A
  B. Class B
  C. Class C
  D. Class D
ANSWER B: If it's working for 180 degrees, it's a Class B amplifier. See figure at question 3F11.

3F14 What is a Class AB amplifier?
  A. Output is present for more than 180 degrees but less than 360 degrees of the signal input cycle
  B. Output is present for exactly 180 degrees of the input signal cycle
  C. Output is present for the entire input signal cycle
  D. Output is present for less than 180 degrees of the input signal cycle
ANSWER A: The Class AB amplifier has some of the properties of both the Class A and the Class B amplifier. It has output for more than 180 degrees of the cycle, but less than 360 degrees of the cycle. See figure at question 3F11.

5 GENERAL RADIOTELEPHONE OPERATOR LICENSE

3F15 What is the distinguishing feature of a Class C amplifier?
- A. Output is present for less than 180 degrees of the input signal cycle
- B. Output is present for exactly 180 degrees of the input signal cycle
- C. Output is present for the entire input signal cycle
- D. Output is present for more than 180 degrees but less than 360 degrees of the input signal cycle

ANSWER A: A Class C amplifier gets to rest a lot! It has great efficiency, but its linearity is not as good as a Class B or Class A amp. Class C amplifiers are best for digital modes. The Class C amplifier output is always less than 180 degrees of the input signal cycle. See figure at question 3F11.

3F16 What class of amplifier is distinguished by the bias being set well beyond cutoff?
- A. Class A
- B. Class B
- C. Class C
- D. Class AB

ANSWER C: If the bias is set well beyond cutoff, the amplifier will only work less than 180 degrees of the cycle, and is a Class C amplifier. Class C amplifiers are very efficient, but not so great on linearity. See figure at question 3F11.

3F17 Which class of amplifier provides the highest efficiency?
- A. Class A
- B. Class B
- C. Class C
- D. Class AB

ANSWER C: The highest efficiency comes out of a Class C amplifier. Since it operates the minimum time out of the cycle, it dissipates less power, but at the same time, produces narrow pulses which can cause loss of linearity.

3F18 Which class of amplifier has the highest linearity and least distortion?
- A. Class A
- B. Class B
- C. Class C
- D. Class AB

ANSWER A: The best amplifier for a pure signal with the highest linearity and least distortion is a Class A amplifier. See figure at question 3F11.

3F19 Which class of amplifier has an operating angle of more than 180 degrees but less than 360 degrees when driven by a sine wave signal?
- A. Class A
- B. Class B
- C. Class C
- D. Class AB

ANSWER D: If it's more than 180, but less than 360, it must be a Class AB amplifier. See figure at question 3F11.

3F20 What is an L-network?
A. A network consisting entirely of four inductors
B. A network consisting of an inductor and a capacitor
C. A network used to generate a leading phase angle
D. A network used to generate a lagging phase angle

ANSWER B: When troubleshooting two-way radio equipment, make absolutely sure you have the technical manual before going out to the job. This will help you spot different types of networks, such as the L-network that has an inductor and capacitor in it.

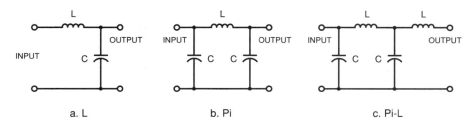

Matching Networks

3F21 What is a pi-network?
A. A network consisting entirely of four inductors or four capacitors
B. A Power Incidence network
C. An antenna matching network that is isolated from ground
D. A network consisting of one inductor and two capacitors or two inductors and one capacitor

ANSWER D: The pi-network consists of three components, looking like the Greek symbol pi (π). The legs could be either capacitors or coils, and the horizontal line at the top could be either a coil or a capacitor.

3F22 What is a pi-L-network?
A. A Phase Inverter Load network
B. A network consisting of two inductors and two capacitors
C. A network with only three discrete parts
D. A matching network in which all components are isolated from ground

ANSWER B: The pi-L-network consists of two coils and two capacitors. Depending on how they are arranged, it could be a high-pass network or a low-pass network.

3F23 Does the L-, pi-, or pi-L-network provide the greatest harmonic suppression?
A. L-network
B. Pi-network
C. Inverse L-network
D. Pi-L-network

ANSWER D: The network with the most components will provide the greatest harmonic suppression. The pi-L-network is found in most worldwide marine SSB sets.

5 GENERAL RADIOTELEPHONE OPERATOR LICENSE

3F24 What are the three most commonly used networks to accomplish a match between an amplifying device and a transmission line?
A. M-network, pi-network and T-network
B. T-network, M-network and Q-network
C. L-network, pi-network and pi-L-network
D. L-network, M-network and C-network

ANSWER C: Worldwide and aeronautical long-range SSB equipment may have three different matching devices between the amplifier output and the transmission line—either pi, L, or pi-L networks.

3F25 How are networks able to transform one impedance to another?
A. Resistances in the networks substitute for resistances in the load
B. The matching network introduces negative resistance to cancel the resistive part of an impedance
C. The matching network introduces transconductance to cancel the reactive part of an impedance
D. The matching network can cancel the reactive part of an impedance and change the value of the resistive part of an impedance

ANSWER D: In older tube sets, you could vary the matching network to cancel the reactive (X) part of an impedance, and change the value of the resistive part (R) of an impedance. Newer transistorized sets have a fixed output, and there is no manual operator control of the fixed matching devices. However, manufacturers have come to the rescue in antenna matching by providing new worldwide SSB radios with built-in automatic antenna tuners.

3F26 Which type of network offers the greater transformation ratio?
A. L-network
B. Pi-network
C. Constant-K
D. Constant-M

ANSWER B: The pi-network is found in many worldwide transceivers. It offers the greatest transformation ratio.

3F27 Why is the L-network of limited utility in impedance matching?
A. It matches a small impedance range
B. It has limited power handling capabilities
C. It is thermally unstable
D. It is prone to self resonance

ANSWER A: The L-network alone can only match to a small impedance range. This is why we always prefer a pi-network.

3F28 What is an advantage of using a pi-L-network instead of a pi-network for impedance matching between the final amplifier of a vacuum-tube type transmitter and a multiband antenna?
A. Greater transformation range
B. Higher efficiency
C. Lower losses
D. Greater harmonic suppression

ANSWER D: Worldwide SSB equipment working into a multi-band antenna system, such as a quad antenna or a private coast station, could generate harmonics. The pi-L-network always gives us greater harmonic suppression.

3F – PRACTICAL CIRCUITS

3F29 Which type of network provides the greatest harmonic suppression?
A. L-network
B. Pi-network
C. Pi-L-network
D. Inverse-Pi network

ANSWER C: The greatest harmonic suppression is provided by a pi-L-network.

3F30 What are the three general groupings of filters?
A. High-pass, low-pass and band-pass
B. Inductive, capacitive and resistive
C. Audio, radio and capacitive
D. Hartley, Colpitts and Pierce

ANSWER A: High-pass filters are found in older TV sets. Low-pass filters are found in marine SSB sets. Band-pass filters may be found in VHF and UHF equipment.

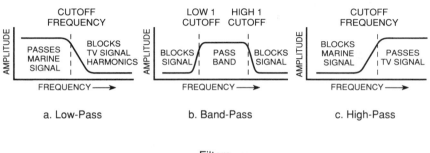

Filters

3F31 What is a constant-K filter?
A. A filter that uses Boltzmann's constant
B. A filter whose velocity factor is constant over a wide range of frequencies
C. A filter whose product of the series- and shunt-element impedances is a constant for all frequencies
D. A filter whose input impedance varies widely over the design bandwidth

ANSWER C: The constant K-filter is *constant for all frequencies* (key words), as opposed to a wide range of frequencies in answer B.

3F32 What is an advantage of a constant-K filter?
A. It has high attenuation for signals on frequencies far removed from the passband
B. It can match impedances over a wide range of frequencies
C. It uses elliptic functions
D. The ratio of the cutoff frequency to the trap frequency can be varied

ANSWER A: The constant K-filter continues to suppress harmonics and spurs far removed from its pass band.

3F33 What is an m-derived filter?
A. A filter whose input impedance varies widely over the design bandwidth
B. A filter whose product of the series- and shunt-element impedances is a constant for all frequencies
C. A filter whose schematic shape is the letter "M"
D. A filter that uses a trap to attenuate undesired frequencies too near cutoff for a constant-k filter.

ANSWER D: The m-derived filter has a tuned *trap* (key word) to attenuate undesired spurious emissions too near the normal cutoff for a constant-K filter.

3F34 What are the distinguishing features of a Butterworth filter?
A. A filter whose product of the series- and shunt-element impedances is a constant for all frequencies
B. It only requires capacitors
C. It has a maximally flat response over its passband
D. It requires only inductors

ANSWER C: The Butterworth filter can be counted on for its flat response over the entire pass band of frequencies. Its cutoff, unfortunately, is not as sharp as other filters.

3F35 What are the distinguishing features of a Chebyshev filter?
A. It has a maximally flat response over its passband
B. It allows ripple in the passband
C. It only requires inductors
D. A filter whose product of the series- and shunt-element impedances is a constant for all frequencies

ANSWER B: It's a man's name who created the design. It sounds like "Chevy" connected to "Shev," but it's "Chebyshev." The Chebyshev filter may allow ripple to pass through on the pass band.

3F36 When would it be more desirable to use an m-derived filter over a constant-K filter?
A. When the response must be maximally flat at one frequency
B. When you need more attenuation at a certain frequency that is too close to the cut-off frequency for a constant-K filter
C. When the number of components must be minimized
D. When high power levels must be filtered

ANSWER B: Occasionally you need to knock down a close-in spur next to your transmitted signal. We would use an m-derived filter because it offers more attenuation at a certain frequency close to its cutoff.

3F37 What are three major oscillator circuits often used in radio equipment?
A. Taft, Pierce and negative feedback
B. Colpitts, Hartley and Taft
C. Taft, Hartley and Pierce
D. Colpitts, Hartley and Pierce

ANSWER D: Colpitts oscillators have a capacitor just like C in its name. Hartley is tapped, and a Pierce oscillator uses a crystal.

3F – PRACTICAL CIRCUITS

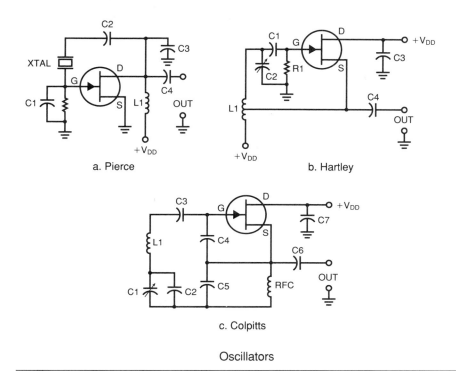

a. Pierce
b. Hartley
c. Colpitts

Oscillators

3F38 How is the positive feedback coupled to the input in a Hartley oscillator?
A. Through a neutralizing capacitor
B. Through a capacitive divider
C. Through link coupling
D. Through a tapped coil

ANSWER D: Hartley is always tapped. The tapped coil provides inductive coupling for positive feedback.

3F39 How is the positive feedback coupled to the input in a Colpitts oscillator?
A. Through a tapped coil
B. Through link coupling
C. Through a capacitive divider
D. Through a neutralizing capacitor

ANSWER C: On the Colpitts oscillator, we use a capacitive divider to provide feedback.

3F40 How is the positive feedback coupled to the input in a Pierce oscillator?
A. Through a tapped coil
B. Through link coupling
C. Through a capacitive divider
D. Through capacitive coupling

ANSWER D: On the Pierce oscillator, we also use capacitive coupling. Think of a quartz crystal as an element sandwiched between two plates. The two plates act a little bit like a capacitor.

3F41 Which of the three major oscillator circuits used in radio equipment utilizes a quartz crystal?
  A. Negative feedback
  B. Hartley
  C. Colpitts
  D. Pierce

ANSWER D: Quartz crystals are found in Pierce oscillators.

3F42 What is the piezoelectric effect?
  A. Mechanical vibration of a crystal by the application of a voltage
  B. Mechanical deformation of a crystal by the application of a magnetic field
  C. The generation of electrical energy by the application of light
  D. Reversed conduction states when a P-N junction is exposed to light

ANSWER A: If you apply a voltage to a quartz crystal, it will vibrate at a specific frequency. This is the reason crystal oscillators are so accurate. They oscillate at the frequency, or harmonics of the frequency, of the crystal, and continue to do so with great accuracy unless the temperature changes outside design limits.

3F43 What is the major advantage of a Pierce oscillator?
  A. It is easy to neutralize
  B. It doesn't require an LC tank circuit
  C. It can be tuned over a wide range
  D. It has a high output power

ANSWER B: The nice thing about a Pierce oscillator is that it doesn't require an inductive-capacitive tank circuit for excellent frequency stability. It operates on one frequency, not variable.

3F44 Which type of oscillator circuit is commonly used in a VFO?
  A. Pierce
  B. Colpitts
  C. Hartley
  D. Negative feedback

ANSWER B: Since the Colpitts oscillator uses a big capacitor in a variable frequency oscillator (VFO), it's the most common oscillator circuit for older VFO radios. Newer radios, even though they say they have a VFO, are really digitally controlled via an optical reader. They just look like they have a big capacitor behind that big tuning dial!

3F45 Why is the Colpitts oscillator circuit commonly used in a VFO?
  A. The frequency is a linear function of the load impedance
  B. It can be used with or without crystal lock-in
  C. It is stable
  D. It has high output power

ANSWER C: The Colpitts oscillator is relatively stable. After you let things warm up, the Colpitts oscillator normally stays put on a single frequency.

3F – PRACTICAL CIRCUITS

3F46 What is meant by the term *modulation*?
A. The squelching of a signal until a critical signal-to-noise ratio is reached
B. Carrier rejection through phase nulling
C. A linear amplification mode
D. A mixing process whereby information is imposed upon a carrier

ANSWER D: When we speak into a microphone, we are putting information out on the air on a radio frequency carrier. In frequency modulation, our modulation alters the carrier frequency; in amplitude modulation, our modulation varies the carrier amplitude.

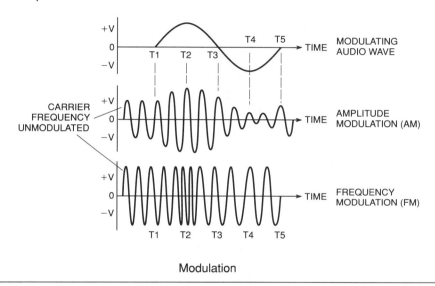

Modulation

3F47 How is an G3E FM-phone emission produced?
A. With a balanced modulator on the audio amplifier
B. With a reactance modulator on the oscillator
C. With a reactance modulator on the final amplifier
D. With a balanced modulator on the oscillator

ANSWER B: The *reactance modulator* (key words) causes the oscillator to vary frequency in accordance with the *phase* modulation. It's the oscillator (key word) that is modulated.

3F48 What is a reactance modulator?
A. A circuit that acts as a variable resistance or capacitance to produce FM signals
B. A circuit that acts as a variable resistance or capacitance to produce AM signals
C. A circuit that acts as a variable inductance or capacitance to produce FM signals
D. A circuit that acts as a variable inductance or capacitance to produce AM signals

ANSWER C: The reactance (key word) modulator is found in an FM transceiver, and varies the inductance or capacitance to produce the FM signal. The key words are *inductance, capacitance* and *FM*. Watch out for answer A!

5 GENERAL RADIOTELEPHONE OPERATOR LICENSE

3F49 What is a balanced modulator?
A. An FM modulator that produces a balanced deviation
B. A modulator that produces a double-sideband, suppressed carrier signal
C. A modulator that produces a single-sideband, suppressed carrier signal
D. A modulator that produces a full carrier signal

ANSWER B: The balanced modulator produces an upper and lower sideband, with the carrier suppressed. Remember, the double sidebands are *balanced* into an upper and lower sideband around a *suppressed* carrier.

3F50 How can a single-sideband phone signal be generated?
A. By driving a product detector with a DSB signal
B. By using a reactance modulator followed by a mixer
C. By using a loop modulator followed by a mixer
D. By using a balanced modulator followed by a filter

ANSWER D: In a single-sideband transceiver, a lower or upper sideband filter removes the unwanted sideband.

3F51 How can a double-sideband phone signal be generated?
A. By feeding a phase modulated signal into a low pass filter
B. By using a balanced modulator followed by a filter
C. By detuning a Hartley oscillator
D. By modulating the plate voltage of a class C amplifier

ANSWER D: For a double-sideband signal, we could modulate the plate voltage of an amplifier for amplitude modulation.

3F52 How is the efficiency of a power amplifier determined?
A. Efficiency = (RF power out / DC power in) × 100%
B. Efficiency = (RF power in / RF power out) × 100%
C. Efficiency = (RF power in / DC power in) × 100%
D. Efficiency = (DC power in / RF power in) × 100%

ANSWER A: The typical efficiency of a power amplifier is approximately half the power output for the amount of power input. The output power is RF power; the input power is DC power. The formula is: Eff(%) = P_{RF}/P_{DC} × 100. You can spot the answer by the key word *out*, followed by the key word *in*.

3F53 For reasonably efficient operation of a transistor amplifier, what should the load resistance be with 12 volts at the collector and 5 watts power output?
A. 100.3 ohms
B. 14.4 ohms
C. 10.3 ohms
D. 144 ohms

ANSWER B: To calculate the optimum load resistance for a transistor amplifier (assuming maximum efficiency of 50%), the following formula is used:

$$R_L = \frac{V_{CC}^2}{2P_O}$$
Where: R_L = Load resistance in **ohms**
V_{CC} = Transistor's collector voltage in **volts**
P_O = Amplifier's power output in **watts**

3F – PRACTICAL CIRCUITS

In this problem, 12 volts (V_{cc}) squared equals 144. Two times P_o is $2 \times 5 = 10$. Dividing 144 by 10 equals 14.4 ohms. The calculator keystrokes are: Clear, 12 $\times$ 12 = 144 ÷ 10 = 14.4.

3F54 What is the fl*ywheel effect?*
A. The continued motion of a radio wave through space when the transmitter is turned off
B. The back and forth oscillation of electrons in an LC circuit
C. The use of a capacitor in a power supply to filter rectified AC
D. The transmission of a radio signal to a distant station by several hops through the ionosphere

ANSWER B: Some of you may remember the old farm machinery that had an external engine flywheel. The flywheel keeps the back and forth oscillations going. The flywheel effect is the way we describe how electrons flow in an LC circuit.

3F55 What order of Q is required by a tank-circuit sufficient to reduce harmonics to an acceptable level?
A. Approximately 120
B. Approximately 12
C. Approximately 1200
D. Approximately 1.2

ANSWER B: A nice tight tank circuit with enough Q to reduce harmonics will have a Q of at least 12. The higher the number, the greater the quality (Q) of the circuit. You might remember that you need "quality above 10" to help you choose the right answer.

3F56 How can parasitic oscillations be eliminated from a power amplifier?
A. By tuning for maximum SWR
B. By tuning for maximum power output
C. By neutralization
D. By tuning the output

ANSWER C: We can reduce parasitic oscillations, and clean up our signal, through the steps of neutralization in a power amplifier.

3F57 What is the process of detection?
A. The process of masking out the intelligence on a received carrier to make an S-meter operational
B. The recovery of intelligence from the modulated RF signal
C. The modulation of a carrier
D. The mixing of noise with the received signal

ANSWER B: Your radio's detector *recovers intelligence* (key words) from a modulated RF signal. You have modulation to put information on a carrier; you have detection to take information off a carrier.

3F58 What is the principle of detection in a diode detector?
A. Rectification and filtering of RF
B. Breakdown of the Zener voltage
C. Mixing with noise in the transition region of the diode
D. The change of reactance in the diode with respect to frequency

5 GENERAL RADIOTELEPHONE OPERATOR LICENSE

ANSWER A: Since a diode only conducts for half of the AC signal, it may be used as a rectifier. By filtering out the radio frequency energy after rectification, detection is accomplished—the modulating signal is what remains.

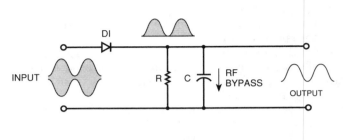

Diode Detector

3F59 What is a product detector?
A. A detector that provides local oscillations for input to the mixer
B. A detector that amplifies and narrows the band-pass frequencies
C. A detector that uses a mixing process with a locally generated carrier
D. A detector used to detect cross-modulation products

ANSWER C: A product detector is found in SSB receivers. It mixes the incoming signal with a beat frequency oscillator signal. The beat frequency oscillator signal is a locally generated carrier that is mixed with the incoming signal.

3F60 How are FM-phone signals detected?
A. By a balanced modulator
B. By a frequency discriminator
C. By a product detector
D. By a phase splitter

ANSWER B: FM signals must be detected differently than AM signals because frequency changes must be detected rather than amplitude changes. We use a reactance modulator to transmit FM. We use a *frequency discriminator* (key words) to detect an FM phone signal.

3F61 What is a frequency discriminator?
A. A circuit for detecting FM signals
B. A circuit for filtering two closely adjacent signals
C. An automatic bandswitching circuit
D. An FM generator

ANSWER A: Anytime you see the words *frequency discriminator* you know that you are dealing with frequency modulation and an FM transceiver.

3F62 What is the mixing process?
A. The elimination of noise in a wideband receiver by phase comparison
B. The elimination of noise in a wideband receiver by phase differentiation
C. Distortion caused by auroral propagation
D. The combination of two signals to produce sum and difference frequencies

ANSWER D: Inside the two-way radio equipment are several stages of mixers that combine two input signals to produce sum and different frequencies. These

3F – PRACTICAL CIRCUITS **5**

two inputs are the desired RF signal and the local oscillator (LO) signal. One of the output signals becomes the intermediate frequency (IF) signal inside the receiver section.

3F63 What are the principal frequencies which appear at the output of a mixer circuit?
A. Two and four times the original frequency
B. The sum, difference and square root of the input frequencies
C. The original frequencies and the sum and difference frequencies
D. 1.414 and 0.707 times the input frequency

ANSWER C: Out of the mixer comes your original two frequencies, and the sum and difference frequencies. The more elaborate the transceiver, the more mixing stages found in the sets. This helps filter out unwanted signals or phantom signals that could cause interference.

3F64 What are the advantages of the frequency-conversion process?
A. Automatic squelching and increased selectivity
B. Increased selectivity and optimal tuned-circuit design
C. Automatic soft limiting and automatic squelching
D. Automatic detection in the RF amplifier and increased selectivity

ANSWER B: Receivers with dual conversion are better than a set with only a single conversion. Triple conversion is better yet. The more stages of conversion, the more stages of increased selectivity and optimal tuned-circuit design.

3F65 What occurs in a receiver when an excessive amount of signal energy reaches the mixer circuit?
A. Spurious mixer products are generated
B. Mixer blanking occurs
C. Automatic limiting occurs
D. A beat frequency is generated

ANSWER A: Some VHF power amplifiers may incorporate a hefty pre-amplifier circuit for use on receiving. It boosts incoming signal levels. The pre-amp turned on could generate spurious mixer products as a result of the high signal level. The spurious products are not really signals that you are trying to receive. If you use a handheld with a power amp, turn the pre-amp off.

3F66 How much gain should be used in the RF amplifier stage of a receiver?
A. As much gain as possible short of self oscillation
B. Sufficient gain to allow weak signals to overcome noise generated in the first mixer stage
C. Sufficient gain to keep weak signals below the noise of the first mixer stage
D. It depends on the amplification factor of the first IF stage

ANSWER B: Very sensitive receiver transistors now give the signal amplifier stages all the gain they need. If you try to add a pre-amplifier on a modern two-way transceiver, you could create more interference than the possible weak-signal amplification would gain. You need enough gain to allow weak signals to overcome noise generated in the first mixer stage.

3F67 Why should the RF amplifier stage of a receiver only have sufficient gain to allow weak signals to overcome noise generated in the first mixer stage?
  A. To prevent the sum and difference frequencies from being generated
  B. To prevent bleed-through of the desired signal
  C. To prevent the generation of spurious mixer products
  D. To prevent bleed-through of the local oscillator

ANSWER C: Too much additional gain by an outside pre-amp could generate spurious mixer products and a terrible-sounding receiver.

3F68 What is the primary purpose of an RF amplifier in a receiver?
  A. To provide most of the receiver gain
  B. To vary the receiver image rejection by utilizing the AGC
  C. To improve the receiver's noise figure
  D. To develop the AGC voltage

ANSWER C: The RF amplifier, factory designed for your receiver, improves the *receiver's noise figure* (key words). Low noise transistors are now available for an improved noise-figure rating.

3F69 What is an i-f amplifier stage?
  A. A fixed-tuned pass-band amplifier
  B. A receiver demodulator
  C. A receiver filter
  D. A buffer oscillator

ANSWER A: The IF (intermediate frequency) amplifier is the next stage after the mixer and is tuned to the incoming intermediate frequency. It is fixed-tuned as a pass-band amplifier.

3F70 What factors should be considered when selecting an intermediate frequency?
  A. Cross-modulation distortion and interference
  B. Interference to other services
  C. Image rejection and selectivity
  D. Noise figure and distortion

ANSWER C: One of the key areas of receiver design is *image rejection* and *selectivity* (key words). Your author prefers a selective receiver any day over one that is slightly more sensitive.

3F71 What is the primary purpose of the first i-f amplifier stage in a receiver?
  A. Noise figure performance
  B. Tune out cross-modulation distortion
  C. Dynamic response
  D. Selectivity

ANSWER D: The first intermediate-frequency stage in an amplifier provides selectivity and image rejection. Look for the answer *selectivity* and *image rejection* (key words) on your examination. In this question, only *selectivity* is given.

3F – PRACTICAL CIRCUITS

3F72 What is the primary purpose of the final i-f amplifier stage in a receiver?
  A. Dynamic response
  B. Gain
  C. Noise figure performance
  D. Bypass undesired signals
ANSWER B: The purpose of the final IF amplifier stage in a receiver is both *gain* (key word) as well as selectivity. The answer for this question is gain.

3F73 What is a *flip-flop* circuit?
  A. A binary sequential logic element with one stable state
  B. A binary sequential logic element with eight stable states
  C. A binary sequential logic element with four stable states
  D. A binary sequential logic element with two stable states
ANSWER D: *"Flip flop"*, two words, two states. A flip-flop circuit is a form of bistable multivibrator—it remains in one of two states until triggered to change to the other state.

3F74 How many bits of information can be stored in a single *flip-flop* circuit?
  A. 1
  B. 2
  C. 3
  D. 4
ANSWER A: One flip-flop circuit can only store a single bit of information. If semiconductor memories were made of flip-flops, one flip-flop would be required for each bit.

3F75 What is a *bistable multivibrator* circuit?
  A. An "AND" gate
  B. An "OR" gate
  C. A flip-flop
  D. A clock
ANSWER C: A bistable multivibrator is a flip-flop. Bistable means two stable states.

3F76 How many output changes are obtained for every two trigger pulses applied to the input of a *bistable T flip-flop* circuit?
  A. No output level changes
  B. One output level change
  C. Two output level changes
  D. Four output level changes
ANSWER C: Two trigger pulses, two output level changes.

3F77 The frequency of an AC signal can be divided electronically by what type of digital circuit?
  A. A free-running multivibrator
  B. An OR gate
  C. A bistable multivibrator
  D. An astable multivibrator
ANSWER C: A bistable multivibrator can be used to divide the frequency of an AC signal.

3F78 What type of digital IC is also known as a *latch*?
A. A decade counter
B. An OR gate
C. A flip-flop
D. An op-amp

ANSWER C: A flip-flop is used in a digital IC circuit that latches onto a bit and holds it until the time when it is triggered to latch onto the next bit.

3F79 How many *flip-flops* are required to divide a signal frequency by 4?
A. 1
B. 2
C. 4
D. 8

ANSWER B: Since a flip-flop has two stable states, you will need two flip-flops to divide a signal frequency by four.

3F80 What is an *astable multivibrator*?
A. A circuit that alternates between two stable states
B. A circuit that alternates between a stable state and an unstable state
C. A circuit set to block either a 0 pulse or a 1 pulse and pass the other
D. A circuit that alternates between two unstable states

ANSWER D: An *astable* multivibrator continuously switches back and forth between two states. It doesn't remain stable for a permanent time in either of its two states.

3F81 What is a *monostable multivibrator*?
A. A circuit that can be switched momentarily to the opposite binary state and then returns after a set time to its original state
B. A "clock" circuit that produces a continuous square wave oscillating between 1 and 0
C. A circuit designed to store one bit of data in either the 0 or the 1 configuration
D. A circuit that maintains a constant output voltage, regardless of variations in the input voltage

ANSWER A: This type of multivibrator may *momentarily* be *monostable* (MO MO). It stays in an original state until triggered to its other state, where it remains for a time usually determined by external components, after which it returns to the original state.

3F82 What is an *AND gate*?
A. A circuit that produces a logic "1" at its output only if all inputs are logic "1"
B. A circuit that produces a logic "0" at its output only if all inputs are logic "1"
C. A circuit that produces a logic "1" at its output if only one input is a logic "1"
D. A circuit that produces a logic "1" at its output if all inputs are logic "0"

ANSWER A: The AND gate has two or more inputs with a single output, and produces a logic "1" at its output only if *all* inputs are logic "1". Remember AND, *all*, and the number 1 and the number 1 looking like the letters ll in "all."

3F – PRACTICAL CIRCUITS

3F83 What is a *NAND gate*?
- A. A circuit that produces a logic "0" at its output only when all inputs are logic "0"
- B. A circuit that produces a logic "1" at its output only when all inputs are logic "1"
- C. A circuit that produces a logic "0" at its output if some but not all of its inputs are logic "1"
- D. A circuit that produces a logic "0" at its output only when all inputs are logic "1"

ANSWER D: The word *NAND* reminds me of the word "naw", meaning no or nothing. The *NAND* gate produces a logic "0" at its output only when all inputs are logic "1". N before AND means negative AND. If an AND gate outputs a "1" when all inputs are "1", a NAND gate outputs a "0", the negative of "1".

3F84 What is an *OR gate*?
- A. A circuit that produces a logic "1" at its output if any input is logic "1"
- B. A circuit that produces a logic "0" at its output if any input is logic "1"
- C. A circuit that produces a logic "0" at its output if all inputs are logic "1"
- D. A circuit that produces a logic "1" at its output if all inputs are logic "0"

ANSWER A: The OR gate will produce a logic "1" at its output if *any* input is, or all inputs are, a logic "1".

3F85 What is a *NOR gate*?
- A. A circuit that produces a logic "0" at its output only if all inputs are logic "0"
- B. A circuit that produces a logic "1" at its output only if all inputs are logic "1"
- C. A circuit that produces a logic "0" at its output if any or all inputs are logic "1"
- D. A circuit that produces a logic "1" at its output if some but not all of its inputs are logic "1"

ANSWER C: The NOR gate produces a logic "0" at its output if *ANY* or all inputs are logic "1". It's a negative OR. With OR gate or NOR gate, spot that word "ANY".

3F86 What is a *NOT gate*?
- A. A circuit that produces a logic "0" at its output when the input is logic "1" and vice versa
- B. A circuit that does not allow data transmission when its input is high
- C. A circuit that allows data transmission only when its input is high
- D. A circuit that produces a logic "1" at its output when the input is logic "1" and vice versa

ANSWER A: The NOT gate will produce a "0" or a "1" at the output—whichever is opposite from the level that is on the input.

3F87 What is a *truth table*?
- A. A table of logic symbols that indicate the high logic states of an op-amp
- B. A diagram showing logic states when the digital device's output is true

C. A list of input combinations and their corresponding outputs that characterizes a digital device's function
D. A table of logic symbols that indicates the low logic states of an op-amp

ANSWER C: We use a "truth table" in digital circuitry to characterize a digital device's function. If you buy the owner's technical manual on a piece of radio gear, you will usually find a page describing the truth table of digital devices used in the equipment.

3F88 In a positive-logic circuit, what level is used to represent a logic 1?
A. A low level
B. A positive-transition level
C. A negative-transition level
D. A high level

ANSWER D: In a positive-logic circuit, the logic "1" is a high level, or the most positive level. Think of the four letters in *high* and the four letters in *plus* (for positive).

3F89 In a positive-logic circuit, what level is used to represent a logic 0?
A. A low level
B. A positive-transition level
C. A negative-transition level
D. A high level

ANSWER A: In a positive-logic circuit, a logic "0" is a low level, or the least positive level.

3F90 In a negative-logic circuit, what level is used to represent a logic 1?
A. A low level
B. A positive-transition level
C. A negative-transition level
D. A high level

ANSWER A: In a negative-logic circuit, the logic "1" is now a low level, or the least negative level.

3F91 In a negative-logic circuit, what level is used to represent a logic 0?
A. A low level
B. A positive-transition level
C. A negative-transition level
D. A high level

ANSWER D: In a negative-logic circuit, the logic "0" is now a high level, or the most negative level.

3F92 What is a *crystal-controlled marker generator*?
A. A low-stability oscillator that "sweeps" through a band of frequencies
B. An oscillator often used in aircraft to determine the craft's location relative to the inner and outer markers at airports
C. A high-stability oscillator whose output frequency and amplitude can be varied over a wide range
D. A high-stability oscillator that generates a series of reference signals at known frequency intervals

ANSWER D: The crystal-controlled marker generator is a free-running oscillator that generates a series of reference signals at known frequency intervals.

3F – PRACTICAL CIRCUITS

3F93 What additional circuitry is required in a 100-kHz *crystal-controlled marker generator* to provide markers at 50 and 25 kHz?
  A. An emitter-follower
  B. Two frequency multipliers
  C. Two flip-flops
  D. A voltage divider
ANSWER C: We would use two flip-flops to take a 100-kHz signal and provide markers at 50 kHz and 25 kHz.

3F94 What is the purpose of a *prescaler circuit*?
  A. It converts the output of a JK flip-flop to that of an RS flip-flop
  B. It multiplies an HF signal so a low-frequency counter can display the operating frequency
  C. It prevents oscillation in a low frequency counter circuit
  D. It divides an HF signal so a low-frequency counter can display the operating frequency
ANSWER D: You will find a prescaler in frequency counters to divide down HF and VHF signals so they can be counted and displayed on a low-frequency counter.

3F95 What does the accuracy of a *frequency counter* depend on?
  A. The internal crystal reference
  B. A voltage-regulated power supply with an unvarying output
  C. Accuracy of the AC input frequency to the power supply
  D. Proper balancing of the power-supply diodes
ANSWER A: Every service shop should have a high frequency receiver that can tune in WWV and WWVH signals at 2.5 MHz, 5 MHz, 10 MHz, 15 MHz, or 20 MHz. If you hear the female voice, that is from WWVH in Hawaii. Either station may be used for frequency calibration of your frequency counter.

3F96 How many states does a decade counter digital IC have?
  A. 6
  B. 10
  C. 15
  D. 20
ANSWER B: Studying this material may seem like it is taking a decade, right? A decade is 10. Maybe a decade of days, or a decade of weeks, but, hopefully, not a decade of months.

3F97 What is the function of a decade counter digital IC?
  A. Decode a decimal number for display on a seven-segment LED display
  B. Produce one output pulse for every ten input pulses
  C. Produce ten output pulses for every input pulse
  D. Add two decimal numbers
ANSWER B: A decade counter digital IC gives one output pulse for every 10 input pulses.

3F98 What are the advantages of using an op-amp instead of LC elements in an audio filter?
A. Op-amps are more rugged and can withstand more abuse than can LC elements
B. Op-amps are fixed at one frequency
C. Op-amps are available in more styles and types than are LC elements
D. Op-amps exhibit gain rather than insertion loss

ANSWER D: An op amp exhibits high gain with negligible insertion loss because its input impedance is also high.

3F99 What determines the gain and frequency characteristics of an op-amp RC active filter?
A. Values of capacitances and resistances built into the op-amp
B. Values of capacitances and resistances external to the op-amp
C. Voltage and frequency of DC input to the op-amp power supply
D. Regulated DC voltage output from the op-amp power supply

ANSWER B: One of the nice features of an op amp is that it will operate as an RC filter—Resistors (R) and capacitors (C) are in a circuit outside the op amp. They are not built into the op amp; they are *external* (key word).

3F100 What are the principle uses of an op-amp RC active filter?
A. Op-amp circuits are used as high-pass filters to block RFI at the input to receivers
B. Op-amp circuits are used as low-pass filters between transmitters and transmission lines
C. Op-amp circuits are used as filters for smoothing power-supply output
D. Op-amp circuits are used as audio filters for receivers

ANSWER D: In radio transceivers, op-amp circuits are used in the audio section of receivers.

3F101 What type of capacitors should be used in an op-amp RC active filter circuit?
A. Electrolytic
B. Disc ceramic
C. Polystyrene
D. Paper dielectric

ANSWER C: The tiny op-amp filter should use polystyrene capacitors because they are very stable and will not vary when the circuit begins to heat up.

3F102 How can unwanted ringing and audio instability be prevented in a multisection op-amp RC audio filter circuit?
A. Restrict both gain and Q
B. Restrict gain, but increase Q
C. Restrict Q, but increase gain
D. Increase both gain and Q

ANSWER A: In order to keep an op amp from going into oscillation, the gain and the Q are restricted and set by the feedback circuit.

3F – PRACTICAL CIRCUITS

3F103 Where should an op-amp RC active audio filter be placed in a receiver?
A. In the IF strip, immediately before the detector
B. In the audio circuitry immediately before the speaker or phone jack
C. Between the balanced modulator and frequency multiplier
D. In the low-level audio stages

ANSWER D: We use an operational amplifier in the low-level audio stages in most mobile two-way radio transceivers.

3F104 What parameter must be selected when designing an audio filter using an op-amp?
A. Bandpass characteristics
B. Desired current gain
C. Temperature coefficient
D. Output-offset overshoot

ANSWER A: What frequencies do you want an op amp to amplify? This will help determine the bandpass characteristics required of the op amp.

3F105 What two factors determine the *sensitivity* of a receiver?
A. Dynamic range and third-order intercept
B. Cost and availability
C. Intermodulation distortion and dynamic range
D. Bandwidth and noise figure

ANSWER D: VHF and UHF transceivers with top sensitivity use low-noise front-end transistors. When choosing a receiver, look closely at the specifications for the bandwidth and the noise figure.

3F106 What is the limiting condition for sensitivity in a communications receiver?
A. The noise floor of the receiver
B. The power-supply output ripple
C. The two-tone intermodulation distortion
D. The input impedance to the detector

ANSWER A: Very weak signals get masked by the background noise of a receiver—the so-called noise floor of the receiver. The sensitivity of a receiver is limited by the noise generated by the devices inside, which contribute to the receiver's noise floor.

3F107 What is the theoretical minimum *noise floor* of a receiver with a 400-hertz bandwidth?
A. –141 dBm
B. –148 dBm
C. –174 dBm
D. –180 dBm

ANSWER B: A receiver with a CW filter switched in at 400-Hz bandwidth will have approximately a –148 dBm minimum noise floor.

5 GENERAL RADIOTELEPHONE OPERATOR LICENSE

3F108 How can *selectivity* be achieved in the front-end circuitry of a communications receiver?
- A. By using an audio filter
- B. By using a preselector
- C. By using an additional RF amplifier stage
- D. By using an additional IF amplifier stage

ANSWER B: A preselector achieves selectivity in the front-end circuitry of a communications receiver. When you change bands, different preselectors are automatically brought on line.

3F109 A receiver selectivity of 2.4 kHz in the IF circuitry is optimum for what type of signals?
- A. CW
- B. SSB voice
- C. Double-sideband AM voice
- D. FSK RTTY

ANSWER B: A good 2.4-kHz filter in the IF circuitry is perfect for SSB voice. A 1.8 kHz filter would even be better, but might make the audio sound pinched.

3F110 What occurs during CW reception if too narrow a filter bandwidth is used in the IF stage of a receiver?
- A. Undesired signals will reach the audio stage
- B. Output-offset overshoot
- C. Cross-modulation distortion
- D. Filter ringing

ANSWER D: While working CW, if you adjust the shaping of your filter network in the IF stage of a receiver, you can only pinch the bandpass so tight. Any further tightening creates filter ringing.

3F111 To find the supply power with only a voltmeter, measure between: (See Figure 3F111 below.)
- A. Z to W voltage times 150,000 divided by 2000
- B. Y to Z voltage squared divided by 2000
- C. W to Y voltage divided by 150
- D. W to X voltage divided by 150,000 times W to Z voltage

ANSWER D: If we measure the voltage with our voltmeter between points W and X, we will read the voltage drop across the 150,000 ohm resistor. Now to solve for current, use $I = E \div R$. We divide the voltage by 150,000, to determine the current for the circuit, remembering that in a series circuit like this, we will have the same amount of current throughout the circuit. Now that we have calculated the current, use $P = E \times I$ to find the power. Multiply the current by the voltage between W and Z, to calculate the total supply power.

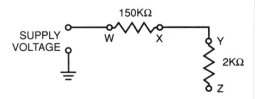

Figure 3F111

3F – PRACTICAL CIRCUITS

3F112 A receiver selectivity of 10 kHz in the IF circuitry is optimum for what type of signals?
   A. SSB voice
   B. Double-sideband AM
   C. CW
   D. FSK RTTY

ANSWER B: A very wide 10-kHz filter would be appropriate for tuning in short-wave broadcast stations on double-sideband AM.

3F113 What degree of selectivity is desirable in the IF circuitry of a single-sideband phone receiver?
   A. 1 kHz
   B. 2.4 kHz
   C. 4.2 kHz
   D. 4.8 kHz

ANSWER B: For SSB, a 2.4-kHz filter bandwidth is the most popular for good fidelity audio.

3F114 What is an undesirable effect of using too wide a filter bandwidth in the IF section of a receiver?
   A. Output-offset overshoot
   B. Undesired signals will reach the audio stage
   C. Thermal-noise distortion
   D. Filter ringing

ANSWER B: By using an AM 10-kHz IF filter for pulling in SSB signals, it would result in undesired signals beyond the normal bandwidth of the desired signal.

3F115 How should the filter bandwidth of a receiver IF section compare with the bandwidth of a received signal?
   A. Filter bandwidth should be slightly greater than the received-signal bandwidth
   B. Filter bandwidth should be approximately half the received-signal bandwidth
   C. Filter bandwidth should be approximately two times the received-signal bandwidth
   D. Filter bandwidth should be approximately four times the received-signal bandwidth

ANSWER A: For best fidelity, filter bandwidth should be slightly greater than the received signal bandwidth. If you are using one of those new SSB base stations, you have plenty of choices!

3F116 What degree of selectivity is desirable in the IF circuitry of an FM-phone receiver?
   A. 1 kHz
   B. 2.4 kHz
   C. 4.2 kHz
   D. 15 kHz

ANSWER D: In your FM equipment, the filter must accommodate a minimum of ±5 kHz, a total of 10-kHz bandwidth. The 15-kHz filter is your best choice.

3F117 How can selectivity be achieved in the IF circuitry of a communications receiver?
A. Incorporate a means of varying the supply voltage to the local oscillator circuitry
B. Replace the standard JFET mixer with a bipolar transistor followed by a capacitor of the proper value
C. Remove AGC action from the IF stage and confine it to the audio stage only
D. Incorporate a high-Q filter

ANSWER D: Circuits with high-Q filters will give you the best selectivity in the IF section.

3F118 What is meant by the *dynamic range* of a communications receiver?
A. The number of kHz between the lowest and the highest frequency to which the receiver can be tuned
B. The maximum possible undistorted audio output of the receiver, referenced to one milliwatt
C. The ratio between the minimum discernible signal and the largest tolerable signal without causing audible distortion products
D. The difference between the lowest-frequency signal and the highest-frequency signal detectable without moving the tuning knob

ANSWER C: The dynamic range of a communications receiver is a *ratio* of the minimum signal to the maximum signal without distortion. Look for the key word *ratio* in the answer to this question.

3F119 What is the term for the ratio between the largest tolerable receiver input signal and the minimum discernible signal?
A. Intermodulation distortion
B. Noise floor
C. Noise figure
D. Dynamic range

ANSWER D: If your client's older two-way radio equipment sounds distorted when receiving extremely strong signals, chances are it has poor dynamic range. Good dynamic range allows the receiver to capture and send to the amplifier stage good, clean audio on extremely strong signal inputs all the way down to the minimum discernible signal.

3F120 What type of problems are caused by poor *dynamic range* in a communications receiver?
A. Cross-modulation of the desired signal and desensitization from strong adjacent signals
B. Oscillator instability requiring frequent retuning, and loss of ability to recover the opposite sideband, should it be transmitted
C. Cross-modulation of the desired signal and insufficient audio power to operate the speaker
D. Oscillator instability and severe audio distortion of all but the strongest received signals

ANSWER A: In a receiver that does not have good dynamic range, cross-modulation of the desired signal can be encountered, and the receiver can drop in sensitivity (be desensitized) by strong adjacent frequency signals.

3F — PRACTICAL CIRCUITS

3F121 The ability of a communications receiver to perform well in the presence of strong signals outside the band of interest is indicated by what parameter?
A. Noise figure
B. Blocking dynamic range
C. Signal-to-noise ratio
D. Audio output

ANSWER B: Look for a receiver that has good "blocking" to keep unwanted signals out of the pass band of the radio band you are working.

3F122 What is meant by the term *noise figure* of a communications receiver?
A. The level of noise entering the receiver from the antenna
B. The relative strength of a received signal 3 kHz removed from the carrier frequency
C. The level of noise generated in the front end and succeeding stages of a receiver
D. The ability of a receiver to reject unwanted signals at frequencies close to the desired one

ANSWER C: For VHF and UHF satellite gear, always look for the lowest noise figure with a hot front-end complement of transistors.

3F123 Which stage of a receiver primarily establishes its *noise figure*?
A. The audio stage
B. The IF strip
C. The RF stage
D. The local oscillator

ANSWER C: It's the RF amplifier stage where we are most concerned about a low noise figure.

3F124 What is an *inverting op-amp circuit*?
A. An operational amplifier circuit connected such that the input and output signals are 180 degrees out of phase
B. An operational amplifier circuit connected such that the input and output signals are in phase
C. An operational amplifier circuit connected such that the input and output signals are 90 degrees out of phase
D. An operational amplifier circuit connected such that the input impedance is held at zero, while the output impedance is high

ANSWER A: On an inverting op amp, the input and output signals are 180 degrees out of phase. Remember *inverting* means a 180° phase shift.

3F125 What is a *noninverting op-amp circuit*?
A. An operational amplifier circuit connected such that the input and output signals are 180 degrees out of phase
B. An operational amplifier circuit connected such that the input and output signals are in phase
C. An operational amplifier circuit connected such that the input and output signals are 90 degrees out of phase
D. An operational amplifier circuit connected such that the input impedance is held at zero while the output impedance is high

ANSWER B: In a *non-inverting* op amp circuit, the input and output signals are *in phase*. *Non-inverting* means *no phase shift*; *inverting* means 180° *phase shift*.

Ideal Operational Amplifier

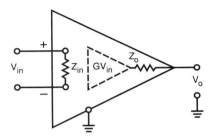

Z_{in} = Infinity
G = Infinity
Z_o = Zero
Bandwith = Infinity
V_o has no offset
(V_o=0 when V_{in}=0)

IC Operational Ampliifer
Even though IC operational amplifiers do not meet all of these specifications, G and Z_{in} are very large. Because G is very large, any small input voltage would drive the output into saturation (for practical supply voltages); therefore, normal operational amplifier operation is with feedback to set the gain. Here's an example for an inverting amplifier:

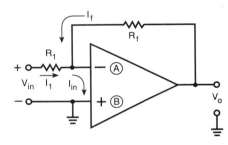

Since Z_{in} is very large, I_{in} = 0

$\therefore I_1 + I_f = 0$

$I_1 = \dfrac{V_{in}}{R_1}$

$I_f = \dfrac{V_o}{R_f}$

$\therefore \dfrac{V_{in}}{R_1} + \dfrac{V_o}{R_f} = 0$

$\dfrac{V_o}{R_f} = -\dfrac{V_{in}}{R_1}$

$\therefore \dfrac{V_o}{V_{in}} = -\dfrac{R_f}{R_1}$

Ⓐ Inverting Input — An input voltage that is more positive on this input will cause the output voltage to be less positive.

Ⓑ Non-Inverting Input — An input voltage that is more positive on this input will cause the output voltage to be more positive.

The gain of an *inverting IC operational amplifier* is:

$$G = -\dfrac{R_f}{R_1}$$

The minus sign means the output is out of phase with the input.

Operational Amplifier Basics

3F126 How does the gain of a theoretically ideal operational amplifier vary with frequency?
A. The gain increases linearly with increasing frequency
B. The gain decreases linearly with increasing frequency
C. The gain decreases logarithmically with increasing frequency
D. The gain does not vary with frequency

ANSWER D: The gain on an *ideal* op amp should not vary with frequency.

3F – PRACTICAL CIRCUITS

3F127 What determines the input impedance in a FET common-source amplifier?
- A. The input impedance is essentially determined by the resistance between the drain and substrate
- B. The input impedance is essentially determined by the resistance between the source and drain
- C. The input impedance is essentially determined by the gate biasing network
- D. The input impedance is essentially determined by the resistance between the source and substrate

ANSWER C: Since the input impedance of the device itself is very high, the input impedance to a FET common source amplifier is determined by the *gate biasing network*.

3F128 What determines the output impedance in a FET common-source amplifier?
- A. The output impedance is essentially determined by the drain resistor
- B. The output impedance is essentially determined by the input impedance of the FET
- C. The output impedance is essentially determined by the drain supply voltage
- D. The output impedance is essentially determined by the gate supply voltage

ANSWER A: The resistance connected to the drain (drain resistor) determines the output impedance of a FET common-source amplifier.

3F129 What is the purpose of a bypass capacitor?
- A. It increases the resonant frequency of the circuit
- B. It removes direct current from the circuit by shunting DC to ground
- C. It removes alternating current by providing a low impedance path to ground
- D. It acts as a voltage divider

ANSWER C: Bypass capacitors shunt AC through a low impedance path to ground.

3F130 What is the purpose of a coupling capacitor?
- A. It blocks direct current and passes alternating current
- B. It blocks alternating current and passes direct current
- C. It increases the resonant frequency of the circuit
- D. It decreases the resonant frequency of the circuit

ANSWER A: Coupling capacitors block direct current, and pass the AC signal. Put an ohmmeter across a capacitor and you read an open circuit (after the capacitor charges), yet holding onto one lead of a capacitor and putting a scope on the other lead produces a signal on the scope.

3F131 What condition must exist for a circuit to oscillate?
- A. It must have a gain of less than 1
- B. It must be neutralized
- C. It must have positive feedback sufficient to overcome losses
- D. It must have negative feedback sufficient to cancel the input

ANSWER C: The circuit must have just enough positive feedback to overcome losses in order to continue to properly oscillate.

3F132 What is the voltage drop across R1? (Refer to Figure 3F132.)
A. 9 volts
B. 7 volts
C. 5 volts
D. 3 volts

ANSWER C: Look at Figure 3F132 and notice that the diode is reverse biased, and will look like an open circuit to this series resistance circuit problem. Current will be the same throughout all of the circuit, and voltage drops will equal the source voltage. Combine R3 and R2 as a total of 300 ohms, R3 + R2 equals the resistance of R1, 300 ohms, so each will drop 5 volts. The voltage drop across R1 equals 5 volts.

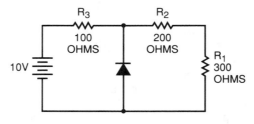

Figure 3F132

3F133 What is the voltage drop across R1? (Refer to Figure 3F133.)
A. 1.2 volts
B. 2.4 volts
C. 3.7 volts
D. 9 volts

ANSWER D: Look at Figure 3F133 and note that the Zener diode will drop 3 volts from the supply 12 volts, and 12 minus 3 volts equals a 9-volt drop at R1.

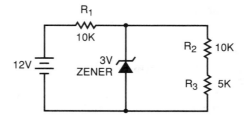

Figure 3F133

3F134 In a properly operating marine transmitter, if the power supply bleeder resistor opens:
A. Short circuit of supply voltage due to overload
B. Regulation would decrease
C. Next stage would fail due to short circuit
D. Filter capacitors might short from voltage surge

ANSWER D: Bleeder resistors assist in voltage regulation in the filter capacitor section of a power supply. If the bleeder resistors should open, the filter capacitors would no longer be protected, and they could short from a voltage surge.

3F – PRACTICAL CIRCUITS

3F135 Which of the following can occur that would least affect this circuit? (Refer to Figure 3F135.)
- A. C1 shorts
- B. C1 opens
- C. C3 shorts
- D. C18 opens

ANSWER D: It is assumed that the circuit is an operating circuit (e.g., Class C operation) and doesn't need base biasing. Based on this, C18 is a bypass capacitor, and if it opens, the circuit would probably continue to function. However, if it shorts, the circuit will not operate properly.

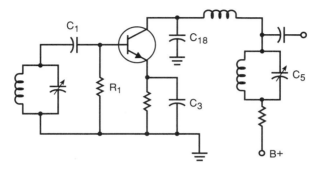

Figure 3F135

3F136 When S1 is closed, lights L1 and L2 go on. What is the condition of both lamps when both S1 and S2 are closed? (Refer to Figure 3F136.)
- A. both lamps stay on
- B. L1 turns off; L2 stays on
- C. both lamps turn off
- D. L1 stays on; L2 turns off

ANSWER D: When both switch 1 and switch 2 are closed, they will provide a positive voltage on the bases of both NPN transistors. Both transistors will conduct, and lamp L1 will be in series and will illuminate. With both switches closed, lamp L2 is shunted by the lower transistor, and there will not be enough current through it to light.

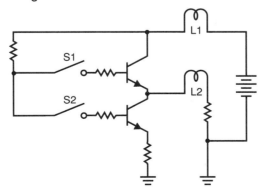

Figure 3F136

3F137 If S1 is closed both lamps light, what happens when S1 and S2 are closed? (Refer again to Figure 3F136.)
A. L1 and L2 are off
B. L1 is on and L2 is flashing
C. L1 is off and L2 is on
D. L1 is on and L2 is off

ANSWER D: This is almost the same question as above, but worded slightly different. The top lamp, L1, in Figure 3F136 will stay illuminated, and L2 will be off due to insufficient current.

3F138 How can we correct the defect, if any, in this voltage doubler circuit? (Refer to Figure 3F138.)
A. Omit C1
B. Reverse polarity signs on load
C. Ground T
D. Reverse polarity on C1

ANSWER B: Look at Figure 3F138, and notice the load polarity—this is not correct. You will normally find the positive leg on top; and if you look at the components within the diagram, you will see that the load polarity has reversed polarity signs.

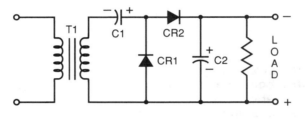

Figure 3F138

3F139 The FCC deleted this question.

Subelement 3G – Signals and Emissions (9 questions)

3G1 What is emission A3C?
A. Facsimile
B. RTTY
C. ATV
D. Slow Scan TV

ANSWER A: Three character signals and emissions may still show up on paperwork going to and from the FCC, or paperwork to local frequency coordinators. When you see the number 3 in the middle of an emission designator, it usually means voice or other analog information. The A means amplitude modulation, and the C is what gives away this answer—C for faCsimile. See chart at 3G70.

3G – SIGNALS AND EMISSIONS

3G2 What type of emission is produced when an amplitude-modulated transmitter is modulated by a facsimile signal?
 A. A3F
 B. A3C
 C. F3F
 D. F3C

ANSWER B: Remember, C for fa*C*simile. Put this together with A for Amplitude modulated and you have the correct answer, 3AC. Most marine SSB sets will accept weather facsimile programs when tied into a computer, and read out FAXed weather charts quite well.

3G3 What is facsimile?
 A. The transmission of tone-modulated telegraphy
 B. The transmission of a pattern of printed characters designed to form a picture
 C. The transmission of printed pictures by electrical means
 D. The transmission of moving pictures by electrical means

ANSWER C: Just like your telephone FAX, radio facsimile will give you *printed pictures by electrical means* (key words).

3G4 What is emission F3C?
 A. Voice transmission
 B. Slow Scan TV
 C. RTTY
 D. Facsimile

ANSWER D: Think of the letter "C" in the word fa*C*simile. F for FM.

3G5 What type of emission is produced when a frequency-modulated transmitter is modulated by a facsimile signal?
 A. F3C
 B. A3C
 C. F3F
 D. A3F

ANSWER A: If it is a facsimile signal then remember the C, but frequency modulated means the new designator will then be F3C, the F for *frequency* modulation.

3G6 What is emission A3F?
 A. RTTY
 B. Television
 C. SSB
 D. Modulated CW

ANSWER B: You will get an F on your exam if all you do is watch television, and not study this book. A3F is amplitude modulated TV.

3G7 What type of emission is produced when an amplitude-modulated transmitter is modulated by a television signal?
 A. F3F
 B. A3F
 C. A3C
 D. F3C

ANSWER B: Television, A3F. A for Amplitude modulated and F for TV.

5 GENERAL RADIOTELEPHONE OPERATOR LICENSE

3G8 What is emission F3F?
A. Modulated CW
B. Facsimile
C. RTTY
D. Television

ANSWER D: Same thing, F for television, but this time F for the frequency modulated part of the TV signal.

3G9 What type of emission is produced when a frequency modulated transmitter is modulated by a television signal?
A. A3F
B. A3C
C. F3F
D. F3C

ANSWER C: Television, F3F—F for FM, F for TV. Since it is frequency modulated, answer A is incorrect.

3G10 How can an FM-phone signal be produced?
A. By modulating the supply voltage to a class-B amplifier
B. By modulating the supply voltage to a class-C amplifier
C. By using a reactance modulator on an oscillator
D. By using a balanced modulator on an oscillator

ANSWER C: They are coming back around. Here's that word again, reactance modulator. An FM phone signal is produced by using a reactance modulator.

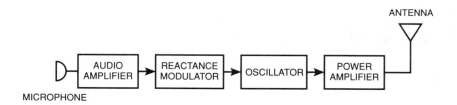

FM Transmitter

3G11 How can a double-sideband phone signal be produced?
A. By using a reactance modulator on an oscillator
B. By varying the voltage to the varactor in an oscillator circuit
C. By using a phase detector, oscillator and filter in a feedback loop
D. By modulating the plate supply voltage to a class C amplifier

ANSWER D: Since we're talking about double-sideband phone, it means AM with both sidebands present. Answer A is wrong because it's talking about FM. Answer D is the correct answer because we're talking about a Class C amplifier, an okay amplifier to plate modulate with a double-sideband phone signal.

3G – SIGNALS AND EMISSIONS

3G12 How can a single-sideband phone signal be produced?
A. By producing a double sideband signal with a balanced modulator and then removing the unwanted sideband by filtering
B. By producing a double sideband signal with a balanced modulator and then removing the unwanted sideband by heterodyning
C. By producing a double sideband signal with a balanced modulator and then removing the unwanted sideband by mixing
D. By producing a double sideband signal with a balanced modulator and then removing the unwanted sideband by neutralization

ANSWER A: On single sideband, the signal is first produced by a balanced modulator with both sidebands present, and a filter then removes the unwanted upper or lower sideband. It is the *filtering* (key word) that removes the unwanted sideband.

3G13 What is meant by the term *deviation ratio*?
A. The ratio of the audio modulating frequency to the center carrier frequency
B. The ratio of the maximum carrier frequency deviation to the highest audio modulating frequency
C. The ratio of the carrier center frequency to the audio modulating frequency
D. The ratio of the highest audio modulating frequency to the average audio modulating frequency

ANSWER B: It's important to set the deviation ratio properly on your FM transceiver to prevent splatter, and a signal bandwidth that is too wide. Deviation ratio is the ratio of the maximum carrier swing to your highest audio modulating frequency when you speak into the microphone.

$$\text{Deviation Ratio} = \frac{\text{Maximum Carrier Frequency Deviation in kHz}}{\text{Maximum Modulation Frequency in kHz}}$$

3G14 In an FM-phone signal, what is the term for the maximum deviation from the carrier frequency divided by the maximum audio modulating frequency?
A. Deviation index
B. Modulation index
C. Deviation ratio
D. Modulation ratio

ANSWER C: Simple division of the maximum audio frequency into the maximum deviation of the carrier frequency will lead you to the deviation ratio. Division produces a ratio.

3G15 What is the deviation ratio for an FM-phone signal having a maximum frequency swing of plus or minus 5 kHz and accepting a maximum modulation rate of 3 kHz?
A. 60
B. 0.16
C. 0.6
D. 1.66

ANSWER D: Use formula at 3G13. Divide the maximum carrier frequency swing by the maximum modulation frequency. 3 kHz modulation frequency into 5 kHz frequency swing gives you a deviation ratio of 1.66, answer D.

3G16 What is the deviation ratio of an FM-phone signal having a maximum frequency swing of plus or minus 7.5 kHz and accepting a maximum modulation rate of 3.5 kHz?
   A. 2.14
   B. 0.214
   C. 0.47
   D. 47

ANSWER A: Divide 3.5 (maximum modulation frequency) into 7.5 (maximum carrier frequency deviation), and you end up with 2.14, your deviation ratio.

3G17 What is meant by the term *modulation index?*
   A. The processor index
   B. The ratio between the deviation of a frequency modulated signal and the modulating frequency
   C. The FM signal-to-noise ratio
   D. The ratio of the maximum carrier frequency deviation to the highest audio modulating frequency

ANSWER B: With modulation index, instead of looking at the highest audio frequency, we look at a specific modulating frequency. This is an important consideration for the amount of deviation on a packet station on FM.

3G18 In an FM-phone signal, what is the term for the ratio between the deviation of the frequency-modulated signal and the modulating frequency?
   A. FM compressibility
   B. Quieting index
   C. Percentage of modulation
   D. Modulation index

ANSWER D: Be careful with this one. Since they don't say the highest modulating frequency, but say only "modulating frequency", they are looking for the modulation index. Notice that there is not any answer that says "deviation ratio" to get you in trouble.

3G19 How does the modulation index of a phase-modulated emission vary with the modulated frequency?
   A. The modulation index increases as the RF carrier frequency (the modulated frequency) increases
   B. The modulation index decreases as the RF carrier frequency (the modulated frequency) increases
   C. The modulation index varies with the square root of the RF carrier frequency (the modulated frequency)
   D. The modulation index does not depend on the RF carrier frequency (the modulated frequency)

ANSWER D: This is one of those answers that has a "no" in it that makes it correct. When you calculate modulation index, you *do not* (key words) look at the actual RF carrier frequency as part of your calculations.

3G – SIGNALS AND EMISSIONS

3G20 In an FM-phone signal having a maximum frequency deviation of 3.0 kHz either side of the carrier frequency, what is the modulation index when the modulating frequency is 1.0 kHz?
- A. 3
- B. 0.3
- C. 3000
- D. 1000

ANSWER A: 3000 divided by 1000 gives you a modulation index of 3.

3G21 What is the modulation index of an FM-phone transmitter producing an instantaneous carrier deviation of 6 kHz when modulated with a 2 kHz modulating frequency?
- A. 6000
- B. 3
- C. 2000
- D. 1/3

ANSWER B: 6 divided by 2 gives us a modulation index of 3.

3G22 What are *electromagnetic waves*?
- A. Alternating currents in the core of an electromagnet
- B. A wave consisting of two electric fields at right angles to each other
- C. A wave consisting of an electric field and a magnetic field at right angles to each other
- D. A wave consisting of two magnetic fields at right angles to each other

ANSWER C: When you think of an electromagnetic radio wave, think of it as a combination of two fields at right angles traveling in waves through the atmosphere at the same time. One field is vertically polarized and the other is horizontally polarized. If the antenna is vertically polarized, the electric field will be vertical, and the magnetic field will be horizontal. See figure at 3G25.

3G23 What is a *wave front*?
- A. A voltage pulse in a conductor
- B. A current pulse in a conductor
- C. A voltage pulse across a resistor
- D. A fixed point in an electromagnetic wave

ANSWER D: When we analyze propagation of signals, we take into account the wave front which is a fixed point at the beginning of an electromagnetic wave.

3G24 At what speed do electromagnetic waves travel in free space?
- A. Approximately 300 million meters per second
- B. Approximately 468 million meters per second
- C. Approximately 186,300 feet per second
- D. Approximately 300 million miles per second

ANSWER A: Radio waves travel at the rate of 300 million meters per second in free space. That's where the number 300 comes from when we convert frequency in MHz to wavelength in meters, or visa versa, wavelength in meters to frequency in MHz.

5 GENERAL RADIOTELEPHONE OPERATOR LICENSE

3G25 What are the two interrelated fields considered to make up an electromagnetic wave?
A. An electric field and a current field
B. An electric field and a magnetic field
C. An electric field and a voltage field
D. A voltage field and a current field

ANSWER B: Radio waves consist of an electric field and a magnetic field traveling together at right angles to each other.

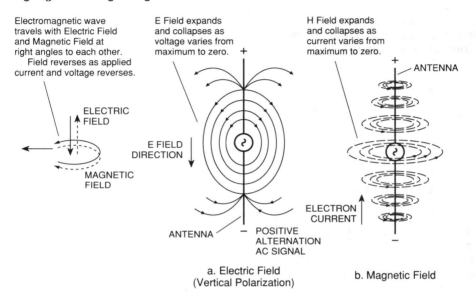

Propagating Electromagnetic Waves

Source: *Antennas – Selection and Installation*, A.J. Evans, ©1986 Master Publishing, Inc., Richardson, Texas

---

3G26 Why do electromagnetic waves not penetrate a good conductor to any great extent?
A. The electromagnetic field induces currents in the insulator
B. The oxide on the conductor surface acts as a shield
C. Because of Eddy currents
D. The resistivity of the conductor dissipates the field

ANSWER C: Radio waves travel best on the outside of a conductor. The reason they do not travel inside a conductor, to any great extent, is because of *Eddy* currents which tend to cancel the inside travel of the radio wave.

3G27 What is meant by referring to electromagnetic waves as horizontally polarized?
A. The electric field is parallel to the earth
B. The magnetic field is parallel to the earth
C. Both the electric and magnetic fields are horizontal
D. Both the electric and magnetic fields are vertical

ANSWER A: When we talk about polarization, we always refer to the electric lines of force, not the magnetic lines of force. On a horizontally polarized wave, the electric field is *parallel to the earth* (key words).

3G – SIGNALS AND EMISSIONS

3G28 What is meant by referring to electromagnetic waves as having circular polarization?
A. The electric field is bent into a circular shape
B. The electric field rotates
C. The electromagnetic wave continues to circle the earth
D. The electromagnetic wave has been generated by a quad antenna

ANSWER B: Some satellite enthusiasts will use circular polarization to improve their transmission and reception through an orbiting satellite. With circular polarization, *the electric field rotates* (key words). It could be right-hand circular, or left-hand circular, depending on the antenna feed system.

3G29 When the electric field is perpendicular to the surface of the earth, what is the polarization of the electromagnetic wave?
A. Circular
B. Horizontal
C. Vertical
D. Elliptical

ANSWER C: If the electric field is perpendicular to the earth, polarization is vertical.

3G30 When the magnetic field is parallel to the surface of the earth, what is the polarization of the electromagnetic wave?
A. Circular
B. Horizontal
C. Elliptical
D. Vertical

ANSWER D: Watch out for this one—here they are asking about the magnetic field which is parallel to the surface of the earth. If the magnetic field is parallel, the *electric field is vertical*, so polarization would be *vertical*.

3G31 When the magnetic field is perpendicular to the surface of the earth, what is the polarization of the electromagnetic field?
A. Horizontal
B. Circular
C. Elliptical
D. Vertical

ANSWER A: If the magnetic field is perpendicular (vertical) to the surface of the earth, the *electric* lines of force would be *horizontal*, and this would give us *horizontal polarization*.

3G32 When the electric field is parallel to the surface of the earth, what is the polarization of the electromagnetic wave?
A. Vertical
B. Horizontal
C. Circular
D. Elliptical

ANSWER B: Here the *electric* field is *parallel*, so the polarization is *horizontal*.

3G33 What is a *sine wave*?
A. A constant-voltage, varying-current wave
B. A wave whose amplitude at any given instant can be represented by a point on a wheel rotating at a uniform speed
C. A wave following the laws of the trigonometric tangent function
D. A wave whose polarity changes in a random manner

ANSWER B: A sine wave is as smooth as a turning wheel, giving us a uniform cycle of energy. The point on a wheel is like a vector rotating through 360°. If the sine of the angle is plotted against degrees rotation, we will plot a sinewave.

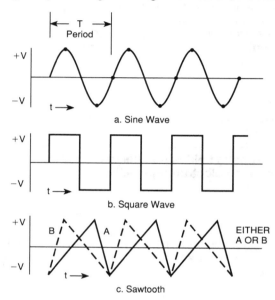

Sine, Square and Sawtooth Waveforms

3G34 If a sine wave begins from above or below the zero axis, how many times will it cross the zero axis in one complete cycle?
A. 180 times
B. 4 times
C. 2 times
D. 360 times

ANSWER C: In a complete cycle, the sine wave begins at the axis, goes up to a maximum, then down through the axis once to a maximum in the opposite direction, and then back to the zero axis for a total of two crossings, completing one cycle.

3G35 How many degrees are there in one complete sine wave cycle?
A. 90 degrees
B. 270 degrees
C. 180 degrees
D. 360 degrees

ANSWER D: One complete sine wave cycle is 360 degrees, just like a wheel.

3G – SIGNALS AND EMISSIONS

3G36 What is the *period* of a wave?
　A. The time required to complete one cycle
　B. The number of degrees in one cycle
　C. The number of zero crossings in one cycle
　D. The amplitude of the wave
ANSWER A: The period of a wave is expressed and measured in time. Just like school periods for classes, and periods of studying this book, the period of the wave is the time required to complete one cycle.

3G37 What is a *square wave*?
　A. A wave with only 300 degrees in one cycle
　B. A wave which abruptly changes back and forth between two voltage levels and which remains an equal time at each level
　C. A wave that makes four zero crossings per cycle
　D. A wave in which the positive and negative excursions occupy unequal portions of the cycle time
ANSWER B: Square waves have abrupt changes back and forth between two voltage levels. The *abrupt change* (key words) of a square wave is not desirable as a CW key-attack and key-release wave form. Square waves on CW generate key clicks because of their abrupt changes. See figure at question 3G33.

3G38 What is a wave called which abruptly changes back and forth between two voltage levels and which remains an equal time at each level?
　A. A sine wave
　B. A cosine wave
　C. A square wave
　D. A rectangular wave
ANSWER C: A square waveform has steep waveform edges, in which the waveform changes rapidly from one voltage level to another. The times that the waveform is at each level are equal. The key words are abruptly changes back and forth; when the waveform does, then it is a square wave. See figure at question 3G33.

3G39 Which sine waves make up a square wave?
　A. 0.707 times the fundamental frequency
　B. The fundamental frequency and all odd and even harmonics
　C. The fundamental frequency and all even harmonics
　D. The fundamental frequency and all odd harmonics
ANSWER D: The square wave is a tough one to get rid of harmonics, because the fundamental frequency and all odd harmonics may be found in a square wave. But only the odd harmonics—1, 3, 5, 7, etc. Square people are odd, aren't they?

3G40 What type of wave is made up of sine waves of the fundamental frequency and all the odd harmonics?
　A. Square wave
　B. Sine wave
　C. Cosine wave
　D. Tangent wave
ANSWER A: If you're odd, you're probably square.

3G41 What is a *sawtooth wave?*
A. A wave that alternates between two values and spends an equal time at each level
B. A wave with a straight line rise time faster than the fall time (or vice versa)
C. A wave that produces a phase angle tangent to the unit circle
D. A wave whose amplitude at any given instant can be represented by a point on a wheel rotating at a uniform speed

ANSWER B: A sawtooth wave has a straight line rise, and an immediate fall, or vice versa. It looks like a saw blade if you look at it on an oscilloscope. See figure at question 3G33.

3G42 What type of wave is characterized by a rise time significantly faster than the fall time (or vice versa)?
A. A cosine wave
B. A square wave
C. A sawtooth wave
D. A sine wave

ANSWER C: The sawtooth wave can be backwards with immediate rise time significantly faster than the fall time. See figure at question 3G33.

3G43 Which sine waves make up a sawtooth wave?
A. The fundamental frequency and all prime harmonics
B. The fundamental frequency and all even harmonics
C. The fundamental frequency and all odd harmonics
D. The fundamental frequency and all harmonics

ANSWER D: The sawtooth wave may be used in frequency multipliers because the fundamental frequency is also rich in all harmonics, both odd and even.

3G44 What type of wave is made up of sine waves at the fundamental frequency and all the harmonics?
A. A sawtooth wave
B. A square wave
C. A sine wave
D. A cosine wave

ANSWER A: If it has *all the harmonics* (key words), the answer is a sawtooth wave.

3G45 What is the meaning of the term *root-mean-square* value of an AC voltage?
A. The value of an AC voltage found by squaring the average value of the peak AC voltage
B. The value of a DC voltage that would cause the same heating effect in a given resistor as a peak AC voltage
C. The value of an AC voltage that would cause the same heating effect in a given resistor as a DC voltage of the same value
D. The value of an AC voltage found by taking the square root of the average AC value

ANSWER C: You sometimes see this as RMS, also called effective value of an AC voltage. Since AC voltage is continually changing in amplitude, we must look at its root mean square value before we can come up with the same heating effect in a given resistor as a constant DC voltage.

3G – SIGNALS AND EMISSIONS 5

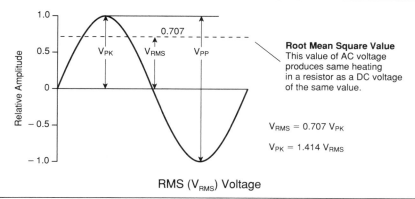

RMS (V_{RMS}) Voltage

3G46 What is the term used in reference to a DC voltage that would cause the same heating in a resistor as a certain value of AC voltage?
- A. Cosine voltage
- B. Power factor
- C. Root mean square
- D. Average voltage

ANSWER C: When we compare a DC voltage to an AC voltage that produces the same heating in a resistor, we need to choose the root-mean-square AC voltage.

3G47 What would be the most accurate way of determining the RMS voltage of a complex waveform?
- A. By using a grid dip meter
- B. By measuring the voltage with a D'Arsonval meter
- C. By using an absorption wavemeter
- D. By measuring the heating effect in a known resistor

ANSWER D: In a very complex wave form, one way to determine its effective value is to measure the heating effect in a known resistor, compared to a DC known voltage.

3G48 What is the RMS voltage at a common household electrical power outlet?
- A. 117-V AC
- B. 331-V AC
- C. 82.7-V AC
- D. 165.5-V AC

ANSWER A: Household alternating current electrical power is rated at 117 VAC RMS.

3G49 What is the peak voltage at a common household electrical outlet?
- A. 234 volts
- B. 165.5 volts
- C. 117 volts
- D. 331 volts

ANSWER B: When you measure house power with a voltmeter, you are measuring RMS. If you were to look at it on an oscilloscope, and look at the electrical peaks, it would be 1.414 times higher than 117 volts, or a total of 165 volts peak.

3G50 What is the peak-to-peak voltage at a common household electrical outlet?
A. 234 volts
B. 117 volts
C. 331 volts
D. 165.5 volts

ANSWER C: If you really want to impress your friends about your house power, tell them that you have 331 volts coming out of the socket. If they bet you don't, tell them you are stating the voltage as peak-to-peak, which is twice as much as peak voltage.

3G51 What is RMS voltage of a 165-volt peak pure sine wave?
A. 233-V AC
B. 330-V AC
C. 58.3-V AC
D. 117-V AC

ANSWER D: If your scope shows 165 volts peak pure sine wave, the RMS value is 117 volts AC, normal RMS house power. Multiply the peak voltage times 0.707 to find the RMS volts.

3G52 What is the RMS value of a 331-volt peak-to-peak pure sine wave?
A. 117-V AC
B. 165-V AC
C. 234-V AC
D. 300-V AC

ANSWER A: 331 volts peak-to-peak divided in half is 165.5 volts peak. Multiply the peak voltage by 0.707, and you end up with house power, 117 volts RMS.

3G53 For many types of voices, what is the ratio of PEP to average power during a modulation peak in a single-sideband phone signal?
A. Approximately 1.0 to 1
B. Approximately 25 to 1
C. Approximately 2.5 to 1
D. Approximately 100 to 1

ANSWER C: When you talk about power output from your SSB worldwide set, you normally refer to power as peak envelope power. Depending on your modulation, the ratio of PEP to average power is about 2.5 to 1. You can remember 2.5 because that's very close to 2.5 kHz of band occupancy for a properly modulated SSB signal.

3G54 In a single-sideband phone signal, what determines the PEP-to-average power ratio?
A. The frequency of the modulating signal
B. The degree of carrier suppression
C. The speech characteristics
D. The amplifier power

ANSWER C: It's not all that easy to determine the ratio of peak envelope power (PEP) to average power in your worldwide SSB set because your speech characteristics may vary from those when another operator is using the same equipment.

3G – SIGNALS AND EMISSIONS

3G55 What is the approximate DC input power to a Class B RF power amplifier stage in an FM-phone transmitter when the PEP output power is 1500 watts?
A. Approximately 900 watts
B. Approximately 1765 watts
C. Approximately 2500 watts
D. Approximately 3000 watts

ANSWER C: You can determine input power by dividing PEP output power by the efficiency of the amplifier. Because efficiency is normally rated in a percentage, divide percentage efficiency by 100 to get the decimal value. In this problem, divide 1500 watts PEP output power by 0.6 because the efficiency of a Class B amplifier is approximately 60 percent. They don't tell you that—you must remember that a Class B amp is a bit more efficient than a class AB amplifier that has an efficiency of 50%.

1500 divided by 0.6 gives a power input of approximately 2500 watts, and for the common Class B amplifier, 2500 watts in will normally provide about 1500 watts out.

3G56 What is the approximate DC input power to a Class C RF power amplifier stage in a RTTY transmitter when the PEP output power is 1000 watts?
A. Approximately 850 watts
B. Approximately 1250 watts
C. Approximately 1667 watts
D. Approximately 2000 watts

ANSWER B: The Class C amplifier is more efficient—sometimes as good as 80 percent. If our output is 1000 watts, our 80 percent efficient amplifier would need only 1250 watts for input power.

You can do this in your head, but double check your work using the following steps: Divide the 80% by 100 to get 0.8. Divide the PEP power of 1000 by 0.8 to get input power of 1250 watts.

3G57 What is the approximate DC input power to a Class AB RF power amplifier stage in an unmodulated carrier transmitter when the PEP output power is 500 watts?
A. Approximately 250 watts
B. Approximately 600 watts
C. Approximately 800 watts
D. Approximately 1000 watts

ANSWER D: The Class AB amplifier is a real work horse, and is one of the most common amps found in worldwide radio shacks. It's 50 percent efficient. If you're putting out 500 watts, your input power is approximately 1000 watts.

Remember, Class C amplifier efficiency is about 80 percent; Class B is approximately 60 percent; and Class AB is approximately 50 percent. These values will not be given on your test, so be sure to have them memorized before the big examination.

3G58 Where is the noise generated which primarily determines the signal-to-noise ratio in a VHF (150 MHz) marine band receiver?
A. In the receiver front end
B. Man-made noise
C. In the atmosphere
D. In the ionosphere

ANSWER A: Back in the early days of marine band on 2 MHz, most noise was generated in the atmosphere and ionosphere, plus plenty onboard. Now that we have marine VHF frequencies, there is little noise that can be detected by an FM receiver, so the only noise that we really need to work with is noise found in the receiver front-end transistorized circuitry.

3G59 In a pulse-width modulation system, what parameter does the modulating signal vary?
A. Pulse duration
B. Pulse frequency
C. Pulse amplitude
D. Pulse intensity

ANSWER A: It is the *duration* of the pulse that the modulating signal varies.

3G60 What is the type of modulation in which the modulating signal varies the duration of the transmitted pulse?
A. Amplitude modulation
B. Frequency modulation
C. Pulse-width modulation
D. Pulse-height modulation

ANSWER C: When the modulating signals varies the *duration* of the transmitted pulse, it is pulse-width modulation.

3G61 In a pulse-position modulation system, what parameter does the modulating signal vary?
A. The number of pulses per second
B. Both the frequency and amplitude of the pulses
C. The duration of the pulses
D. The time at which each pulse occurs

ANSWER D: This question is about a pulse-position modulation system. The position is the time at which each pulse occurs. The modulating signal varies the *time* when pulses occur.

3G62 Why is the transmitter peak power in a pulse modulation system much greater than its average power?
A. The signal duty cycle is less than 100%
B. The signal reaches peak amplitude only when voice modulated
C. The signal reaches peak amplitude only when voltage spikes are generated within the modulator
D. The signal reaches peak amplitude only when the pulses are also amplitude modulated

ANSWER A: Since pulse modulation is not a continuous burst of power, the duty cycle is always less than 100 percent.

3G – SIGNALS AND EMISSIONS

3G63 What is one way that voice is transmitted in a pulse-width modulation system?
- A. A standard pulse is varied in amplitude by an amount depending on the voice waveform at that instant
- B. The position of a standard pulse is varied by an amount depending on the voice waveform at that instant
- C. A standard pulse is varied in duration by an amount depending on the voice waveform at that instant
- D. The number of standard pulses per second varies depending on the voice waveform at that instant

ANSWER C: Remember *pulse-width* and *duration*. A standard pulse is varied *in duration* by an amount depending on the modulating waveform of the voice signal.

3G64 The International Organization for Standardization has developed a seven-level reference model for a packet-radio communications structure. What level is responsible for the actual transmission of data and handshaking signals?
- A. The physical layer
- B. The transport layer
- C. The communications layer
- D. The synchronization layer

ANSWER A: The data and handshaking signals are in the *physical layer*.

3G65 The International Organization for Standardization has developed a seven-level reference model for a packet-radio communications structure. What level arranges the bits into frames and controls data flow?
- A. The transport layer
- B. The link layer
- C. The communications layer
- D. The synchronization layer

ANSWER B: It is the *link layer* that arranges the bits into frames, and also controls data flow.

3G66 What is one advantage of using the ASCII code, with its larger character set, instead of the Baudot code?
- A. ASCII includes built-in error-correction features
- B. ASCII characters contain fewer information bits than Baudot characters
- C. It is possible to transmit upper and lower case text
- D. The larger character set allows store-and-forward control characters to be added to a message

ANSWER C: ASCII code allows upper and lower case text to be transmitted; Baudot does not.

3G67 What is the duration of a 45-baud Baudot RTTY data pulse?
- A. 11 milliseconds
- B. 40 milliseconds
- C. 31 milliseconds
- D. 22 milliseconds

ANSWER D: The very slow 45-baud RTTY data pulse is 22 milliseconds long.

5 GENERAL RADIOTELEPHONE OPERATOR LICENSE

3G68 What is the duration of a 45-baud Baudot RTTY start pulse?
A. 11 milliseconds
B. 22 milliseconds
C. 31 milliseconds
D. 40 milliseconds
ANSWER B: The 45-baud RTTY start pulse is also 22 milliseconds long.

3G69 What is the duration of a 45-baud Baudot RTTY stop pulse?
A. 11 milliseconds
B. 18 milliseconds
C. 31 milliseconds
D. 40 milliseconds
ANSWER C: The 45-baud stop pulse is slightly longer than the data and start pulse—31 milliseconds.

3G70 What is the necessary bandwidth of a 170-hertz shift, 45-baud Baudot emission F1B transmission?
A. 45 Hz
B. 249 Hz
C. 442 Hz
D. 600 Hz
ANSWER B: The formula for bandwidth for a Baudot transmission is:

    BW = baud rate + (1.2 × frequency shift).

In this problem the baud rate is 45 plus 1.2 × 170. This works out to be 249 Hz, the correct answer B.

First symbol – type of main modulation
(Note: Modulation is the process of varying the ratio to convey information)
- N – Emission of an unmodulated carrier
- A – Amplitude Modulation
- J – Single sideband, suppressed carrier (the emission you are authorized on ten meters between 28.3 and 28.5 MHz)
- F – Frequency modulation
- G – Phase modulation
- P – Sequence of unmodulated pulses

Second symbol – Nature of signals modulating the main carrier
- 0 – No modulating signal
- 1 – Single channel digital information, no modulation
- 2 – Single channel digital information with modulation
- 3 – Single channel carrying analog information
- 7 – Two or more channels carrying digital information
- 8 – Two or more channels carrying analog information
- 9 – Combination of digital and analog information

Third symbol – Type of Information to be transmitted
- N – No information transmitted
- A – Telegraphy to be received by human hearing
- B – Telegraphy to be received by automatic equipment
- C – Facsimile, transmission of pictures by radio
- D – Data transmission, telemetry, telecommand
- E – Telephony (voice information)
- F – Video, television
- W – Combination of above

ITU Emissions

3G – SIGNALS AND EMISSIONS

3G71 What is the necessary bandwidth of a 170-hertz shift, 45-baud Baudot emission J2B transmission?
- A. 45 Hz
- B. 249 Hz
- C. 442 Hz
- D. 600 Hz

ANSWER B: Use the same formula as in question 3G70:

BW = 45 plus 1.2 × 170 = 249 Hz, which is answer B.

3G72 What is the necessary bandwidth of a 170-hertz shift, 74-baud Baudot emission F1B transmission?
- A. 250 Hz
- B. 278 Hz
- C. 442 Hz
- D. 600 Hz

ANSWER B: Use the same formula as in question 3G70:

BW = 74 plus 1.2 × 170 = 278 Hz, which is answer B.

3G73 What is the necessary bandwidth of a 170-hertz shift, 74-baud Baudot emission J2B transmission?
- A. 250 Hz
- B. 278 Hz
- C. 442 Hz
- D. 600 Hz

ANSWER B: Use the same formula as in question 3G70:

BW = 74 plus 1.2 × 170 = 278 Hz, which is answer B.

3G74 What is the necessary bandwidth of a 1000-hertz shift, 1200-baud ASCII emission F1D transmission?
- A. 1000 Hz
- B. 1200 Hz
- C. 440 Hz
- D. 2400 Hz

ANSWER D: Now the question is back to digital bandwidth. Use the formula in question 3G70. Take your baud rate at 1200 and add it to (1.2 × 1000). This gives you 1200 plus 1200, or 2400 Hz which matches answer D.

3G75 What is the necessary bandwidth of a 4800-hertz frequency shift, 9600-baud ASCII emission F1D transmission?
- A. 15.36 kHz
- B. 9.6 kHz
- C. 4.8 kHz
- D. 5.76 kHz

ANSWER A: Digital bandwidth again—use the formula at question 3G70. In this problem, the baud rate is 9600 and the frequency shift is 4800. Bandwidth = 9600 + (1.2 × 4800) = 15360. The answer is 15360 Hz, and when the decimal point is moved three places to the left, the bandwidth works out to be 15.36 kHz, matching answer A.

5 GENERAL RADIOTELEPHONE OPERATOR LICENSE

3G76 What is the necessary bandwidth of a 4800-hertz frequency shift, 9600-baud ASCII emission J2D transmission?
  A. 15.36 kHz
  B. 9.6 kHz
  C. 4.8 kHz
  D. 5.76 kHz

ANSWER A: Digital bandwidth again; use the same formula again. Same calculation: Bandwidth = 9600 + (1.2 × 4800) = 15360. You get the same answer of 15.36 kHz.

3G77 What is the necessary bandwidth of a 170-hertz shift, 110-baud ASCII emission F1B transmission?
  A. 304 Hz
  B. 314 Hz
  C. 608 Hz
  D. 628 Hz

ANSWER B: Use the formula at question 3G70. The calculations are: Bandwidth = 110 + (1.2 × 170) = 314 Hz.

3G78 What is the necessary bandwidth of a 170-hertz shift, 110-baud ASCII emission J2B transmission?
  A. 304 Hz
  B. 314 Hz
  C. 608 Hz
  D. 628 Hz

ANSWER B: Same digital transmission, same formula as question 3G77. Even though the the emission changed from F1B to J2B, the bandwidth solution is the same.

3G79 What is the necessary bandwidth of a 170-hertz shift, 300-baud ASCII emission F1D transmission?
  A. 0 Hz
  B. 0.3 kHz
  C. 0.5 kHz
  D. 1.0 kHz

ANSWER C: Same digital transmission, same formula, same solution: In words this time, bandwidth equals the baud rate of 300 plus the product of 1.2 times the frequency shift of 170 Hz. Bandwidth works out to be 504 Hz, and moving the decimal point three places to the left gives 0.5 kHz.

3G80 What is the necessary bandwidth for a 170-hertz shift, 300-baud ASCII emission J2D transmission?
  A. 0 Hz
  B. 0.3 kHz
  C. 0.5 kHz
  D. 1.0 kHz

ANSWER C: No change here from the last question, other than the emission, which does not effect the calculation of the bandwidth. It's the same: Bandwidth = 300 + (1.2 × 170) = 504 Hz = 0.504 kHz.

3G – SIGNALS AND EMISSIONS

3G81 What is *amplitude compandored single sideband*?
- A. Reception of single sideband with a conventional CW receiver
- B. Reception of single sideband with a conventional FM receiver
- C. Single sideband incorporating speech compression at the transmitter and speech expansion at the receiver
- D. Single sideband incorporating speech expansion at the transmitter and speech compression at the receiver

ANSWER C: Compandoring a single-sideband signal will compress it at the transmitter, and the receiver will then expand it at the other end of the circuit. ACSB may soon become popular on land mobile VHJF frequencies where it's now still considered an experimental emission. ACSB compression provides more efficient use of transmitter power; the expansion provides better signal-to-noise ratio at the receiver.

3G82 What is meant by *compandoring*?
- A. Compressing speech at the transmitter and expanding it at the receiver
- B. Using an audio-frequency signal to produce pulse-length modulation
- C. Combining amplitude and frequency modulation to produce a single-sideband signal
- D. Detecting and demodulating a single-sideband signal by converting it to a pulse-modulated signal

ANSWER A: We compress the speech to compandor our transmitted signal. It is expanded again at the receiver. That's where the name compandoring comes from—COMpression and exPANDing.

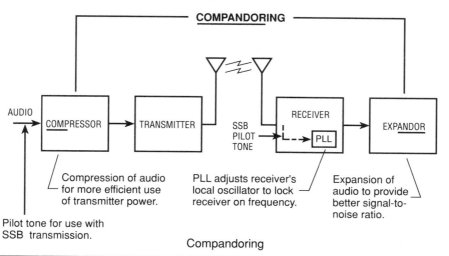

Compandoring

3G83 What is the purpose of a *pilot tone* in an amplitude compandored single-sideband system?
- A. It permits rapid tuning of a mobile receiver.
- B. It replaces the suppressed carrier at the receiver.
- C. It permits rapid change of frequency to escape high-powered interference.
- D. It acts as a beacon to indicate the present propagation characteristic of the band.

ANSWER A: The pilot tone, used only for ACSB transmissions, allows an automatic mobile receiver to lock onto a signal. This allows rapid tuning of the mobile receiver in the automatic mode.

3G84 What is the approximate frequency of the *pilot tone* in an amplitude compandored single-sideband system?
 A. 1 kHz
 B. 5 MHz
 C. 455 kHz
 D. 3 kHz

ANSWER D: The pilot tone is a 3000 Hz (3 kHz) signal. After the pilot tone is used for tuning the receiver, it is filtered out so it will not be heard in the speaker.

3G85 How many more voice transmissions can be packed into a given frequency band for amplitude-compandored single-sideband systems over conventional FM-phone systems?
 A. 2
 B. 4
 C. 8
 D. 16

ANSWER B: In the VHF land mobile radio service, ACSB gives us four channels in the place of one conventional, narrow-band, FM channel. Although there is no tremendous use of ACSB on VHF 150-MHz land mobile frequencies, there promises to be plenty of ACSB up on the newly created 220-222-MHz land mobile band.

3G86 What term describes a wide-bandwidth communications system in which the RF carrier varies according to some predetermined sequence?
 A. Amplitude compandored single sideband
 B. SITOR
 C. Time-domain frequency modulation
 D. Spread spectrum communication

ANSWER D: Spread spectrum communications are becoming more popular. You probably don't realize it, but your 900-MHz cordless phone uses spread spectrum communications. The frequency hops in a pre-arranged sequence, eliminating interference and eavesdropping. The new global positioning system (GPS) at 1500 MHz uses spread-spectrum techniques.

3G87 What is the term used to describe a *spread spectrum communications system* where the center frequency of a conventional carrier is altered many times per second in accordance with a pseudo-random list of channels?
 A. Frequency hopping
 B. Direct sequence
 C. Time-domain frequency modulation
 D. Frequency compandored spread spectrum

ANSWER A: Spread spectrum communications alters the center frequency of an RF carrier in psuedo-random manner, causing the frequency to "hop" around from channel to channel. This is termed *frequency hopping*.

3G – SIGNALS AND EMISSIONS

3G88 What term is used to describe a *spread spectrum communications system* in which a very fast binary bit stream is used to shift the phase of an RF carrier?
- A. Frequency hopping
- B. Direct sequence
- C. Binary phase-shift keying
- D. Phase compandored spread spectrum

ANSWER B: The term "direct sequence" is used to describe a system where a fast binary stream shifts the phase of an RF carrier.

3G89 What is the term for the amplitude of the maximum positive excursion of a signal as viewed on an oscilloscope?
- A. Peak-to-peak voltage
- B. Inverse peak negative voltage
- C. RMS voltage
- D. Peak positive voltage

ANSWER D: The maximum positive excursion for the amplitude of a signal on an oscilloscope is called the peak positive voltage.

3G90 What is the term for the amplitude of the maximum negative excursion of a signal as viewed on an oscilloscope?
- A. Peak-to-peak voltage
- B. Inverse peak positive voltage
- C. RMS voltage
- D. Peak negative voltage

ANSWER D: AC signals alternate from positive to negative. The maximum negative excursion is called the peak negative voltage.

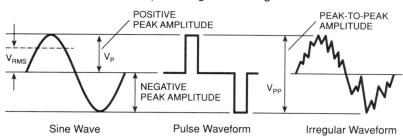

V_P = Peak voltage
V_{PP} = Peak-to-peak voltage
V_{RMS} = Root-mean-square voltage

$V_P = \dfrac{V_{PP}}{2}$ $V_{PP} = 2V_P$
$V_{RMS} = 0.707\ V_P$
$V_P = 1.414\ V_{RMS}$

Peak, Peak-to-Peak and RMS Voltages

3G91 What is the easiest voltage amplitude dimension to measure by viewing a pure sine wave signal on an oscilloscope?
- A. Peak-to-peak voltage
- B. RMS voltage
- C. Average voltage
- D. DC voltage

5 GENERAL RADIOTELEPHONE OPERATOR LICENSE

ANSWER A: Looking at a pure sine wave signal on a scope, it is easy to identify the peak-to-peak voltage measured from the maximum positive excursion of the signal to the maximum negative excursion of the signal.

3G92 What is the relationship between the peak-to-peak voltage and the peak voltage amplitude in a symmetrical wave form?
    A. 1:1
    B. 2:1
    C. 3:1
    D. 4:1

ANSWER B: The relationship between peak-to-peak voltage and peak voltage is 2:1. It is important to realize that the waveform must be symmetrical—as in a sine wave. See figure at question 3G90.

3G93 What input-amplitude parameter is valuable in evaluating the signal-handling capability of a Class A amplifier?
    A. Peak voltage
    B. Average voltage
    C. RMS voltage
    D. Resting voltage

ANSWER A: In a class A amplifier, peak voltage provides a double check on the capability and linearity of that amplifier. Many times, distortion occurs when the amplifier is handling a peak voltage.

3G94 If 480 kHz is radiated from a 1/4 wavelength antenna, what is the 7th harmonic?
    A. 3.360 MHz
    B. 840 KHz
    C. 3350 kHz
    D. 480 kHz

ANSWER A: Disregard the electrical wavelength of the antenna. Multiple 7 × 480 = 3360 kHz. To change kHz to MHz, remember KLM (kilohertz-left-megahertz), and convert kilohertz to megahertz by moving the decimal point 3 places to the left.

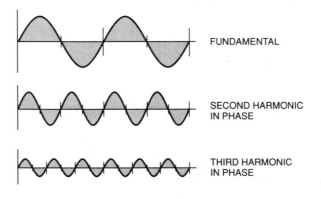

Fundamental Radio Wave and Harmonics

236

3H – ANTENNAS AND FEED LINES

3G95 What is the seventh harmonic of a 100-MHz quarter wavelength antenna?
  A. 14.28 MHz.
  B. 107 MHz.
  C. 149 MHz.
  D. 700 MHz.
ANSWER D: The seventh harmonic of 100 MHz is simply 7 × 100 = 700 MHz.

3G96 What is the seventh harmonic of 2182 kHz when the transmitter is connected to a half-wave Hertz antenna?
  A. 2182 kHz
  B. 15.27 MHz
  C. 311.7 kHz
  D. 7.64 MHz
ANSWER B: Here again, disregard the length of the antenna. The harmonics are actually generated within the transmitter section. It's just that some antennas will radiate harmonics more than others. 7 × 2182 = 15,274 kHz = 15.27 MHz.

3G97 What is the fifth harmonic frequency of a transmitter operating on 480 kHz with a 1/4 wavelength antenna?
  A. 2.4 MHz
  B. 240 MHz
  C. 600 kHz
  D. 1.2 MHz
ANSWER A: One more time, disregard the length of the antenna. Multiply the frequency by the harmonics: 480 kHz × 5 = 2400 kHz, (KLM) 2.4 MHz. Just keep in mind that a multi-band trap antenna is prone to making these harmonics sound stronger over the airwaves.

Subelement 3H – Antennas and Feed Lines   (5 questions)

3H1 What is meant by the term *antenna gain?*
  A. The numerical ratio relating the radiated signal strength of an antenna to that of another antenna
  B. The ratio of the signal in the forward direction to the signal in the back direction
  C. The ratio of the amount of power produced by the antenna compared to the output power of the transmitter
  D. The final amplifier gain minus the transmission line losses (including any phasing lines present)
ANSWER A: We normally rate antenna gain figures in dB. This is the numerical ratio relating the radiated signal strength of your antenna to that of another antenna down the street. Generally, the higher gain antenna will get better signal reports.

3H2 What is the term for a numerical ratio which relates the performance of one antenna to that of another real or theoretical antenna?
  A. Effective radiated power
  B. Antenna gain

C. Conversion gain
D. Peak effective power

ANSWER B: Antenna gain is an important consideration when choosing an antenna system.

3H3 What is meant by the term *antenna bandwidth*?
A. Antenna length divided by the number of elements
B. The frequency range over which an antenna can be expected to perform well
C. The angle between the half-power radiation points
D. The angle formed between two imaginary lines drawn through the ends of the elements

ANSWER B: As a ship station on worldwide frequencies switches to lower bands, such as the 4-MHz, 6-MHz, or 8-MHz band, antenna bandwidth becomes important. The associated automatic antenna tuner will probably seek a new LC setting with frequency excursions as little as 50 kHz.

3H4 What is the wavelength of a shorted stub used to absorb even harmonics?
A. 1/2 wavelength
B. 1/3 wavelength
C. 1/4 wavelength
D. 1/8 wavelength

ANSWER C: We would use a 1/4 wavelength shorted stub to absorb even harmonics for a specific frequency antenna system.

3H5 What is a *trap antenna*?
A. An antenna for rejecting interfering signals
B. A highly sensitive antenna with maximum gain in all directions
C. An antenna capable of being used on more than one band because of the presence of parallel LC networks
D. An antenna with a large capture area

ANSWER C: For worldwide frequencies aboard a boat, 2 MHz through 22 MHz, a multi-band trap antenna is a good performer. This will allow the ship station to use a 50-ohm SSB transceiver without an auxiliary antenna tuner. The trap antenna consists of parallel resonant LC networks to trap out the higher bands from getting up to the top of the antenna. These traps also provide loading for the lower bands. Trap antennas use parallel LC networks which prevent the signal from traveling beyond the trap.

3H6 What is an advantage of using a trap antenna?
A. It has high directivity in the high-frequency bands
B. It has high gain
C. It minimizes harmonic radiation
D. It may be used for multiband operation

ANSWER D: The trap antenna may be used for multiband operation, from a single feed line. There are trap dipoles, trap verticals, and trap beam antennas for multiband operation.

3H – ANTENNAS AND FEED LINES

3H7 What is a disadvantage of using a trap antenna?
A. It will radiate harmonics
B. It can only be used for single band operation
C. It is too sharply directional at lower frequencies
D. It must be neutralized

ANSWER A: Trap antennas have one big disadvantage—on older equipment they can radiate harmonics. Since marine band harmonics relate to bands twice the operating frequency, you can see that the second harmonic of 8 MHz ends up on sensitive frequencies at 16 MHz from a trap antenna. If your ship station is using a 50-ohm trap antenna, pay particular attention to unintentional harmonics that trap antennas are well known for!

3H8 What is the principle of a trap antenna?
A. Beamwidth may be controlled by non-linear impedances
B. The traps form a high impedance to isolate parts of the antenna
C. The effective radiated power can be increased if the space around the antenna "sees" a high impedance
D. The traps increase the antenna gain

ANSWER B: The trap antenna will use loading coils and outside capacitive coil sleeves to develop a high-impedance network to isolate longer parts of the antenna to those frequencies requiring a shorter radiating element.

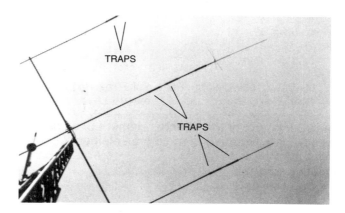

Yagi Antenna with Traps

3H9 What is a *parasitic element* of an antenna?
A. An element polarized 90 degrees opposite the driven element
B. An element dependent on the antenna structure for support
C. An element that receives its excitation from mutual coupling rather than from a transmission line
D. A transmission line that radiates radio-frequency energy

ANSWER C: You can build a beam antenna on a broomstick. Even though the elements are not electrically connected, they receive their excitation from *mutual coupling* (key words).

5 GENERAL RADIOTELEPHONE OPERATOR LICENSE

3H10 How does a parasitic element generate an electromagnetic field?
A. By the RF current received from a connected transmission line
B. By interacting with the earth's magnetic field
C. By altering the phase of the current on the driven element
D. By currents induced into the element from a surrounding electric field

ANSWER D: Radio frequency currents are induced into the parasitic elements from a surrounding electric field.

3H11 How does the length of the reflector element of a parasitic element beam antenna compare with that of the driven element?
A. It is about 5% longer
B. It is about 5% shorter
C. It is twice as long
D. It is one-half as long

ANSWER A: The beam antenna reflector element is normally 5 percent longer than the driven element. The reflector helps form the front lobe of the radiation pattern.

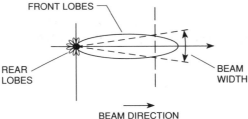

Directional Radiation Pattern of a Yagi Beam

Source: *Antennas – Selection and Installation*, A.J. Evans, ©1986 Master Publishing, Inc., Richardson, Texas

---

3H12 How does the length of the director element of a parasitic element beam antenna compare with that of the driven element?
A. It is about 5% longer
B. It is about 5% shorter
C. It is one-half as long
D. It is twice as long

ANSWER B: The director of a beam antenna is approximately 5 percent shorter than the driven element. This is why most beams have a tapered look when you view them from the ground. If receiving, the shorter end always points toward the transmitting stations.

3H13 What is meant by the term *radiation resistance* for an antenna?
A. Losses in the antenna elements and feed line
B. The specific impedance of the antenna
C. An equivalent resistance that would dissipate the same amount of power as that radiated from an antenna
D. The resistance in the trap coils to received signals

ANSWER C: Radiation resistance is an important term because it compares the power radiated from an antenna to the equivalent resistance that would dissipate the same amount of power as heat. The higher the radiation resistance of an antenna, the better its performance. Don't confuse radiation resistance with simple DC resistance—they are exactly opposite.

3H – ANTENNAS AND FEED LINES

3H14 What are the factors that determine the radiation resistance of an antenna?
A. Transmission line length and height of antenna
B. The location of the antenna with respect to nearby objects and the length/diameter ratio of the conductors
C. It is a constant for all antennas since it is a physical constant
D. Sunspot activity and the time of day

ANSWER B: Radiation resistance of a mobile antenna may be severely compromised if the antenna is mounted too low. The location of the antenna with respect to nearby objects, as well as the length to diameter ratio of the conductors, is very important.

3H15 What is a *driven element* of an antenna?
A. Always the rearmost element
B. Always the forwardmost element
C. The element fed by the transmission line
D. The element connected to the rotator

ANSWER C: The driven element of a Yagi antenna is the element fed by the *transmission line* (key words).

3H16 What is the usual electrical length of a driven element in a HF beam antenna?
A. 1/4 wavelength
B. 1/2 wavelength
C. 3/4 wavelength
D. 1 wavelength

ANSWER B: The usual electrical length of a driven element on a high-frequency beam antenna is like a dipole, one-half wavelength long.

3H17 What is the term for an antenna element which is supplied power from a transmitter through a transmission line?
A. Driven element
B. Director element
C. Reflector element
D. Parasitic element

ANSWER A: The transmission line supplies power from the transmitter to the driven element.

3H18 How is antenna "efficiency" computed?
A. Efficiency = (radiation resistance/transmission resistance) × 100%
B. Efficiency = (radiation resistance/total resistance) × 100%
C. Efficiency = (total resistance/radiation resistance) × 100%
D. Efficiency = (effective radiated power/transmitter output) × 100%

ANSWER B: The efficiency of your antenna is an important consideration when you plan to upgrade your antenna system. Efficiency equals the radiation resistance divided by the total resistance, multiplied by 100 percent. You want to keep your ohmic resistance as low as possible to make your antenna more efficient. Generally, bigger antenna elements have greater radiation resistance, lower ohmic resistance, higher efficiency, and increased bandwidth. Bigger is always better!

3H19 What is the term for the ratio of the radiation resistance of an antenna to the total resistance of the system?
A. Effective radiated power
B. Radiation conversion loss
C. Antenna efficiency
D. Beamwidth

ANSWER C: The ratio of the radiation resistance of an antenna to the total resistance of the system times 100 is the percent antenna efficiency.

3H20 What is included in the total resistance of an antenna system?
A. Radiation resistance plus space impedance
B. Radiation resistance plus transmission resistance
C. Transmission line resistance plus radiation resistance
D. Radiation resistance plus ohmic resistance

ANSWER D: You want the *ohmic resistance* (key words) as small as possible for improved antenna efficiency.

3H21 How can the antenna efficiency of a HF grounded vertical antenna be made comparable to that of a half-wave antenna?
A. By installing a good ground radial system
B. By isolating the coax shield from ground
C. By shortening the vertical
D. By lengthening the vertical

ANSWER A: Shipboard vertical antennas are 1/4 wavelength long, and use a mirror image counterpoise to compare to that of a half-wave antenna. A good ground radial system, interconnected to bonded through-hulls, will give the 1/4 wavelength radiator good performance, and the good ground will also lower the noise floor and lower the take-off angle, which leads to longer range results.

3H22 Why does a half-wave antenna operate at very high efficiency?
A. Because it is non-resonant
B. Because the conductor resistance is low compared to the radiation resistance
C. Because earth-induced currents add to its radiated power
D. Because it has less corona from the element ends than other types of antennas

ANSWER B: Half-wave antennas operate at high efficiency because their ohmic conductor resistance is low compared to their radiation resistance. For marine applications, insulated stainless steel stays supporting the mast may make good antenna elements with extremely low ohmic resistance.

3H23 What is a *folded dipole* antenna?
A. A dipole that is one-quarter wavelength long
B. A ground plane antenna
C. A dipole whose ends are connected by another one-half wavelength piece of wire
D. A fictional antenna used in theoretical discussions to replace the radiation resistance

ANSWER C: The folded dipole is a fun one to build out of television twin-lead cable. As shown in the diagram, the twin-lead folded dipole is constructed by using a half-wavelength piece of twin lead. Scrape the insulation from the wire

3H – ANTENNAS AND FEED LINES

ends and twist the wires together on each end. In the center of one of the leads in the twin-lead, cut the lead and scrape some of the insulation off to expose bare wires on the end of each quarter-wave piece. Feed these wire with standard 300-ohm twin-lead, or connect a balun transformer to the bare wires to permit 50-ohm or 75-ohm coax cable to feed the folded dipole.

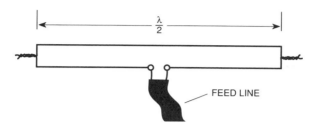

Folded Dipole Antenna Made from 300-ohm Twin-lead

3H24 How does the bandwidth of a folded dipole antenna compare with that of a simple dipole antenna?
A. It is 0.707 times the simple dipole bandwidth
B. It is essentially the same
C. It is less than 50% that of a simple dipole
D. It is greater

ANSWER D: The advantage of a folded dipole over a simple dipole is *greater bandwidth* (key words). The bandwidth is greater because twice the added half-wavelength of wire connecting the driven-element dipole makes the folded dipole look like a "thick" antenna, and thus gives it a wider bandwidth.

3H25 What is the input terminal impedance at the center of a folded dipole antenna?
A. 300 ohms
B. 72 ohms
C. 50 ohms
D. 450 ohms

ANSWER A: The impedance at the center of a folded dipole may be a whopping 300 ohms. You'll need an impedance-matching transformer to feed it with 50-ohm coax; however, ordinary TV twin-lead would match very well.

3H26 What is the meaning of the term *velocity factor* of a transmission line?
A. The ratio of the characteristic impedance of the line to the terminating impedance
B. The index of shielding for coaxial cable
C. The velocity of the wave on the transmission line multiplied by the velocity of light in a vacuum
D. The velocity of the wave on the transmission line divided by the velocity of light in a vacuum

ANSWER D: Radio waves travel in free space at 300 million meters per second. But in coaxial cable transmission lines, and other types of feed lines, the radio waves move slower. Velocity factor is the velocity of the radio wave on the transmission line divided by the velocity in free space.

5 GENERAL RADIOTELEPHONE OPERATOR LICENSE

3H27 What is the term for the ratio of actual velocity at which a signal travels through a line to the speed of light in a vacuum?
- A. Velocity factor
- B. Characteristic impedance
- C. Surge impedance
- D. Standing wave ratio

ANSWER A: Comparing the velocity of radio signals through a piece of feed line to the speed of light in a vacuum is called the velocity factor of the feed line.

3H28 What is the velocity factor for non-foam dielectric 50- or 75-ohm flexible coaxial cable such as RG-8, 11, 58 and 59?
- A. 2.70
- B. 0.66
- C. 0.30
- D. 0.10

ANSWER B: Most marine grade coax cable has a velocity factor of 0.66. This is not stated on the outside jacket. You may need to check the manufacturer's coax cable specification sheet to determine the velocity factor.

3H29 What determines the velocity factor in a transmission line?
- A. The termination impedance
- B. The line length
- C. Dielectrics in the line
- D. The center conductor resistivity

ANSWER C: There are some great low-loss coaxial cable types out there in radio land with different dielectrics separating the center conductor and the outside shield. It's the dielectric that makes a big difference in the velocity factor.

3H30 Why is the physical length of a coaxial cable transmission line shorter than its electrical length?
- A. Skin effect is less pronounced in the coaxial cable
- B. RF energy moves slower along the coaxial cable
- C. The surge impedance is higher in the parallel feed line
- D. The characteristic impedance is higher in the parallel feed line

ANSWER B: Radio frequency energy moves slightly *slower* (key word) inside coaxial cable than it would in free space.

3H31 What would be the physical length of a typical coaxial transmission line which is electrically one-quarter wavelength long at 14.1 MHz?
- A. 20 meters
- B. 3.51 meters
- C. 2.33 meters
- D. 0.25 meters

ANSWER B: When you work on antenna phasing harnesses, you will need to calculate the velocity factor for a piece of coax in order to know where to place the connectors. You must be accurate down to a fraction of an inch! You can calculate the physical length of a coax cable which is electrically one-quarter wavelength long using the formula:

3H – ANTENNAS AND FEED LINES

   $L \text{ (in feet)} = \dfrac{984\lambda V}{f}$ Where: L = Antenna length in **feet**
λ = Wavelength in **meters**
V = Velocity factor
f = Frequency in **MHz**

For this problem, multiply 984 × 0.25, because the coax is one quarter wavelength. Then multiply the result by 0.66, the velocity factor. Remember, 0.66 is the velocity factor for coax. We are assuming that value for these problems. Divide the total by 14.1, the frequency in MHz. The length comes out to about 11.51 feet. The answer needs to be in meters, so divide 11.5 by 3.28 (3.28 feet = 1 meter). The result is 3.51 meters. (For a quick approximation, divide 11.5 by 3. The result is 3.83, which gets close enough to pick an answer.)

3H32 What would be the physical length of a typical coaxial transmission line which is electrically one-quarter wavelength long at 7.2 MHz?
  A. 10.5 meters
  B. 6.88 meters
  C. 24 meters
  D. 50 meters

ANSWER B: The coax cable is again one-quarter wavelength, so multiple 984 × 0.25, and the result by 0.66, the velocity factor. We again will assume this value. They won't give you 0.66 on the exam. You must know it. Now divide these three multiplied figures by the frequency, 7.2 MHz, and you end up with 22.55 feet of coax. Dividing by 3 to get approximate length of 7.5 meters, leads you to 6.88 meters of answer B.

3H33 What is the physical length of a parallel antenna feedline which is electrically one-half wavelength long at 14.10 MHz? (assume a velocity factor of 0.82.)
  A. 15 meters
  B. 24.3 meters
  C. 8.7 meters
  D. 70.8 meters

ANSWER C: In this problem, the coax is a half-wavelength long, and the velocity factor is given as 0.82. No problem—multiply 984 × 0.5 × 0.82, and divide by 14.1, the frequency in MHz. The answer is 28.61 feet. This works out to be about 8.7 meters long. There actually are 3.28 feet to a meter.

3H34 What is the physical length of a twin lead transmission feedline at 36.5 MHz? (assume a velocity factor of 0.80.)
  A. Electrical length times 0.8
  B. Electrical length divided by 0.8
  C. 80 meters
  D. 160 meters

ANSWER A: This problem is slightly different—all they want to know is the physical length of a feed line at 36.5 MHz with a velocity factor of 0.8. They don't ask anything about how many wavelengths long, so simply take electrical length and multiply it times 0.8. Be careful, times 0.8! Your answer is A.

3H35 In a half-wave antenna, where are the *current* nodes?
A. At the ends
B. At the center
C. Three-quarters of the way from the feed point toward the end
D. One-half of the way from the feed point toward the end

ANSWER A: On a half-wave dipole, current nodes are areas of minimum current on the tip ends of a dipole. Current is maximum in the center. See figure at question 3H36.

3H36 In a half-wave antenna, where are the *voltage* nodes?
A. At the ends
B. At the feed point
C. Three-quarters of the way from the feed point toward the end
D. One-half of the way from the feed point toward the end

ANSWER B: For a voltage node in a half-wave dipole, it is just opposite of current—the voltage node of minimum voltage is at the feed point where current is a maximum.

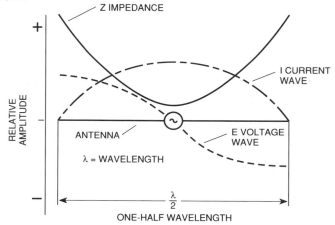

Voltage, Current and Impedance on a Half-wave Dipole

3H37 At the ends of a half-wave antenna, what values of current and voltage exist compared to the remainder of the antenna?
A. Equal voltage and current
B. Minimum voltage and maximum current
C. Maximum voltage and minimum current
D. Minimum voltage and minimum current

ANSWER C: There is maximum voltage on the tip ends of a dipole, but minimum current. This is why you can sometimes spark a lead pencil (be careful) off of the tip ends of a dipole.

3H38 At the center of a half-wave antenna, what values of voltage and current exist compared to the remainder of the antenna?
A. Equal voltage and current
B. Maximum voltage and minimum current
C. Minimum voltage and minimum current
D. Minimum voltage and maximum current

3H – ANTENNAS AND FEED LINES

ANSWER D: At the center of a halfwave dipole, voltage is minimum, but current is maximum. Since current at the feed point is maximum, we always want to insure a tight feed point connection.

3H39 What happens to the base feed point of a fixed-length mobile antenna as the frequency of operation is lowered?
  A. The resistance decreases and the capacitive reactance decreases
  B. The resistance decreases and the capacitive reactance increases
  C. The resistance increases and the capacitive reactance decreases
  D. The resistance increases and the capacitive reactance increases

ANSWER B: As we operate lower on the bands, the base feed point resistance decreases (lower decreases), and the capacitive reactance (X_c) increases. You can visualize the correct answer by thinking the lower you operate in frequency, the more coil (inductive reactance, X_L) you will need to offset and equal the added capacitive reactance, which increases.

3H40 Why should an HF mobile antenna loading coil have a high ratio of reactance to resistance?
  A. To swamp out harmonics
  B. To maximize losses
  C. To minimize losses
  D. To minimize the Q

ANSWER C: When you see those giant open-air coils halfway up the mast on a mobile whip antenna, you are looking at high Q coils which minimize losses. The bigger the coil, the lower the losses.

Mobile Antenna Loading Coil

3H41 Why is a loading coil often used with an HF mobile antenna?
  A. To improve reception
  B. To lower the losses
  C. To lower the Q
  D. To tune out the capacitive reactance

ANSWER D: Since X_L must equal X_C in order for a whip antenna to achieve resonance, we need to add a loading coil on an antenna for lower frequency operation to tune out capacitive reactance.

3H42 For a shortened vertical antenna, where should a loading coil be placed to minimize losses and produce the most effective performance?
   A. Near the center of the vertical radiator
   B. As low as possible on the vertical radiator
   C. As close to the transmitter as possible
   D. At a voltage node

ANSWER A: *Center loading* (key words) is one of the best ways to produce an effective high-frequency mobile antenna system. Also, the larger the coil, the lower the losses.

3H43 What happens to the bandwidth of an antenna as it is shortened through the use of loading coils?
   A. It is increased
   B. It is decreased
   C. No change occurs
   D. It becomes flat

ANSWER B: The shorter your mobile whip, the more loading you need, and bandwidth is decreased. Also, your efficiency usually goes down—so for high efficiency, and more bandwidth, go for a longer whip with less loading.

3H44 Why are self-resonant antennas popular in many applications?
   A. They are very broad banded
   B. They have high gain in all azimuthal directions
   C. They are the most efficient radiators
   D. They require no calculations

ANSWER C: If you can approach the natural quarter-wave, half-wave, or full-wave resonance of an antenna system, it will be the most efficient radiator. Anytime you add a lot of loading, efficiency drops.

3H45 What is an advantage of using top loading in a shortened HF vertical antenna?
   A. Lower Q
   B. Greater structural strength
   C. Higher losses
   D. Improved radiation efficiency

ANSWER D: Helical top-loaded whips have *improved radiation efficiency* (key words) over shorter whips with big fat loading coils in the center. While they are not as aerodynamic on the top of your mobile unit, the top helical-loaded whip has improved radiation efficiency.

3H46 What is an *isotropic radiator*?
   A. A hypothetical, omnidirectional antenna
   B. In the northern hemisphere, an antenna whose directive pattern is constant in southern directions
   C. An antenna high enough in the air that its directive pattern is substantially unaffected by the ground beneath it
   D. An antenna whose directive pattern is substantially unaffected by the spacing of the elements

3H – ANTENNAS AND FEED LINES

ANSWER A: The isotropic radiator radiates equally well in all directions. It is a hypothetical antenna, used as a standard. The isotropic radiator is at least 2 dB down from the simple dipole antenna that is many times also used as a standard.

3H47 When is it useful to refer to an *isotropic radiator*?
A. When comparing the gains of directional antennas
B. When testing a transmission line for standing wave ratio
C. When (in the northern hemisphere) directing the transmission in a southerly direction
D. When using a dummy load to tune a transmitter

ANSWER A: When the gain of directional antennas and collinear arrays are compared, they are usually compared to an isotropic radiator.

3H48 What theoretical reference antenna provides a comparison for antenna measurements?
A. Quarter-wave vertical
B. Yagi-Uda array
C. Bobtail curtain
D. Isotropic radiator

ANSWER D: The isotropic radiator is a hypothetical antenna used for a theoretical reference.

3H49 What purpose does an *isotropic radiator* serve?
A. It is used to compare signal strengths (at a distant point) of different transmitters
B. It is used as a reference for antenna gain measurements
C. It is used as a dummy load for tuning transmitters
D. It is used to measure the standing-wave-ratio on a transmission line

ANSWER B: Marine antenna manufacturers may advertise their VHF antenna gain as compared to an isotropic radiator. This elevates the gain figure, and makes their antennas more competitive. However, in the land mobile industry, most antenna specifications meet EIA (Electronics Industry of America) guidelines, and they are usually several dB less than the same type of marine antenna. Most land mobile VHF and UHF collinear array gain figures are compared to a dipole (dBd).

3H50 How much gain does a 1/2-wavelength dipole have over an *isotropic radiator*?
A. About 1.5 dB
B. About 2.1 dB
C. About 3.0 dB
D. About 6.0 dB

ANSWER B: The halfwave dipole has about 2.1 dB advantage in gain over an isotropic radiator.

3H51 How much gain does an antenna have over a 1/2-wavelength dipole when it has 6 dB gain over an *isotropic radiator*?
A. About 3.9 dB
B. About 6.0 dB
C. About 8.1 dB
D. About 10.0 dB

ANSWER A: If an antenna has 6 dB gain over an isotropic radiator, it will have 2.1 dB less over a dipole, resulting in a true 3.9 (6 - 2.1) dBd—that little "d" at the end means over a dipole. dBi means gain over an isotropic radiator.

3H52 How much gain does an antenna have over a 1/2-wavelength dipole when it has 12 dB gain over an *isotropic radiator*?
 A. About 6.1 dB
 B. About 9.9 dB
 C. About 12.0 dB
 D. About 14.1 dB

ANSWER B: An antenna with 12 dBi gain is only 9.9 dBd. Since the dipole has a 2.1 dB gain over an isotropic antenna, the dBi of the antenna in the question is reduced to 9.9 dBd when referenced to the dipole.

3H53 What is the antenna pattern for an *isotropic radiator*?
 A. A figure-8
 B. A unidirectional cardioid
 C. A parabola
 D. A sphere

ANSWER D: The isotropic radiator radiates in all directions, like a sphere.

3H54 What type of directivity pattern does an *isotropic radiator* have?
 A. A figure-8
 B. A unidirectional cardioid
 C. A parabola
 D. A sphere

ANSWER D: The isotropic antenna (radiator) radiates in a pattern that is equal at any point the same distance from the antenna in all directions. Connecting the points would form a perfect sphere.

3H55 What factors determine the receiving antenna gain required at a station in earth operation?
 A. Height, transmitter power and antennas of satellite
 B. Length of transmission line and impedance match between receiver and transmission line
 C. Preamplifier location on transmission line and presence or absence of RF amplifier stages
 D. Height of earth antenna and satellite orbit

ANSWER A: If you are planning to set up a shore station to operate through, or receive, satellites, you will need antenna height, and plenty of transmitting power.

3H56 What factors determine the EIRP required by a station in earth operation?
 A. Satellite antennas and height, satellite receiver sensitivity
 B. Path loss, earth antenna gain, signal-to-noise ratio
 C. Satellite transmitter power and orientation of ground receiving antenna
 D. Elevation of satellite above horizon, signal-to-noise ratio, satellite transmitter power

ANSWER A: The effective isotropic radiated power (EIRP) required by a station in earth operation is dependent on the type of satellite antennas and their height, and the satellite receiver sensitivity.

3H57 What factors determine the EIRP required by a station in telecommand operation?
  A. Path loss, earth antenna gain, signal-to-noise ratio
  B. Satellite antennas and height, satellite receiver sensitivity
  C. Satellite transmitter power and orientation of ground receiving antenna
  D. Elevation of satellite above horizon, signal-to-noise ratio, satellite transmitter power

ANSWER B: VHF or UHF frequencies are used for telecommand operation, so satellite antennas and height, plus satellite receiver sensitivity, are important factors for determining EIRP.

3H58 How does the gain of a parabolic dish type antenna change when the operating frequency is doubled?
  A. Gain does not change
  B. Gain is multiplied by 0.707
  C. Gain increases 6 dB
  D. Gain increases 3 dB

ANSWER C: From 1270 MHz on up, you may wish to use a parabolic dish antenna for microwave work. If you double the frequency, the gain of the dish increases by 6 dB, a 4 times increase.

Parabolic Antenna

3H59 What happens to the beamwidth of an antenna as the gain is increased?
  A. The beamwidth increases geometrically as the gain is increased
  B. The beamwidth increases arithmetically as the gain is increased
  C. The beamwidth is essentially unaffected by the gain of the antenna
  D. The beamwidth decreases as the gain is increased

ANSWER D: The parabolic dish concentrates that beam width as gain increases. The higher the gain of a dish, the narrower the beam width. Beam width *decreases* as gain *increases*.

5　GENERAL RADIOTELEPHONE OPERATOR LICENSE

3H60　What is the beamwidth of a symmetrical pattern antenna with a gain of 20 dB as compared to an isotropic radiator?
A. 10.1 degrees
B. 20.3 degrees
C. 45.0 degrees
D. 60.9 degrees

ANSWER B: Beam width (in degrees) is equal to a fixed value of 203 divided by the square root of 10, raised to a power represented by the antenna gain, in dB, divided by 10:

$$\text{Beamwidth} = \frac{203}{(\sqrt{10})^x} \text{ in degrees} \qquad X = \frac{\text{Antenna gain in dB}}{10}$$

In this first question, the antenna gain is 20 dB, and $X = 20 \div 10 = 2$

$$\text{Beamwidth} = \frac{203}{\sqrt{10}^2} = \frac{203}{10} = 20.3 \text{ degrees}$$

The beamwidth for this antenna is 20.3° matching answer B.

3H61　What is the beamwidth of a symmetrical pattern antenna with a gain of 30 dB as compared to an isotropic radiator?
A. 3.2 degrees
B. 6.4 degrees
C. 37 degrees
D. 60.4 degrees

ANSWER B: Let's use the same formula as the previous question. In this question the antenna gain is 30 dB and $X = 30 \div 10 = 3$; therefore,

$$\text{Beamwidth} = \frac{203}{\sqrt{10}^3} = \frac{203}{10\sqrt{10}} = \frac{203}{31.63} = 6.42°$$

The beamwidth for this antenna is 6.42°, matching answer B.

3H62　What is the beamwidth of a symmetrical pattern antenna with a gain of 15 dB as compared to an isotropic radiator?
A.　72 degrees
B.　52 degrees
C.　36.1 degrees
D.　3.61 degrees

ANSWER C: Same formula as question 3H60. The gain is 15 dB and $X = 15 \div 10 = 1.5$. The beamwidth is:

$$\text{Beamwidth} = \frac{203}{\sqrt{10}^{1.5}} = \frac{203}{3.16^{1.5}} = \frac{203}{5.62} = 36.1°$$

You may not know how to find $3.16^{1.5}$. What you can do is estimate. 3.16^2 is 10; therefore, the value you are looking for is between 3.16 and 10. Let's call it Y; therefore, $Y = 3.16 \times Z$. Z is somewhere between 0 and 3.16. Since the power is 1.5, you could choose half of 3.16 or 1.58. However, the power function is logarithmic so 0.5 is weighted past the linear one-half point. Let's choose $Z = 1.7$, therefore, $Y = 3.16 \times 1.7 = 5.37$. With 5.37, the beamwidth = $203 \div 5.37 = 37.8°$. In this question, you would be close enough to choose the correct answer C of 36.1°.

3H – ANTENNAS AND FEED LINES

3H63 What is the beamwidth of a symmetrical pattern antenna with a gain of 12 dB as compared to an isotropic radiator?
A. 34.8 degrees
B. 45.0 degrees
C. 58.0 degrees
D. 51.0 degrees

ANSWER D: Same formula, but now the gain is 12 dB and $X = 12 \div 10 = 1.2$.

$$\text{Beamwidth} = \frac{203}{\sqrt{10}^{1.2}} = \frac{203}{3.16^{1.2}} = \frac{203}{3.98} = 51°$$

If you estimate $3.16^{1.2} = 3.16 \times 1.2 = 3.79$, the beamwidth would be 53.6°. This would get you close enough to choose answer D, 51 degrees.

Let's review what's been done in the last four questions. A big 20 dB dish will provide a tight pattern of 20 degrees—20/20. A giant 30 dB dish will narrow the beam down to about 6.5 degrees. But a much smaller 15 dB dish will widen the beam to a rather broad pattern of 36 degrees, and a 12 dB dish even a wider pattern of 51 degrees. You can almost visualize these patterns.

3H64 How is circular polarization produced using linearly-polarized antennas?
A. Stack two yagis, fed 90 degrees out of phase, to form an array with the respective elements in parallel planes
B. Stack two yagis, fed in phase, to form an array with the respective elements in parallel planes
C. Arrange two yagis perpendicular to each other, with the driven elements in the same plane, fed 90 degrees out of phase
D. Arrange two yagis perpendicular to each other, with the driven elements in the same plane, fed in phase

ANSWER C: To get circular polarization out of a pair of Yagis, arrange the Yagis perpendicular to each other, with the driven elements in the same plane, and fed 90 degrees out of phase.

3H65 Why does an antenna system for *earth operation* (for communications through a satellite) need to have rotators for both azimuth and elevation control?
A. In order to point the antenna above the horizon to avoid terrestrial interference
B. Satellite antennas require two rotators because they are so large and heavy
C. In order to track the satellite as it orbits the earth
D. The elevation rotator points the antenna at the satellite and the azimuth rotator changes the antenna polarization

ANSWER C: You need both an azimuth as well as an elevation rotor control to track the satellites up in orbit. This will be useful if you plan to track orbiting weather satellites at 137 MHz for weather facsimile reception.

5 GENERAL RADIOTELEPHONE OPERATOR LICENSE

3H66 What term describes a method used to match a high-impedance transmission line to a lower impedance antenna by connecting the line to the driven element in two places, spaced a fraction of a wavelength on each side of the driven element center?
   A. The gamma matching system
   B. The delta matching system
   C. The omega matching system
   D. The stub matching system

ANSWER B: The delta matching system is a popular way to change the impedance of the transmission line to the impedance of the antenna input.

3H67 What term describes an unbalanced feed system in which the driven element is fed both at the center of that element and a fraction of a wavelength to one side of center?
   A. The gamma matching system
   B. The delta matching system
   C. The omega matching system
   D. The stub matching system

ANSWER A: A gamma match is used with coax cable that forms an unbalanced feed system going to a tube that acts as a capacitor in series with the connection point.

3H68 What term describes a method of antenna impedance matching that uses a short section of transmission line connected to the antenna feed line near the antenna and perpendicular to the feed line?
   A. The gamma matching system
   B. The delta matching system
   C. The omega matching system
   D. The stub matching system

ANSWER D: When a short section of transmission line is connected to the antenna feed line near the antenna, it is called stub matching.

3H69 What kind of impedance does a 1/8-wavelength transmission line present to a generator when the line is shorted at the far end?
   A. A capacitive reactance
   B. The same as the characteristic impedance of the line
   C. An inductive reactance
   D. The same as the input impedance to the final generator stage

ANSWER C: If a 1/8-wavelength transmission line is shorted at the far end, it will look like an inductive reactance. See figure at question 3H70.

3H70 What kind of impedance does a 1/8-wavelength transmission line present to a generator when the line is open at the far end?
   A. The same as the characteristic impedance of the line
   B. An inductive reactance
   C. A capacitive reactance
   D. The same as the input impedance of the final generator stage

ANSWER C: If that 1/8-wavelength transmission line is open at the far end, it will look like a capacitive reactance.

3H – ANTENNAS AND FEED LINES

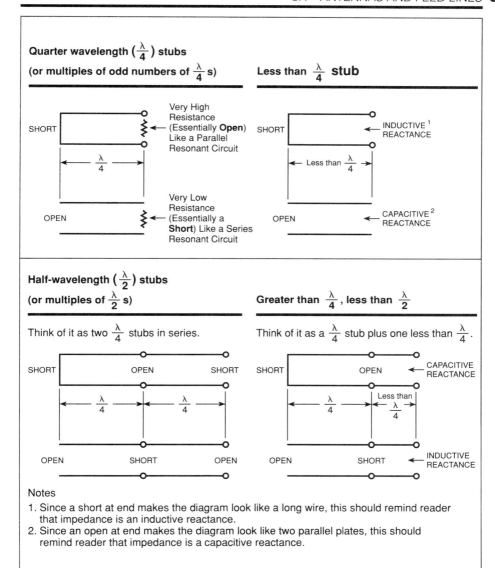

Notes
1. Since a short at end makes the diagram look like a long wire, this should remind reader that impedance is an inductive reactance.
2. Since an open at end makes the diagram look like two parallel plates, this should remind reader that impedance is a capacitive reactance.

Impedance of Matching Stubs

3H71 What kind of impedance does a 1/4-wavelength transmission line present to a generator when the line is shorted at the far end?
A. A very high impedance
B. A very low impedance
C. The same as the characteristic impedance of the transmission line
D. The same as the generator output impedance

ANSWER A: On a quarter-wavelength transmission line, shorting it at the far end will look like a very high impedance. See above figure (at question 3H70).

3H72 What kind of impedance does a 1/4-wavelength transmission line present to a generator when the line is open at the far end?
 A. A very high impedance
 B. A very low impedance
 C. The same as the characteristic impedance of the line
 D. The same as the input impedance to the final generator stage
ANSWER B: If the far end of a quarter-wavelength transmission line is open, it will look like a very low impedance. See figure at question 3H70.

3H73 What kind of impedance does a 3/8-wavelength transmission line present to a generator when the line is shorted at the far end?
 A. The same as the characteristic impedance of the line
 B. An inductive reactance
 C. A capacitive reactance
 D. The same as the input impedance to the final generator stage
ANSWER C: On a 3/8-wavelength transmission line, if it is shorted at the far end, it looks like a capacitive reactance. See figure at question 3H70.

3H74 What kind of impedance does a 3/8-wavelength transmission line present to a generator when the line is open at the far end?
 A. A capacitive reactance
 B. The same as the characteristic impedance of the line
 C. An inductive reactance
 D. The same as the input impedance to the final generator stage
ANSWER C: If that same 3/8-wavelength transmission line is now open at the far end, it will look like an inductive reactance. See figure at question 3H70.

3H75 What kind of impedance does a 1/2-wavelength transmission line present to a generator when the line is shorted at the far end?
 A. A very high impedance
 B. A very low impedance
 C. The same as the characteristic impedance of the line
 D. The same as the output impedance of the generator
ANSWER B: If a half-wavelength transmission line is shorted at the far end, it will have a very low impedance. See figure at question 3H70.

3H76 What kind of impedance does a 1/2-wavelength transmission line present to a generator when the line is open at the far end?
 A. A very high impedance
 B. A very low impedance
 C. The same as the characteristic impedance of the line
 D. The same as the output impedance of the generator
ANSWER A: If that half-wavelength transmission line is open at the far end, it will have a very high impedance. See figure at question 3H70.

3H77 What is the term used for an equivalent resistance which would dissipate the same amount of energy as that radiated from an antenna?
 A. Space resistance
 B. Loss resistance
 C. Transmission line loss
 D. Radiation resistance

3H – ANTENNAS AND FEED LINES **5**

ANSWER D: Radiation resistance is that equivalent resistance which would dissipate the same amount of energy as that radiated from the antenna.

3H78 Why is the value of the radiation resistance of an antenna important?
A. Knowing the radiation resistance makes it possible to match impedances for maximum power transfer
B. Knowing the radiation resistance makes it possible to measure the near-field radiation density from a transmitting antenna
C. The value of the radiation resistance represents the front-to-side ratio of the antenna
D. The value of the radiation resistance represents the front-to-back ratio of the antenna

ANSWER A: If we know the radiation resistance of a specific antenna, it will allow us to match impedances to that antenna for maximum power transfer. Most coaxial cables have a characteristic impedance of 50 to 52 ohms, and this is what is commonly used in land mobile, air, and marine installations.

3H79 Adding parasitic elements to an antenna will:
A. Decrease its directional characteristics
B. Decrease its sensitivity
C. Increase its directional characteristics
D. Increase its sensitivity

ANSWER C: Adding parasitic elements to a directional antenna for worldwide marine use, or VHF aeronautical use, will increase its directional characteristics.

3H80 What ferrite rod device prevents the formation of reflected waves on a waveguide transmission line?
A. Reflector
B. Isolator
C. Wave-trap
D. SWR refractor

ANSWER B: At microwave frequencies, the ferrite rod isolator prevents reflected waves on a waveguide transmission line.

3H81 Frequencies most affected by knife-edge refraction are:
A. Low and medium frequencies
B. High frequencies
C. Very high and ultra high frequencies
D. 100 kHz. to 3.0 MHz.

ANSWER C: VHF and UHF signals can be bent over mountains, and around and over buildings with a beam antenna and the effect of knife-edge refraction.

3H82 When measuring I and V along a 1/2-wave Hertz antenna where would you find the points where I and V are maximum and minimum?
A. V and I are high at the ends
B. V and I are high in the middle
C. V and I are uniform throughout the antenna
D. V is maximum at both ends, I is maximum in the middle

ANSWER D: Voltage is always maximum at the ends of a halfwave dipole. Current is maximum in the middle and minimum at the ends. Remember volts and the letter "S" on its side at the dipole, and current like the setting sun on the dipole. See figure at question 3H36.

3H83 To increase the resonant frequency of a 1/4 wavelength antenna:
A. Add a capacitor
B. Lower capacitor value
C. Cut antenna
D. Add an inductor

ANSWER A: We add capacitance to raise the resonant frequency of a 1/4-wavelength antenna. Answer C is close, but a better answer would be "trim antenna" rather than to just cut it.

3H84 Why are concentric transmission lines sometimes filled with nitrogen?
A. Reduces resistance at high frequencies
B. Prevent water damage underground
C. Keep moisture out and prevent oxidation
D. Reduce microwave line losses

ANSWER C: In high power commercial land radio stations, we sometimes will back fill concentric transmission line with dry nitrogen to drive out moisture and to prevent oxidation on the center conductor.

3H85 A vertical 1/4 wave antenna receives signals:
A. In the microwave band
B. In one vertical direction
C. In one horizontal direction
D. Equally from all horizontal directions

ANSWER D: While this answer sounds strange, the 1/4 wavelength vertical antenna receives its energy from other vertical antennas in all horizontal directions around the antenna. You have to think about this for a few seconds to get the picture—it receives in all directions—all horizontal directions.

3H86 Which of the following represents the best standing wave ratio (SWR)?
A. 1:1
B. 1:1.5
C. 1:3
D. 1:4

ANSWER A: A 1:1 is the best SWR.

3H87 What is the purpose of *stacking elements* on an antenna?
A. Sharper directional pattern
B. Increased gain
C. Improved bandpass
D. All of the above

ANSWER D: A collinear antenna is normally used for land mobile and marine applications. The collinear antenna has stacked elements inside a white radome. We normally do not use a collinear antenna in the aviation service because it might not pick up well at high elevations.

3H – ANTENNAS AND FEED LINES

3H88 The FCC deleted this question.

3H89 On a half-wave Hertz antenna:
A. Voltage is maximum at both ends and current is maximum at the center of the antenna
B. Current is maximum at both ends and voltage in the center
C. Voltage and current are uniform throughout the antenna
D. Voltage and current are high at the ends

ANSWER A: On the halfwave dipole, voltage is maximum at both ends, and current is maximum at the center. See figure at question 3H36.

3H90 What type of antenna is designed for minimum radiation?
A. Dummy antenna
B. Quarter-wave antenna
C. Half-wave antenna
D. Directional antenna

ANSWER A: We use a dummy antenna when working on two-way radio equipment to keep it from radiating out on the airwaves.

3H91 What is the outcome when you stack antennas at various angles?
A. A more omni-directional reception
B. A more uni-directional reception
C. An overall reception signal increase
D. Both A and C

ANSWER D: When we stack antennas at various angles, it becomes more omni-directional and overall reception increases.

3H92 Adding parasitic elements to a quarter-wavelength antenna will:
A. Reduce its directional characteristics
B. Increase its directional characteristics
C. Increase its sensitivity
D. Both B and C

ANSWER D: Adding parasitic elements in front of and behind a quarter-wavelength antenna will increase its directional characteristics. Parasitic elements also add to the sensitivity of the system *in one direction*; therefore, the answer D because both B and C are correct.

3H93 Ignoring line losses, voltage at a point on a transmission line without standing waves is:
A. Equal to the product of the line current and impedance
B. Equal to the product of the line current and power factor
C. Equal to the product of the line current and the surge impedance
D. Zero at both ends

ANSWER C: Measuring voltage on a transmission line is the product of the line current times the surge impedance.

3H94 Stacking antenna elements:
A. Will suppress odd harmonics
B. Decrease signal to noise ratio
C. Increases sensitivity to weak signals
D. Increases selectivity

ANSWER C: Stacking antenna elements in a collinear array will increase the sensitivity to extremely weak signals down close to the horizon.

3H95 What allows microwaves to pass in only one direction?
A. RF emitter
B. Ferrite isolator
C. Capacitor
D. Varactor-triac

ANSWER B: We use a ferrite isolator to channel microwaves to pass in only one direction to protect the receiver section hooked up to the same isolator.

3H96 What would be added to make a receiving antenna more directional?
A. Inductor
B. Capacitor
C. Parasitic elements
D. Height

ANSWER C: Adding parasitic antenna elements to a receiving antenna will make it more directional.

3H97 Nitrogen is placed in transmission lines to:
A. Improve the 'skin-effect' of microwaves
B. Reduce arcing in the line
C. Reduce the standing wave ratio of the line
D. Prevent moisture from entering the line

ANSWER D: Nitrogen inside a transmission line prevents moisture from getting into the line to protect the center conductor from oxidation.

3H98 Neglecting line losses, the voltage at any point along a transmission line, having no standing waves, will be equal to the:
A. Transmitter output
B. Product of the line voltage and the surge impedance of the line
C. Product of the line current and the surge impedance of the line
D. Product of the resistance and surge impedance of the line

ANSWER C: To calculate voltage in a transmission line, multiply line current by surge impedance.

3H99 Adding a capacitor in series with a Marconi antenna:
A. Increases the antenna circuit resonant frequency
B. Decreases the antenna circuit resonant frequency
C. Blocks the transmission of signals from the antenna
D. Increases the power handling capacity of the antenna

ANSWER A: Adding a capacitor in series with a Marconi antenna will raise the antenna's resonant frequency. Adding an inductor will lower the antenna's resonant frequency.

3H100 An antenna is carrying an unmodulated AM signal, when 100% modulation is applied, the antenna current:
A. Goes up 50%
B. Goes down one half
C. Stays the same
D. Goes up 22.5%

ANSWER D: You can calculate 100 percent modulated antenna current from the following formula:

 % increase in modulated = $\left(\sqrt{1 + \frac{m^2}{2}} - 1\right) \times 100$

Where m = a value between 0 and 1 representing the % modulation, e.g., 1 = 100%

Out in the field, there may be many different types of modulation, so this formula is an approximate way to calculate 100 percent modulated antenna current. Typical is 20 percent to 25 percent.

3H101 An excited 1/2-wavelength antenna produces:
A. Residual fields
B. An electromagnetic field only
C. Both electromagnetic and electrostatic fields
D. An electro flux field sometimes

ANSWER C: All antennas radiate two fields—an electromagnetic field and an electrostatic field that are 90 degrees cross-polarized. It is the electrostatic field that determines antenna polarization.

3H102 An antenna which intercepts signals equally from all horizontal directions is:
A. Parabolic
B. Vertical loop
C. Horizontal Marconi
D. Vertical 1/4 wave

ANSWER D: An antenna that receives signals equally well from all horizontal directions is the vertical quarter-wavelength dipole.

3H103 Loop-antenna:
A. Is bi-directional
B. Is usually vertical
C. Is more often used as a receiving antenna
D. Any of the above

ANSWER D: The loop antenna may be used for direction-finding on a radio receiver. It is bi-directional, and usually mounted vertical.

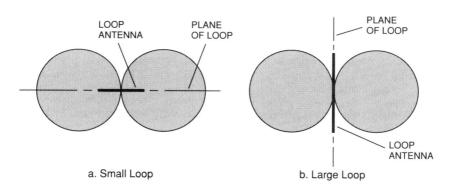

a. Small Loop　　　　b. Large Loop

Radiation Patterns of Loop Antennas

3H104 Referred to the fundamental frequency a shorted stub line attached to the transmission line to absorb even harmonics could have a wavelength of:
 A. 1.41 wavelength
 B. 1/2 wavelengths
 C. 1/4 wavelengths
 D. 1/6 wavelengths
ANSWER C: The 1/4-wavelength shorted stub line is attached to a transmitter to absorb even harmonics on a single frequency antenna system.

3H105 Nitrogen gas in concentric RF transmission lines is used to:
 A. Keep moisture out
 B. Prevent oxidation
 C. Act as insulator
 D. Both A and B
ANSWER D: Nitrogen in a RF line acts as an insulator and helps drive out and keep out moisture. Moisture would cause oxidation.

3H106 "Stacking" elements on an antenna:
 A. Makes for better reception
 B. Makes for poorer reception
 C. Decreases antenna current
 D. Decreases directivity
ANSWER A: Stacking elements in a collinear array increases reception.

3H107 The FCC deleted this question.

3H108 The parasitic elements on a receiving antenna:
 A. Increase its directivity
 B. Decrease its directivity
 C. Have no effect on its impedance
 D. Make it more nearly omnidirectional
ANSWER A: Adding parasitic elements on a receiving antenna will increase its directivity. This is similar to parasitic elements on a Yagi beam antenna.

3H109 The FCC deleted this question.

3H110 The resonant frequency of a Hertz antenna can be lowered by:
 A. Lowering the frequency of the transmitter
 B. Placing a condenser in series with the antenna
 C. Placing a resistor in series with the antenna
 D. Placing an inductance in series with the antenna
ANSWER D: If we add an inductance (coil) in series with a Hertz antenna, it will lower its resonant frequency.

3H111 Parasitic elements are useful in a receiving antenna for:
 A. Increasing directivity
 B. Increasing selectivity
 C. Increasing sensitivity
 D. Both A and C
ANSWER D: Parasitic elements will add both sensitivity in one direction and directivity to an antenna system.

3H – ANTENNAS AND FEED LINES

3H112 The FCC deleted this question.

3H113 In regards to shipboard satellite dish antenna systems, azimuth is:
A. Vertical aiming of the antenna
B. Horizontal aiming of the antenna
C. 0 - 90 degrees
D. North to east

ANSWER B: Azimuth is the horizontal aiming of the antenna. A radar antenna has 360-degrees azimuth when it is rotating.

3H114 What is the effect of adding a capacitor in series to an antenna?
A. Resonant frequency will decrease
B. Resonant frequency will increase
C. Resonant frequency will remain same
D. Electrical length will be longer

ANSWER B: Adding a capacitor in series with an antenna will raise the resonant frequency.

3H115 If a transmission line has a power loss of 6 dB per 100 feet, what is the power at the feed point to the antenna at the end of a 200 foot transmission line fed by a 100 watt transmitter?
A. 70 watts
B. 50 watts
C. 25 watts
D. 6 watts

ANSWER D: You can do this one in your head. If there is 6 dB loss for 100 feet of coax, there will be 12 dB loss for 200 feet. 12 dB loss is more than 10 dB which is easy to remember as a multiplier of 10. Knowing that your loss is greater than 10 times, 6 watts is the only answer that will agree with 12 dB.

3H116 Waveguides are:
A. Used exclusively in high-frequency power supplies
B. Ceramic couplers attached to antenna terminals
C. High-pass filters used at low radio frequencies
D. Hollow metal conductors used to carry high-frequency current

ANSWER D: The key words are *hollow metal conductors*. None of the other answers is even close, but answer D has an unfortunate wording by using "high-frequency" current. It does not refer to the designated HF band of 3-30 MHz. We normally use waveguides at frequencies above 500 MHz, mostly above 1 GHz.

3H117 Which of the following represents the best SWR?
A. 1:1
B. 1:2
C. 1:15
D. 2:1

ANSWER A: The best SWR here is 1:1.

5 GENERAL RADIOTELEPHONE OPERATOR LICENSE

3H118 A 520-kHz signal is fed to a 1/2-wave Hertz antenna. The fifth harmonic will be:
- A. 2.65 MHz.
- B. 2650 kHz.
- C. 2600 kHz.
- D. 104 kHz.

ANSWER C: 5 × 520 = 2600 kHz. This is near 2638 kHz, one of the early AM medium-frequency, ship-to-ship channels.

3H119 When a capacitor is connected in series with a Marconi antenna:
- A. An inductor of equal value must be added
- B. No change occurs to antenna
- C. Antenna open circuit stops transmission
- D. Antenna resonant frequency increases

ANSWER D: If you add a capacitor in series with a Marconi antenna, it will raise the resonant frequency.

3H120 How do you increase the electrical length of an antenna?
- A. Add an inductor in parallel
- B. Add an inductor in series
- C. Add a capacitor in series
- D. Add a resistor in series

ANSWER B: To lower the resonant frequency of a Marconi antenna, you will need to add inductance in series. This is the same as increasing the electrical length of the antenna.

3H121 A coaxial cable has 7 dB of reflected power when the input is 5 watts. What is the output of the transmission line?
- A. 5 watts
- B. 2.5 watts
- C. 1.25 watts
- D. 1 watt

ANSWER D: This one you can do in your head. 6 dB reflection is a 4 times loss in this circuit. Since it's a little bit more than a 4 times loss at 5 watts, you would only have 1 watt getting out of the transmission line into the antenna circuit.

3H122 What is the 7th harmonic of 450 kHz when fed through a 1/4-wavelength vertical antenna?
- A. 3150 Hz
- B. 3150 MHz
- C. 787.5 kHz
- D. 3.15 MHz

ANSWER D: Disregard the length of the antenna. 7 × 450 kHz = 3150 kHz; and if you move the decimal point 3 places to the left to calculate MHz, the answer is 3.15 MHz.

3H123 What is the 5th harmonic of a 450 kHz transmitter carrier fed to a 1/4-wave antenna?
- A. 562.5 MHz
- B. 1125 kHz
- C. 2250 MHz
- D. 2.25 MHz

ANSWER D: Here again, disregard the length of the antenna. 5 × 450 kHz = 2250 kHz; and moving the decimal point 3 places to the left (KLM), you end up with 2.25 MHz.

3H124 Waveguide construction:
A. Should not use silver plating
B. Should not use copper
C. Should have short vertical runs
D. Should not have long horizontal runs

ANSWER D: With any waveguide installation, you should not have any long runs.

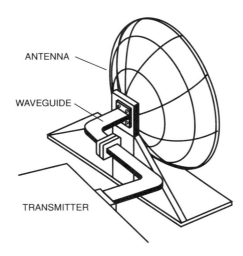

Waveguide Between Transmitter and Antenna

3H125 The FCC deleted this question.

3H126 To lengthen an antenna electrically, add a:
A. Coil
B. Resistor
C. Battery
D. Conduit

ANSWER A: To lengthen an antenna electrically, add an inductor, such as a coil at the base.

3H127 How do you electrically decrease the length of an antenna?
A. Add an inductor in series
B. Add a capacitor in series
C. Add an inductor in parallel
D. Add a resistor in series

ANSWER B: To electrically shorten an antenna, add a capacitor in series at the base.

5 GENERAL RADIOTELEPHONE OPERATOR LICENSE

3H128 If the length of an antenna is changed from 1.5 feet to 1.6 feet its resonant frequency will:
- A. Decrease
- B. Increase
- C. Be 6.7% higher
- D. Be 6% lower

ANSWER A: Lower, longer, giving the antenna better performance at a decrease in frequency. Answer D cannot be verified unless you know more details about how the antenna is loaded or matched.

3H129 To couple energy into and out of a waveguide:
- A. Use wide copper sheeting
- B. Use an LC circuit
- C. Use capacitive coupling
- D. Use a thin piece of wire as an antenna

ANSWER D: With the transmitter turned off, and the waveguide feed assembly out of circuit, take a look inside and spot the tiny thin piece of wire that is acting as the antenna element. This will couple energy into and out of a waveguide.

3H130 An isolator:
- A. Acts as a buffer between a microwave oscillator coupled to a waveguide
- B. Acts as a buffer to protect a microwave oscillator from variations in line load changes
- C. Shields UHF circuits from RF transfer
- D. Both A and B

ANSWER D: The isolator in an antenna circuit will act as a buffer between the microwave oscillator coupled to the waveguide, and it will also shield UHF circuits from radio frequency transfer.

3H131 A high SWR creates losses in transmission lines. A high standing wave ratio might be caused by:
- A. Improper turns ratio between primary and secondary in the plate tank transformer
- B. Screen grid current flow
- C. An antenna electrically too long for its frequency
- D. An impedance mismatch

ANSWER D: An impedance mismatch in the antenna or a deformed transmission line could lead to high SWR.

3H132 A properly installed shunt-fed, 1/4-wave Marconi antenna:
- A. Has zero resistance to ground
- B. Has high resistance to ground
- C. Should be cut to 1/2 wave
- D. Should not be shunt-fed

ANSWER A: The perfect Marconi antenna, DC shunt-fed, should have zero resistance to ground.

3H – ANTENNAS AND FEED LINES

3H133 When a capacitor is connected in series with a Marconi antenna:
A. An inductor of equal value must be added
B. No change occurs to antenna
C. Antenna open circuit stops transmission
D. Antenna resonant frequency increases

ANSWER D: Adding capacitance to an antenna will raise its resonant frequency.

3H134 When excited by RF, a half-wave antenna will radiate:
A. A space wave
B. A ground wave
C. Electromagnetic fields
D. Both electromagnetic and electrostatic fields

ANSWER D: When an antenna emits radio waves, it emits electromagnetic fields and electrostatic fields at right angles to each other.

3H135 Waveguides are:
A. Used exclusively in high frequency power supplies
B. Ceramic couplers attached to antenna terminals
C. High pass filters used at low radio frequencies
D. Hollow metal conductors used to carry high frequency current

ANSWER D: This question is identical to question 3H116. Key words are *hollow metal conductors*.

3H136 A 520-kHz signal is fed to a 1/2-wave Hertz antenna. The fifth harmonic will be:
A. 2.65 MHz
B. 2650 kHz
C. 2600 kHz
D. 1300 kHz

ANSWER C: This question is identical to question 3H118. 5 × 520 kHz = 2600 kHz. Disregard the length of the antenna.

3H137 The voltage produced in a receiving antenna is:
A. Out of phase with the current if connected properly
B. Out of phase with the current if cut to 1/3 wavelength
C. Variable depending on the station's SWR
D. Always proportional to the received field strength

ANSWER D: The voltage produced in a receiving antenna is always proportional to the received field strength at the antenna. Electromagnetic radio waves induce a voltage in the receiving antenna which produces a current in the antenna and input circuit to which the antenna is connected.

3H138 A properly connected transmission line:
A. Is grounded at the transmitter end
B. Is cut to a harmonic of the carrier frequency
C. Is cut to an even harmonic of the carrier frequency
D. Has a SWR as near as 1:1 as possible

ANSWER D: A properly connected transmission line whose impedance matches the load will have an SWR as near as 1:1 as possible.

3H139 Conductance takes place in a waveguide:
A. By interelectron delay
B. Through electrostatic field reluctance
C. In the same manner as a transmission line
D. Through electromagnetic and electrostatic fields in the walls of the waveguide

ANSWER D: In a waveguide, the conductance takes place in the walls of the waveguide through electromagnetic and electrostatic fields.

3H140 In regards to a shunt-fed 1/4-wavelength Marconi antenna:
A. DC resistance of the antenna to ground is zero
B. RF resistance from antenna feed point to ground is zero
C. Harmonic radiation is zero under all conditions
D. It must be grounded at both feed and far ends

ANSWER A: In a shunt-fed, 1/4-wavelength Marconi antenna, DC resistance of the antenna to ground is zero.

3H141 The FCC deleted this question.

3H142 If a 3/4-wavelength transmission is shorted at one end, the impedance at the open will be:
A. Zero
B. Infinite
C. Decreased
D. Increased

ANSWER B: With a 3/4-wavelength transmission line shorted at one end, impedance at the opposite end, open end, will always be high. Refer to the figure at 3H70 and consider the 3/4-wavelength transmission line as a 1/2-wavelength line with a 1/4 wavelength added to it. A short at the end of the 1/2-wavelength reflects a short to the 1/4-wavelength. A short at the end of the 1/4-wavelength reflects a high impedance at the open end.

3H143 A dummy antenna is a:
A. Non-directional receiver antenna
B. Wide bandwidth directional receiver antenna
C. Transmitter test antenna designed for minimum radiation
D. Transmitter non-directional narrow-band antenna

ANSWER C: A dummy antenna is one that is designed for minimal radiation and output.

6

Taking the License Examinations

ABOUT THIS BOOK AND THIS CHAPTER

This book is intended to be a complete guide to the FCC's new Commercial Radio Operator examination program — with emphasis on the *General Radiotelephone Operator License* (GROL). The GROL is, by far, the most popular of all commercial radio operator licenses. The GROL requires examinations taken from the Element 1 question pool and the Element 3 question pool. Element 1 question pool is found in Chapter 4, and Element 3 question pool is found in Chapter 5. Passing examinations on Element 1 and Element 3 question pools are also prerequisites for many of the other commercial licenses (See *Tables 3-1* and *3-2.*). Element 1 (radio law) is a requirement for the Marine Radio Operator Permit, all radiotelegraph certificates and both of the new Global Maritime Distress and Safety System (GMDSS) licenses. Element 3 (the technical questions) is also needed for the GMDSS Maintainer's license.

This book covers everything you need to know about the requirements to obtain any commercial license; however, only Element 1 and Element 3 question pools are included. Therefore, if you are striving to pass an examination for a license other than GROL or MROP, additional question pools will be required. Contact your nearest COLEM for direction.

This chapter tells who is qualified to give examinations, how to find an examination location, how the examination will be given, and what happens after you complete an examination.

WHICH LICENSE DO I NEED?

The U.S. government in 1984 discontinued the requirement that an electronic technician hold a commercial radio operator license when installing, maintaining and repairing radio transmitting equipment in some of the domestic radio services. In its place, they substituted a program of industry technical certification. Chapter 2 details the requirements for the various types of licenses, permits and endorsements. Refer to it for the specific license requirements to make sure you are qualified and prepared for the examination.

WHO GIVES EXAMINATIONS

The Commercial Operator License Examination (COLE) System is the name of the Federal Communications Commission's newly privatized commercial radio operator testing program. The objective of this program is to meet the need of prospective licensees for frequent examination opportunities at convenient locations and to meet the need of industry for examinations that reflect state-of-the-art technology and contemporary operating conditions, at minimal expense to the government.

A Commercial Operator License Examination Manager (COLEM) is an organization that has been approved by the FCC to prepare and administer the various examination elements that must be passed by persons in order to be eligible to apply to the FCC for commercial radio operator licenses. They are independent contractors and each has an extensive nationwide commercial radio operator testing network.

EXAMINATION LOCATIONS

Up until 1992, the GROL examination was given twice a year at 25 different FCC field offices. It cost $35.00 to take the examination and there was a registration deadline a month ahead of the test date.

Now that commercial radio operator testing has been privatized, there are hundreds of locations across the United States where you can be administered commercial radio operator examinations. All approved examiners are required to be affiliated with one of the nine FCC-approved COLEMs. The examination location and time is determined by the examiners. The best way to find a nearby examination location is to simply telephone the COLEM's main office. There is a detailed list of the COLEMs in the Appendix giving telephone numbers, contact person, and examination availability. Most have toll-free "800" telephone numbers and their operators can give you the local telephone number of the nearest test center. Most COLEMs sponsor monthly or quarterly examinations, but some have them available daily on a walk-in basis. Many technical schools and colleges are also affiliated with a COLEM and many will administer the required tests as part of your tuition.

There is probably an examination site within a hour's travel time of your home! We strongly recommend that you telephone ahead to the examiners and let them know you are coming, since space at some test sessions may be limited. This will assure that a space will be reserved for you.

EXAMINATION FEES

You can expect to pay two fees for your new FCC license: an *Examination Fee* to the COLEM examiners and a *Regulatory Fee* to the FCC. The Regulatory Fee, which is sent to the FCC, is paid by a check attached to the license application. This cost is exactly the same regardless of which COLEM organization gives you the required tests. Another fee, payable only when you renew or replace your license, is called an *Application Fee*. These fees are detailed in Chapter 2.

THE QUESTION POOLS AND YOUR EXAMINATION

Until the fall of 1993, the questions contained in the commercial radio operator tests were not readily known. Many publishers printed study manuals with review or sample questions, but they were not the word-for-word questions that you would be asked in an actual examination. This has now changed. The process (called the *Question Pool System*) now is similar to taking the written portion of a state automobile driver's license test. You obtain a license preparation manual such as this book, you study the questions, and you pass the test.

There is one, and only one, question pool for each written element. There no longer will be any "secret" questions appearing on any radio operator examinations. Everything — including the schematic diagrams — was (or will be) released by the FCC and no question, nor any of the multiple-choice answers, may be used in a question set administered to an examinee unless it appears in the question pool of the most recent release. The questions and answers in Chapters 4 (Element 1) and 5 (Element 3) in this book are the same questions and answers that will appear on your examination. An explanation of the correct answer is included to help you understand and remember the answer. About ten percent of the questions in a pool are asked in any one question set. Even the numerical values in a question that requires mathematics or a formula will be the same in your examination if that particular question is asked.

The FCC will periodically update the questions that appear in every question pool and it is important that you study the current version. Maintenance of the various question pools for the written examination is handled by the FCC's Aviation and Marine Branch. Since every COLEM must use exactly the same question pool to construct their examinations, the examinations of one COLEM are no easier or harder than those of another.

The remaining question pools are scheduled for release by the FCC to the public during 1994. Although they usually contain more, FCC regulations (§13.215) require that each question pool must contain at least five times the number of questions required for a single examination. Copies of just-released question pools — including the multiple choices and answers — will be available shortly after release from National Radio Examiners, P.O. Box 565206, Dallas, Texas 75356, (800) 669-9594.

TAKING THE EXAMINATION

You will be asked for positive identification before you are administered any commercial radio examination. One of these identification documents should be a state driver's license with your photograph on it. This is to prevent anyone from fraudulently taking the examination for another person. You will probably be asked to pay the examination fee before the tests are administered. The examination fee is due whether you pass or fail.

Examiners for each COLEM will administer the examinations in basically the same way. Some administer tests at a personal computer while others simply hand out a prepared written examination and you check the appropriate multiple-choice answers. You will answer each question by selecting one of four possible answers. All commercial radio examiners are required by the FCC to use exactly the same question pools, multiple choices, schematic diagrams and answers. Thus, if you study Chapters 4 and 5 of this book, there will be no surprises when you are administered the Element 1 and 3 examinations for the Marine Radio Operator Permit (MROP) or the General Radiotelephone Operator License (GROL).

The Marine Radio Operator Permit requires passing only Element 1. You must correctly answer at least 75% (18) of the 24 questions selected from the Element 1 question pool of 169 questions.

The 100-question examination for the General Radiotelephone Operator License consists of two written multiple-choice examinations; one from Element 1 and one from Element 3. You must correctly answer at least 75% (18) of the 24 questions selected from the Element 1 question pool of 169 questions, and you must correctly answer at least 75% (57) of the 76 questions selected from the Element 3 question pool of 720 questions.

You must abide by the instructions of the examiners. The amount of time that you will be allowed to answer the questions is at the discretion of the examiners administering the tests. As a general guideline, you will be given about one hour for the Element 1 examination and about two hours for the Element 3 examination.

HINTS ON PASSING THE TEST

- Be sure to get a good night's sleep the night before. Don't stay up half the night studying!
- Be at the exam room early.
- Review the explanations of some of the most difficult answers. Try remembering the hints on the answers.
- Examinees are generally allowed to bring a calculator with them into the exam room, but the examiners may ask to clear its memory.
- You won't be allowed to leave the exam room once the examination starts — so make your bathroom stop beforehand.
- Read every question completely before trying to answer it. If you don't know the answer right away, go on to the next question. When you finish the first pass, go back to the ones you skipped over. Don't leave any question unanswered; mark your best guess.
- On non-math questions, try eliminating choices that are obviously wrong. You have a better chance of guessing the right answer if you can reduce the number of choices.
- Try working the math problems in a number of ways until you come up with one of the answers.
- An interesting and very effective way to study for the GROL license is with a personal computer. Some COLEMs; e.g., National Radio Examiners, have low-price interactive computer software that allows you to study for the GROL and take sample examinations right at your IBM or IBM-compatible PC keyboard.

GRADING THE EXAMINATION

Your examiners will grade your examination(s) as soon as you turn in your answer sheet(s). They will immediately tell you your score and whether you passed the examination. They usually will not tell you which questions you missed. If you took your examinations at a computer keyboard, the PC will print out a scored answer sheet.

At the option of the examiner, you may retake a failed test after paying another examination fee. The rules (§13.209(c)) require that a different question set be re-administered to the examinee. Due to time constraints, however, examiners may ask you to appear at their next scheduled testing session.

Some applicants who pass Elements 1 and 3 will want to take the Element 8 examination for the radar endorsement. This will take another hour and consists of 50 questions on the proper operation and servicing of ship radar equipment. If you pass, this endorsement will be placed on your GROL. Check with your COLEM on the availability of the Element 8 examination.

PROOF-OF-PASSING CERTIFICATE (PPC)

Once an applicant has passed any of the examination elements, the examiners will complete a *Proof-of-Passing Certificate* — or PPC, as it is called. A PPC is documented evidence that an applicant has passed a test. An examinee is eligible to submit an application to the FCC for a commercial radio operator license once he/she has accumulated the necessary original PPCs — or other credit document (either a commercial or amateur radio operator license) — indicating that he/she has completed the necessary examination requirements for a license. For example, a photocopy of a Marine Radio Operator Permit (MROP) will grant Element 1 credit toward the General Radiotelephone Operator License (GROL), or any or the radiotelegraph or GMDSS licenses. Likewise, a copy of a GROL confers Element 1 and 3 credit toward the GMDSS Maintainer's License. An Amateur Extra Class operator license grants Telegraphy Element 1 and 2 credit toward the Second and Third Class Radiotelegraph Operator's Certificate. (See § 13.9(c)(2).)

Just because you qualified for a lower class permit or license than you wanted does not mean you have to submit an application for the lower class permit or license. For example, let's say you wanted a GROL. You took the examinations for Element 1 and Element 3. You passed Element 1 but failed Element 3. You do not have to submit an application for a MROP. You will receive a PPC for Element 1 which you may hold until you pass Element 3, then you can use the Element 1 PPC as credit towards the GROL.

FILLING OUT THE FCC FORM 756

Either the applicant — or his agent — may submit FCC Form 756 *Applications for Commercial Radio Operator Licenses* to the FCC. It is an advantage to use the COLEM who coordinated your examinations as your agent. There are all sorts of pitfalls and errors that can be made in this area! You'll get your application back if it is not submitted to the correct address with the proper fees and attachments.

FCC Form 756 may be obtained without charge from the FCC's forms supply house by telephoning (202) 632-FORM. In addition, most COLEM examiners maintain supplies of application forms for examinees. It is very important that Form 756 applications be legible. All applicants should print with blue or black ink (not pencil) using block letter capitals. A properly completed FCC Form 756 is shown in the Appendix. Use only an FCC Form 756 that has been issued on or after 10/93 (October, 1993.) Previous issues are not accepted since they are not compatible with the new COLEM system.

RECEIVING YOUR LICENSE

It normally takes six to eight weeks to receive your commercial radio operator license from the Federal Communications Commission in Gettysburg, Pennsylvania. Other than the restricted permit, applicants are not authorized to perform any operating, installation, service or maintenance functions which require a commercial license until they actually receive the license document. If it has been more than three months, check on your application's status by telephoning the FCC's Consumer Assistance hotline at (717) 337-1212.

TESTING THE HANDICAPPED

Applicants who are afflicted with physical handicaps which would impair their ability to perform as a radio operator under emergency conditions may still be examined and licensed. Applicants are not denied a license solely on the basis of being disabled. The radio examiners are required to "accommodate" the disabled examinee by using special testing methods. Any required special equipment must be supplied by the disabled examinee.

A handicapped person who qualifies for a license employing special accommodative procedures and is unable to accomplish emergency duties involving the safety of life or property will have a special restrictive endorsement added to his/her license.

Applicants who submit an Amateur Extra Class license for Telegraphy Element 1 and 2 examination credit without passing the amateur Element 1(C) examination in a normal manner must check "Yes" to question No. 11 on the FCC Form 756 and explain that they received a handicap exemption for the 20 words-per-minute Morse code test. A restrictive endorsement will be added to their license.

Appendix

COLEMS – COMMERCIAL OPERATOR LICENSE EXAMINATION MANAGER ORGANIZATIONS

COLEM	Examination Availability

National Radio Examiners Division, The W5YI Group, Inc.
P.O. Box 565206
Dallas, TX 75356-5206
Voice: (800) 669-9594
Voice: (817) 461-6443
Fax: (817) 548-9594
	All elements are available on a monthly or quarterly basis, based on demand, at more than 250 test centers in all states.
	Fee: $35.00 per license
	Contact: Frederick O. Maia
	  2000 E. Randol Mill Rd., Suite 608A
	  Arlington, TX 76011
	  MCI Mail: 351-1297; Internet: 3511297@mcimail.com

Drake Training and Technologies 8800 Queen Avenue South Bloomington, MN 55431 Voice: (800) 401-EXAM Fax: (612) 921-7248	All elements are available on a daily basis at over 200 locations in all states except Maine and at over 300 locations worldwide. Evening, weekend, and holiday appointments are available. Fee: $60.00 per examination Contact: Julie Johnson
Electronic Technicians Association International, Inc. (ETAI) 602 North Jackson Street Greencastle, IN 46135 Voice: (317) 653-4301 Voice: (317) 653-8262 Fax: (317) 653-8262	All elements are available at test sites throughout all states. Also at stateside and overseas U.S. military installations (DANTES). Call for schedule information. Fee: $35.00 - $75.00 Contact: Anne Voiles
Elkins Institute, Inc. P.O. Box 797666 Dallas, TX 75379 Voice: (800) 944-1603 Fax: (214) 732-0244	All written examinations are available at test sites throughout all states. Scheduled and "by appointment" examinations are available. Fee: $50.00 for first element; $25.00 for each additional element taken at same sitting. Contact: Ed Lyda
International Society of Certified Electronics Technicians (ISCET) 2708 West Berry Street Fort Worth, TX 76109 Voice: (817) 921-9101 Fax: (817) 921-3741	All elements are available by appointment from 360 examiners in 47 states, Guam, and some foreign countries. (Alaska, Vermont and Wyoming are not available). Fee: $25.00 - $75.00 per element Contact: Dept. 19
National Association of Business and Educational Radio, Inc. (NABER) 1501 Duke Street Alexandria, VA 22314 Registration Voice:(800) 869-1100 Fax: (612) 832-1290	Written elements 1, 3, 7, and 9 are available at 95 test centers nationwide five days a week. Fee: $63.00 - $120.00 Contact: FCC Technician Testing Center Voice: (800) 759-0300 Fax: (703) 836-1608
Sea School 5905 4th Street N. St. Petersburg, FL 33703 Voice: (800) 237-8663 Fax: (813) 522-3155	All elements are available by appointment in 83 coastal cities. Fee: $25.00 - $55.00 Contact: Len Wahl
Sylvan KEE Systems 9135 Guilford Road Columbia, MD 21046 Voice: (800) 967-1100 Fax: (410) 880-8714	All elements are available seven days a week, walk-in or scheduled appointment (except holidays) at over 110 computerized testing centers in 35 states. Fee: $50.00 - $75.00 Contact: National Registration Center

The National Association of Radio Telecommunications Engineers, Inc. (NARTE)
P.O. Box 678
Medway, MA 02053
Voice: (508) 533-8333
Fax: (508) 533-3815
	All elements available by appointment quarterly at NARTE test centers at 120 U.S. universities and colleges. Also available at U.S. and some overseas military bases (DANTES).
	Fee: $40.00 per examination per sitting
	Contact: Susan Stillwell
	  11 Awl Street
	  Medway, MA 02053

GENERAL RADIOTELEPHONE OPERATOR LICENSE

FCC FIELD OFFICE ADDRESSES

ALASKA
FCC – Anchorage Field Office EIC
6721 West Raspberry Road
Anchorage, Alaska 99502-1896
Tel. (907) 243-2153

ARIZONA
FCC – Douglas Field Office
P.O. Box 6
Douglas, Arizona 85608-0006
Tel. (602) 364-8414

CALIFORNIA
FCC – Los Angeles Field Office
Cerritos Corporate Tower
1800 Studebaker Road, Room #660
Cerritos, California 90701-3684
Tel. (310) 809-2096

FCC – San Diego Field Office
4542 Ruffner Street, Room 370
San Diego, California 92111-2216
Tel. (619) 467-0549

FCC – San Francisco Field Office
3777 Depot Road
Hayward, California 94545-1914
Tel. (510) 732-9046

FCC – Livermore Field Office
P.O. Box 311
Livermore, California 94551-03111
Tel. (415) 447-3614

COLORADO
FCC – Denver Field Office
165 South Union Blvd., Suite 860
Lakewood, Colorado 80228-2213
Tel. (303) 969-6497/8

FLORIDA
FCC – Vero Beach Field Office
P.O. Box 1730
Vero Beach, Florida 32961-1730
Tel. (407) 778-3755

FCC – Miami Field Office
Rochester Building, Room 310
8390 N.W. 53rd Street
Miami, Florida 33166-4668
Tel. (305) 526-7420

FCC – Tampa Field Office
2203 N. Lois Avenue, Room 1215
Tampa, Florida 33607-2356
Tel. (813) 228-2872

GEORGIA
FCC – Atlanta Field Office
3575 Koger Blvd., Room 290
Koger Center-Gwinnett
Duluth, Georgia 30136-4958
Tel. (404) 279-4621

FCC – Powder Springs Field Office
P.O. Box 85
Powder Springs, Georgia 30073-0085
Tel. (404) 943-5420

HAWAII
FCC – Honolulu Field Office
P.O. Box 1030
Waipahu, Hawaii 96797-1030
Tel. (8080 677-3318

ILLINOIS
FCC – Chicago Field Office
Park Ridge Office Center #306
1550 Northwest Highway
Park Ridge, Illinois 60068-1460
Tel. (312) 353-0195

LOUISIANA
FCC – New Orleans Field Office
800 W. Commerce Road, Room 505
New Orleans, Louisiana 70123-3333
Tel. (504) 589-2095

MAINE
FCC – Belfast Field Office
P.O. Box 470
Belfast, Maine 04915-0470
Tel. (207) 338-4088

MARYLAND
FCC – Baltimore Field Office
1017 Federal Bldg.
31 Hopkins Plaza
Baltimore, Maryland 21201-2802
Tel. (410) 962-2728

FCC – Laurel Field Office
P.O. Box 250
Columbia, Maryland 21045-9998
Tel. (301) 725-3474

MASSACHUSETTS
FCC – Boston Field Office
1 Batterymarch Park
Quincy, Massachusetts 02169-7495
Tel. (617) 770-4023

MICHIGAN
FCC – Allegan Field Office
P.O. Box 89
Allegan, Michigan 49010-9437
Tel. (616) 673-2063

FCC – Detroit Field Office
24897 Hathaway Street
Farmington Hills, Michigan 48335-1952
Tel. (313) 226-6078

MINNESOTA
FCC – St. Paul Field Office
2025 Sloan Place
Suite 31
Maplewood, Minnesota 55117-2058
Tel. (612) 774-5180

MISSOURI
FCC – Kansas City Field Office
8800 E. 63rd Street, Room 320
Kansas City, Missouri 64133-4895
Tel. (816) 353-3773

NEBRASKA
FCC – Grand Island Field Office
P.O. Box 1588
Grand Island, Nebraska 68802-1588
Tel. (308) 381-4721

NEW YORK
FCC – Buffalo Field Office
1307 Federal Bldg.
111 West Huron Street
Buffalo, New York 14202-2398
Tel. (716) 846-4511

FCC – New York Field Office
201 Varick Street
New York, New York 10014-4870
Tel. (212) 620-3437/8

OREGON
FCC – Portland Field Office
1782 Federal Bluilding
1220 S.W. Third Avenue
Portland, Oregon 97204-2898
Tel. (503) 326-4114/5

PENNSYLVANIA
FCC – Philadelphia Field Office
One Oxford Valley Office Bldg. #404
2300 E. Lincoln Highway
Langhorne, Pennsylvania 19047-1859
Tel. (215) 752-1324

PUERTO RICO
FCC – San Juan Field Office
U.S. Federal Building, Room 747
Hato Rey, Puerto Rico 00918-1731
Tel. (809) 766-5567

TEXAS
FCC – Dallas Field Office
9330 LBJ Freeway, Room 1170
Dallas, Texas 75243-3429
Tel. (214) 235-3369

FCC – Houston Field Office
1225 North Loop West, Room 900
Houston, Texas 77008-1775
Tel. (713) 861-6200

FCC – Kingsville Field Office
P.O. Box 632
Kingsville, Texas 78363-0632
Tel. (512) 592-2531

VIRGINIA
FCC – Norfolk Field Office
1200 Communications Circle
Virginia Beach, Virginia 23455-3725
Tel.(804) 441-6472

WASHINGTON
FCC – Ferndale Field Office
1330 Loomis Trail Road
Custer, Washington 98240-9303
Tel. (206) 354-4892

FCC – Seattle Field Office
11410 N.E. 122nd Way, Suite 312
Kirkland, Washington 98034-6927
Tel. (206) 821-9037

APPENDIX – PART 13

Federal Communications Commission - Rules and Regulations – Complete
47 C.F.R. – PART 13 – COMMERCIAL RADIO OPERATOR RULES

GENERAL

§ 13.1   Basis and purpose.
§ 13.3   Definitions.
§ 13.5   Licensed commercial radio operators required.
§ 13.7   Classification of operator licenses and endorsements.
§ 13.9   Eligibility and application for new license or endorsement.
§ 13.11   Holding more than one commercial radio operator license.
§ 13.13   Application for renewed or modified license.
§ 13.15   License term.
§ 13.17   Replacement license.
§ 13.19   Operator's responsibility.

EXAMINATION SYSTEM

§ 13.201   Qualifying for a commercial operator license or endorsement.
§ 13.203   Examination elements.
§ 13.207   Preparing an examination.
§ 13.209   Examination procedures.
§ 13.211   Commercial radio operator license examination.
§ 13.213   COLEM qualifications.
§ 13.215   Question pools.
§ 13.217   Records.

GENERAL PROVISIONS

§ 13.1 Basis and purpose.

(a) Basis. The basis for the rules contained in this Part is the Communications Act of 1934, as amended, and applicable treaties and agreements to which the United States is a party.

(b) Purpose. The purpose of the rules in this Part is to prescribe the manner and conditions under which commercial radio operators are licensed by the Commission.

§ 13.3 Definitions.

The definitions of terms used in Part 13 are:

(a) *COLEM.* Commercial operator license examination manager.

(b) *Commercial radio operator.* A person holding a license or licenses specified in § 13.7(b).

(c) *GMDSS.* Global Maritime Distress and Safety System.

(d) *FCC.* Federal Communications Commission.

(e) *International Morse code.* A dot-dash code as defined in International Telegraph and Telephone Consultative Committee (CCITT) Recommendation F.1 (1984), Division B, I. Morse code.

(f) *ITU.* International Telecommunication Union.

(g) *PPC.* Proof-of-Passing Certificate.

(h) *Question pool.* All current examination questions for a designated written examination element.

(i) *Question set.* A series of examination questions on a given examination selected from the current question pool.

(j) *Radio Regulations.* The latest ITU Radio Regulations to which the United States is a party.

§ 13.5 Licensed commercial radio operator required.

Rules that require FCC station licensees to have certain transmitter operating, maintenance, and repair duties performed by a commercial radio operator are contained in Parts 23, 73, 74, 80, and 87 of this Chapter.

§ 13.7 Classification of operator licenses and endorsements.

(a) Commercial radio operator licenses issued by the FCC are classified in accordance with the Radio Regulations of the ITU.

(b) There are nine types of commercial radio operator licenses, certificates and permits (licenses). The license's ITU classification, if different from its name, is given in parenthesis.

(1) First Class Radiotelegraph Operator's Certificate.

(2) Second Class Radiotelegraph Operator's Certificate.

(3) Third Class Radiotelegraph Operator's Certificate (radiotelegraph operator's special certificate).

(4) General Radiotelephone Operator License (radiotelephone operator's general certificate).

(5) Marine Radio Operator Permit (radiotelephone operator's restricted certificate).

(6) Restricted Radiotelephone Operator Permit (radiotelephone operator's restricted certificate).

(7) Restricted Radiotelephone Operator Permit-Limited Use (radiotelephone operator's restricted certificate).

(8) GMDSS Radio Operator's License (general operator's certificate).

(9) GMDSS Radio Maintainer's License (technical portion of the first-class radio electronic certificate).

(c) There are six license endorsements affixed by the FCC to provide special authorizations or restrictions. Endorsements may

277

GENERAL RADIOTELEPHONE OPERATOR LICENSE

be affixed to the license (s) indicated in parenthesis.

(1) Ship Radar Endorsement (First and Second Class Radiotelegraph Operator's Certificates, General Radiotelephone Operator License, GMDSS Radio Maintainer's License).

(2) Six Months Service Endorsement (First and Second Class Radiotelegraph Operator's License).

(3) Restrictive endorsements relating to physical handicaps, English language or literacy waivers, or other matters (all licenses).

(4) Marine Radio Operator Permits shall bear the following endorsement: This Permit does not authorize the operation of AM, FM or TV broadcast stations.

(5) General Radiotelephone Operator Licenses issued after December 31, 1985, shall bear the following endorsement: This license confers authority to operate licensed radio stations in the Aviation, Marine and International Fixed Public Radio Service only. This authority is subject to: any endorsement placed upon this license; FCC orders, rules, and regulation; United States statutes; and the provisions of any treaties to which the United States is a party. This license does not confer any authority to operate broadcast stations. It is not assignable or transferable.

(6) (i) If a person is afflicted with an uncorrected physical handicap which would clearly prevent the performance of all or any part of the duties of a radio operator, under the license for which application is made, at a station under emergency conditions involving the safety of life or property, that person still may be issued the license if found qualified. Such a license shall bear a restrictive endorsement as follows:

This license is not valid for the performance of any operating duties, other than installation, service and maintenance duties, at any station licensed by the FCC which is required, directly or indirectly, by any treaty, statute or rule or regulation pursuant to statute, to be provided for safety purposes.

(ii) In the case of a license that does not require an examination in technical radio matters, the endorsement specified in (i) above will be modified by deleting the reference therein to installation, service, and maintenance duties.

(iii) In any case where an applicant who normally would receive or has received a commercial radio operator license bearing the endorsement prescribed by paragraph (i) above, indicates a desire to operate a station falling within the prohibited terms of the endorsement, the applicant may request in writing that such endorsement not be placed upon, or be removed from his or her license, and may submit written comments or statements from other parties in support thereof.

(iv) An applicant who shows that he has performed satisfactorily the duties of a radio operator at a station required to be provided for safety purposes during a period when he or she was afflicted by uncorrected physical handicaps of the same kind and to the same degree as the physical handicaps shown by his or her current application shall not be deemed to be within the provisions of paragraph (i) above.

(d) A Restricted Radiotelephone Operator Permit-Limited Use issued by the FCC to an aircraft pilot who is not legally eligible for employment in the United States is valid only for operating radio stations on aircraft.

(e) A Restricted Radiotelephone Operator Permit-Limited Use issued by the FCC to a person under the provision of § 303 (1)(2) of the Communications Act of 1934, as amended, is valid only for the operation of radio stations for which that person is the station licensee.

§ 13.9 Eligibility and application for new license or endorsement.

(a) If found qualified, the following persons are eligible to apply for commercial radio operator licenses:

(1) Any person legally eligible for employment in the United States.

(2) Any person, for the purpose of operating aircraft radio stations, who holds:

(i) United States pilot certificates; or

(ii) Foreign aircraft pilot certificates which are valid in the United States, if the foreign government involved has entered into a reciprocal agreement under which such foreign government does not impose any similar requirement relating to eligibility for employment upon United States citizens.

(3) Any person who holds a FCC radio station license, for the purpose of operating that station.

(4) Notwithstanding any other provision of the FCC's rules, no person shall be eligible to be issued a commercial radio operator license when:

(i) the person's commercial radio operator license is suspended, or

(ii) the person's commercial radio operator license is the subject of an ongoing suspension proceeding, or

(iii) the person is afflicted with complete deafness or complete muteness or complete inability for any other reason to transmit correctly and to receive correctly by telephone spoken messages in English.

(b) (1) Each application for a new General Radiotelephone Operator License, Marine Radio Operator Permit, First Class Radiotelegraph Operator's Certificate, Second Class Radiotelegraph Operator's Certificate, Third Class Radiotelegraph Operator's Certificate, Ship Radar Endorsement, Six Months Service Endorsement, GMDSS Radio Operator's License or GMDSS Radio Maintainer's License must be made on FCC Form 756.

(2) Each application for a Restricted Radiotelephone Operator Permit must be made on FCC form 753.

(3) Each application for a Restricted Radiotelephone Operator Permit-Limited Use must be made on FCC Form 755.

(c) Each application for a new General Radiotelephone Operator License, Marine Radio Operator Permit, First Class Radiotelegraph Operator's Certificate, Second Class Radiotelegraph Operator's Certificate, Third Class Radiotelegraph Operator's Certificate, Ship Radar Endorsement, GMDSS Radio Operator's License or GMDSS Radio Maintainer's License must be accompanied by the required fee, if any, and submitted to the address specified in Part 1 of the rules. The application must include:

(1) An original PPC(s) from a COLEM(s) showing that the applicant has passed the necessary examination element(s) within the previous 365 days, and

(2) A copy of the applicable license(s) when the applicant claims credit for an examination element. The FCC will give credit only as specified below to an applicant holding any of the following documents:

(i) An unexpired (or within the grace period) FCC-issued commercial radio operator license: the written examination and telegraphy Element(s) required to obtain the license held.

(ii) An unexpired (or within the grace period) FCC-issued Amateur Extra Class operator license: Telegraphy Elements 1 and 2.

(d) Each application for a new six month radiotelegraph endorsement must be submitted to the address specified in Part 1 of the rules. The application must include documentation showing that:

(1) The applicant was employed as a radio operator on board a ship or ships of the United States for a period totaling at least six months.

(2) The ships were equipped with a radio station complying with the provisions of Part II of Title III of the Communications Act, or the ships were owned and operated by the United States Government and equipped with radio stations.

(3) The ships were in service during the applicable six-months period and no portion of any in-port period included in the qualifying six months period exceeded seven days.

(4) The applicant held a FCC-issued First or Second Class Radiotelegraph Operator's Certificate during the entire six-months qualifying period and

(5) The applicant holds a radio officer's license issued by the U.S. Coast Guard at the time the six-months endorsement is requested

(e) No person shall alter, duplicate for fraudulent purposes, or fraudulently obtain or attempt to obtain an operator license. No person shall use a license issued to another or a license that he or she knows to be altered, duplicated for fraudulent purposes, or fraudulently obtained. No person shall obtain or attempt to obtain, or assist another person to obtain or attempt to obtain, an operator license by fraudulent means.

§ 13.11 Holding more than one commercial radio operator license.

(a) An eligible person may hold more than one commercial operator license except as follows:

(1) No person may hold two or more unexpired radiotelegraph operator's certificates at the same time;

(2) No person may hold any class of radiotelegraph operator's certificate and a Marine Radio Operator Permit;

(3) No person may hold any class of radiotelegraph operator's certificate and a Restricted Radiotelephone Operator Permit.

(b) Each person who is not legally eligible for employment in the United States, and certain other persons who were issued permits prior to September 13, 1982, may hold two Restricted Radiotelephone Operator Permits simultaneously when each permit authorizes the operation of a particular station or class of stations.

§ 13.13 Application for a renewed or modified license.

(a) Each application to renew a First Class Radiotelegraph Operator's Certificate, Second Class Radiotelegraph Operator's Certificate, Third Class Radiotelegraph Operator's Certificate, GMDSS Radio Operator's License, or GMDSS Radio Maintainer's License must be made on FCC Form 756. The application must be accompanied by the original document or a legible photocopy unless it has been lost, mutilated, or destroyed. If the license has been lost, mutilated, or destroyed, submit a written explanation. The application must be accompanied by the appropriate fee and submitted to the address specified in Part 1 of the rules.

GENERAL RADIOTELEPHONE OPERATOR LICENSE

(b) A licensee may submit an application for renewal of an unexpired license during the last year of the license term. If a license expires, application for renewal may be made during a grace period of five years after the expiration date without having to retake the required examinations. The application must be accompanied by the required fee and submitted to the address specified in Part 1 of the rules. During the grace period, the expired license is not valid. A license renewed during the grace period will be effective as of the date of the renewal. Licensees who fail to renew their license within the grace period must apply for a new license and take the required examination(s).

(c) Each application involving a change in operator class must be made on FCC Form 756. Each application for a commercial operator license involving a change in operator class must be accompanied by the required fee, if any, and submitted to the address specified in Part 1 of the rules. The application must include:

(1) An original PPC(s) from a COLEM(s) showing that the applicant has passed the necessary examination elements within the past 365 days, and

(2) A copy of the applicable license(s) when the applicant claims credit for an examination element. The FCC will give credit only as specified below to an applicant holding any of the following documents:

(i) An unexpired (or within the grace period) FCC-issued commercial radio operator license: the written examination and telegraphy Element(s) required to obtain the license held

(ii) An unexpired (or within the grace period) FCC-issued Amateur Extra Class operator license: Telegraphy Elements 1 and 2.

(d) The holder of a First Class Radiotelegraph Operator's Certificate, Second Class Radiotelegraph Operator's Certificate, Third Class Radiotelegraph Operator's Certificate, General Radiotelephone Operator License, GMDSS Radio Operator's License, or GMDSS Radio Maintainer's License whose name is legally changed may obtain a modified license by filing a FCC Form 756 with a written explanation. The application must be accompanied by the required fee and submitted to the address specified in Part 1 of the rules.

(e) A licensee who has made application for a renewed or modified operator license or permit may exhibit a photocopy of their license in lieu of the original document.

§ 13.15 License Term.

(a) Commercial radio operator licenses are normally valid for a term of five years from the date of issuance, except as provided in paragraph (b) of this section.

(b) General Radiotelephone Operator Licenses, Restricted Radiotelephone Operator Permits, and Restricted Radiotelephone Operator Permits-Limited Use are normally valid for the lifetime of the holder. The terms of all Restricted Radiotelephone Operator Permits issued prior to November 15, 1953, and valid on that date, are extended to the lifetime of the operator.

§ 13.17 Replacement license.

(a) Each licensee or permittee whose original document is lost, mutilated, or destroyed must request a replacement. The application must be accompanied by the required fee and submitted to the address specified in Part 1 of the rules.

(b) Each application for a replacement General Radiotelephone Operator License, Marine Radio Operator Permit, First Class Radiotelegraph Operator's Certificate, Second Class Radiotelegraph Operator's Certificate, Third Class Radiotelegraph Operator's Certificate, GMDSS Radio Operator's License, GMDSS Radio Maintainer's License, must be made on FCC Form 756 and must include a written explanation as to the circumstances involved in the loss, mutilation, or destruction of the original document.

(c) Each application for a replacement Restricted Radiotelephone Operator Permit must be on FCC Form 753.

(d) Each application for a replacement Restricted Radiotelephone operator Permit-Limited Use must be on FCC Form 755.

(e) A licensee who has made application for a replacement license may exhibit a copy of the application submitted to the FCC or a photocopy of the license in lieu of the original document.

§ 13.19 Operator's responsibility.

(a) The operator responsible for maintenance of a transmitter may permit other persons to adjust that transmitter in the operator's presence for the purpose of carrying out tests or making adjustments requiring specialized knowledge or skill, provided that he or she shall not be relieved thereby from responsibility for the proper operation of the equipment.

(b) In every case where a station operating log or service and maintenance log is required, the operator responsible for the station operation or maintenance shall make the required entries in the station log. If no station log is required, the operator responsible for the service or maintenance duties which may affect the proper operation of the station shall sign and date an entry in the station maintenance records giving:

(1) Pertinent details of all service and maintenance work performed by the operator or conducted under his or her supervision;

(2) His or her name and address; and

(3) The class, serial number and expiration date of the license:

(c) When the operator is on duty and in charge of transmitting systems, or performing service, maintenance or inspection functions, the license or permit document, or a photocopy thereof, must be posted or in the operator's personal possession, and available for inspection upon request by a FCC representative.

(d) The operator on duty and in charge of transmitting systems, or performing service, maintenance or inspection functions, shall not be subject to the requirements of paragraph (b) of this section at a station, or stations of one licensee at a single location, at which the operator is regularly employed and at which his or her license, or a photocopy, is posted.

EXAMINATION SYSTEM

§ 13.201 Qualifying for a commercial operator license or endorsement.

(a) To be qualified to hold any commercial radio operator license, an applicant must have a satisfactory knowledge of FCC rules and must have the ability to send correctly and receive correctly spoken messages in the English language.

(b) An applicant must pass an examination for the issuance of a new commercial radio operator license, other than the Restricted Radiotelephone Operator Permit and the Restricted Radiotelephone Operator Permit-Limited Use, and for each change in operator class. An applicant must pass an examination for the issuance of a new Ship Radar Endorsement. Each applicant for the class of license or endorsement specified below must pass, or otherwise receive credit for, the corresponding examination elements:

(1) First Class Radiotelegraph Operator's Certificate.
 (i) Telegraphy Elements 3 & 4;
 (ii) Written Elements 1, 5, and 6:
 (iii) Applicant must be at least 21 years old;
 (iv) Applicant must have one year of experience in sending and receiving public correspondence by radiotelegraph at a public coast station, a ship station, or both.

(2) Second Class Radiotelegraph Operator's Certificate.
 (i) Telegraphy elements 1 and 2;
 (ii) Written Elements 1, 5, and 6.

(3) Third Class Radiotelegraph Operator's Certificate.
 (i) Telegraphy Elements 1 and 2;
 (ii) Written Elements 1 and 5.

(4) General Radiotelephone Operator License: Written Elements 1 and 3.

(5) Marine Radio Operator Permit: Written Element 1.

(6) GMDSS Radio Operator's License: Written Elements 1 and 7.

(7) GMDSS Radio Maintainer's License: Written Elements 1, 3, and 9.

(8) Ship Radar Endorsement: Written Element 8.

§ 13.203 Examination elements.

(a) A written examination (written Element) must prove that the examinee possesses the operational and technical qualifications to perform the duties required by a person holding that class of commercial radio operator license. Each written examination must be comprised of a question set as follows:

(1) Element 1 (formerly Elements 1 and 2): Basic radio law and operating practice with which every maritime radio operator should be familiar. 24 questions concerning provisions of laws, treaties, regulations, and operating procedures and practices generally followed or required in communicating by means of radiotelephone stations. The minimum passing score is 18 questions answered correctly.

(2) Element 3: General Radio telephone. 76 questions concerning electronic fundamentals and techniques required to adjust, repair, and maintain radio transmitters and receivers at stations licensed by the FCC in the aviation, maritime, and international fixed public radio services. The minimum passing score is 57 questions answered correctly.

(3) Element 5: Radiotelegraph operating practice. 50 questions concerning radio operating procedures and practices generally followed or required in communicating by means of radiotelegraph stations primarily other than in the maritime mobile services of public correspondence. The minimum passing score is 38 questions answered correctly.

(4) Element 6: Advanced radiotelegraph. 100 questions concerning technical, legal and other matters applicable to the operation of all classes of radiotelegraph stations, including operating procedures and practices in the maritime mobile services of public correspondence, and associated matters such as radio navigational aids, message traffic routing and accounting, etc. The minimum passing score is 75 questions answered correctly.

GENERAL RADIOTELEPHONE OPERATOR LICENSE

(5) Element 7: GMDSS radio operating practices. 76 questions concerning GMDSS radio operating procedures and practice sufficient to show detailed practical knowledge of the operation of all GMDSS subsystems and equipment; ability to send and receive correctly by radio telephone and narrow-band direct-printing telegraphy; detailed knowledge of the regulations applying to direct-printing telegraphy; detailed knowledge of the regulations applying to direct-printing telegraphy; detailed knowledge of the regulations applying to radio communications, knowledge of the documents relating to charges for radio communications and knowledge of those provisions of the International Convention for the Safety of Life at Sea which relate to radio; sufficient knowledge of English to be able to express oneself satisfactorily both orally and in writing; knowledge of and ability to perform each function listed in Section 80.1081; and knowledge covering the requirements set forth in IMO Assembly Resolution on Training for Radio Personnel (GMDSS), Annex 3. The minimum passing score is 57 questions answered correctly.

(6) Element 8: Ship radar techniques. 50 questions concerning specialized theory and practice applicable to the proper installation, servicing and maintenance of ship radar equipment in general use for marine navigational purposes. The minimum passing score is 38 questions answered correctly.

(7) Element 9: GMDSS radio maintenance practices and procedures. 50 questions concerning the requirements set forth in IMO Assembly on Training for Radio Personnel (GMDSS), Annex 5 and IMO Assembly on Radio Maintenance Guidelines for the Global Maritime Distress and Safety System related to Sea Areas A3 and A4. The minimum passing score is 38 questions answered correctly.

(b) A telegraphy examination (telegraphy Elements) must prove that the examinee has the ability to send correctly by hand and to receive correctly by ear texts in the international Morse code at not less than the prescribed speed, using all the letters of the alphabet, numerals 0-9, period, comma, question mark, slant mark, and prosigns $\overline{AR}$, $\overline{BT}$ and $\overline{SK}$.

(1) Telegraphy Element 1: 16 code groups (CG) per minute.

(2) Telegraphy Element 2: 20 Plain Language (PL) words per minute.

(3) Telegraphy Element 3: 20 code groups (CG) per minute.

(4) Telegraphy Element 4: 25 Plain Language (PL) words per minute.

§ 13.207 Preparing an examination.

(a) Each telegraphy message and each written question set administered to an examinee for a commercial radio operator license must be provided by a COLEM.

(b) Each question set administered to an examinee must utilize questions taken from the applicable Element question pool. The COLEM may obtain the written question sets from a supplier or other COLEM.

(c) A telegraphy examination must consist of a plain language text or code group message sent in the international Morse code at no less than the prescribed speed for a minimum of five minutes. The message must contain each required telegraphy character at least once. No message known to the examinee may be administered in a telegraphy examination. Each five letters of the alphabet must be counted as one word or one code group. Each numeral, punctuation mark, and prosign must be counted as two letters of the alphabet. The COLEM may obtain the telegraphy message from a supplier or other COLEM.

§ 13.209 Examination procedures.

(a) Each examination for a commercial radio operator license must be administered at a location and a time specified by the COLEM. The COLEM is responsible for the proper conduct and necessary supervision of each examination. The COLEM must immediately terminate the examination upon failure of the examinee to comply with its instructions.

(b) Each examinee, when taking an examination for a commercial radio operator license, shall comply with the instructions of the COLEM.

(c) No examination that has been compromised shall be administered to any examinee. Neither the same telegraphy message nor the same question set may be re-administered to the same examinee.

(d) Passing a telegraphy examination.

(1) To pass a receiving telegraphy examination, an examinee is required to receive correctly the message by ear, for a period of 1 minute without error at the rate of speed specified in Section 13.203 for the class of license sought.

(2) To pass a sending telegraphy examination, an examinee is required to send correctly for a period of 1 minute at the rate of speed prescribed in Section 13.203 (b) for the class of license sought.

(e) Passing a telegraphy receiving examination is adequate proof of an examinee's ability to both send and receive telegraphy. The COLEM, however, may also include a sending segment in a telegraphy examination.

APPENDIX – PART 13

(f) The COLEM is responsible for determining the correctness of the examinee's answers. When the examinee does not score a passing grade on an examination element, the COLEM must inform the examinee of the grade.

(g) When the examinee is credited for all examination elements required for the commercial operator license sought, the examinee may apply to the FCC for the license.

(h) No applicant who is eligible to apply for any commercial radio operator license shall, by reason of any physical handicap, be denied the privilege of applying and being permitted to attempt to prove his or her qualification (by examination if examination is required) for such commercial radio operator license in accordance with procedures established by the COLEM.

(i) The COLEM must accommodate an examinee whose physical disabilities require a special examination procedure. The COLEM may require a physician's certification indicating the nature of the disability before determining which, if any, special procedures are appropriate to use. In the case of a blind examinee, the examination questions may be read aloud and the examinee may answer orally. A blind examinee wishing to use this procedure must make arrangements with the COLEM prior to the date the examination is desired.

(j) The FCC may:

(1) Administer any examination element itself.

(2) Readminister any examination element previously administered by a COLEM, either itself or by designating another COLEM to readminister the examination element.

(3) Cancel the commercial operator license(s) of any licensee who fails to appear for re-administration of an examination when directed by the FCC, or who fails any required element that is re-administered. In case of such cancellation, the person will be issued an operator license consistent with completed examination elements that have not been invalidated by not appearing for, or by failing, the examination upon readministration.

§ 13.211 Commercial radio operator license examination.

(a) Each session where an examination for a commercial radio operator license is administered must be managed by a COLEM or the FCC.

(b) Each examination for a commercial radio operator license must be administered as determined by the COLEM.

(c) The COLEM may limit the number of candidates at any examination.

(d) The COLEM may prohibit from the examination area items the COLEM determines could compromise the integrity of an examination or distract examinees.

(e) Within 10 days of completion of the examination element (s), the COLEM must provide the results of the examination to the examinee and the COLEM must issue a PPC to an examinee who scores a passing grade on an examination element.

(f) A PPC is valid for 365 days from the date it is issued.

§ 13.213 COLEM qualifications.

No entity may serve as a COLEM unless it has entered into a written agreement with the FCC. In order to be eligible to be a COLEM, then entity must:

(a) Agree to abide by the terms of the agreement;

(b) Be capable to serving as a COLEM;

(c) Agree to coordinate examinations for one or more types of commercial radio operator licenses and/or endorsements;

(d) Agree to assure that, for any examination, every examinee eligible under these rules is registered without regard to race, sex, religion, national origin or membership (or lack thereof) in any organization;

(e) Agree to make any examination records available to the FCC, upon request.

(f) Agree not to administer an examination to an employee, relative, or relative of an employee.

§ 13.215 Question pools.

The question pool for each written examination element will be composed of questions acceptable to the FCC. Each question pool must contain at least 5 times the number of questions required for a single examination. The FCC will issue public announcements detailing the questions in the pool for each element. COLEMs must use only the most recent question pool made available to the public when preparing a question set for a written examination element.

§ 13.217 Records.

Each COLEM recovering fees from examinees must maintain records of expenses and revenues, frequency of examinations administered, and examination pass rates. Records must cover the period from January 1 to December 31 of the preceding year and must be submitted as directed by the Commission. Each COLEM must retain records for 1 year and the records must be made available to the FCC upon request.

GENERAL RADIOTELEPHONE OPERATOR LICENSE

Federal Communications Commission - Rules and Regulations – Excerpts from

47 C.F.R. – PART 23 – INTERNATIONAL FIXED PUBLIC RADIOCOMMUNICATION SERVICES

§ 23.1 Definitions.

Assigned frequency. The frequency coinciding with the center of an authorized bandwidth of emission.

Authorized bandwidth. The maximum bandwidth authorized to be used by a station as specified in the station license. This shall be occupied bandwidth or necessary bandwidth, whichever is greater.

Authorized reference frequency. A frequency having a fixed and specific position with respect to the assigned frequency.

Authorized service. The transmission of public correspondence to a point of communication subject to such special provisions as may be contained in the license of the station.

Fixed public service. A radiocommunication service carried on between fixed stations open to public correspondence.

Fixed public press service. A limited radiocommunication service carried on between point-to-point telegraph stations, consisting of transmissions by fixed stations open to limited public correspondence, of news items, or other material related to or intended for publication by press agencies, newspapers, or for public dissemination.

Fixed station. Includes all apparatus used in rendering the authorized service at a particular location under a single instrument of authorization.

Frequency tolerance. The maximum permissible departure by the center frequency of the frequency band occupied by an emission from the assigned frequency or by the carrier, or suppressed carrier, from the reference frequency.

International fixed public radiocommunication service. A fixed service, the stations of which are open to public correspondence and which, in general, is intended to provide radiocommunication between any one of the 50 states or any other U.S. possession and any other point. Communications solely between Alaska, or any one of the contiguous 48 states (including the District of Columbia), and either Canada or Mexico are not deemed to be in the international fixed public radiocommunication service when such radiocommunications are transmitted on frequencies above 72 MHz.

International fixed public control service. A fixed service carried on for the purpose of communicating between transmitting stations, receiving stations, message centers or control points in the international fixed public radiocommunication service.

Occupied bandwidth. The frequency bandwidth such, that, below and above its upper frequency limits, the mean powers radiated are each equal to 0.5 percent of the total mean power radiated by a given emission.

Point-to-point telegraph station. A fixed station authorized for radiotelegraph communication.

Point-to-point telephone station. A fixed station authorized for radiotelephone communication.

Point of communication. A specific location designated in the license to which a station is authorized to communicate for the transmission of public correspondence.

Radiotelegraph. Includes types NØN, A1A, A2A, A3C, F1B, F2B, and F3C emission.

Radiotelephone. With respect to operation on frequencies below 30 MHz, means a system of radiocommunication for the transmission of speech or, in some cases, other sounds by means of amplitude modulation including double sideband (A3E), single sideband (R3E, H3E, J3E) or independent sideband (B3E) transmission.

§ 23.13 Types of emission.

Stations in the international fixed public radiocommunication services may be authorized to use any of the types of emission or combinations thereof - as well as new types which may be developed: *Provided,* That harmful interference to adjacent operations in not caused, *And provided further,* That the intelligence to be transmitted will use the bandwidth requested to a degree of efficiency compatible with the current state of the art.

§ 23.18 Authorization of power.

(a) *Authorized power.* ...is peak-envelope-power for transmitters having full, unkeyed carrier, single sideband or independent sideband emissions and mean power for transmitters having other emissions.

(b) *Use of minimum power.* In the interest of avoiding interference to other operations, all stations shall radiate only as much power as is necessary to ensure a satisfactory service.

§ 23.19 Use of directional antennas

Insofar as practicable, directional antennas, of type consistent with the current state of art, shall be used on all circuits for both transmitting and receiving.

§ 20.20 Assignment of frequencies

(a) Only those frequencies which are in accordance with the international and United States Table of Allocations (§2.106) may be authorized for use by stations in the Fixed Public and Fixed Public Press Services. Selection of specific frequencies within such bands shall be made by the applicant. Its availability for assignment as requested will be determined by a study of the probabilities of interference to and from existing services.

§23.24 Correspondents and points of communication.

Each instrument of authorization issued for fixed public or fixed public press service shall authorize communication to the points of communication and to the organizations, agencies, or persons specified therein only.

§ 23.29 License period and expiration time.

Licenses for stations operating in the fixed public radiocommunications services will be issued for a period of 10 years unless otherwise stated in the instrument of authorization.

§ 23.37 Station identification.

Every radiotelegraph or radiotelephone station in the International Fixed Public or Fixed Public Press Service shall transmit the identifying call sign or other approved identification signal on each of its assigned frequencies below 30 MHz on which energy is being transmitted. The call sign shall be transmitted at the beginning and end of each period of use of the frequency.

§ 23.46 Operators, class required and general duties.

(a) The operation and control of all transmitting apparatus licensed at a station in the international fixed public radiocommunication services shall be carried on only by a person holding a valid operator license issued by the Commission:

(b) Classes of operator licenses required:

(1) Radiotelegraph stations: Radiotelegraph first or second class license except;

(i) A Third Class Radiotelegraph Operator's Certificate or higher is required if manual Morse code keying is used for the purposes of identification or for sending service messages and no public correspondence is transmitted.

(2) Radiotelephone stations: General Radiotelephone Operator License.

§ 23.47 Station records.

(a) Station records shall be kept in an orderly manner, and in such detail that the data required is readily available.

§ 23.54 Use of double sideband radiotelephone

Use of double sideband radiotelephone transmissions, on frequencies below 30 MHz, shall not be transmitted except in cases where the foreign correspondent is unable to receive single sideband.

Federal Communications Commission - Rules and Regulations – Excerpts from

47 C.F.R. – PART 73 – RADIO BROADCAST SERVICES

Subpart H – Rules Applicable to All Broadcast Stations

§ 73.1860 Transmitter duty operators.

(a) Each AM, FM or TV broadcast station must have at least one person holding a commercial radio operator license or permit (any class unless otherwise endorsed) on duty in charge of the transmitter during all periods of broadcast operation. The operator must be on duty at the transmitter location, a remote control point, an ATS monitor and alarm point, or a position where extension meters are installed.

(b) The transmitter operator must be able to observe the required transmitter and monitor metering to determine deviations from normal indications.

§ 73.1870 Chief operators.

(a) The licensee of each AM, FM, or TV broadcast station must designate a person holding a commercial radio operator license or permit (any class unless endorsed) to serve as the station's chief operator. At times when the chief operator is unavailable or unable to act (e.g., vacations, sickness), the licensee shall designate another licensed operator as the acting chief operator on a temporary basis.

(b)(3) The designation of the chief operator must be in writing with a copy of the designation posted with the operator license.

GENERAL RADIOTELEPHONE OPERATOR LICENSE

Federal Communications Commission - Rules and Regulations – Excerpts from
47 C.F.R. – PART 80 – STATIONS IN THE MARITIME SERVICES

Subpart A - General Information
§80.5 Definitions.

Alaska-public fixed station. A fixed station in Alaska which is open to public correspondence and is licensed by the Commission for radio communication with Alaska-Private fixed stations on paired channels.

Alaska-private fixed station. A fixed station in Alaska which is licensed by the Commission for radio communication within Alaska and with associated ship stations, on single frequency channels. Alaska-private fixed stations are also eligible to communicate with Alaska-public fixed stations on paired channels.

Associated ship unit. A portable VHF transmitter for use in the vicinity of the ship station with which it is associated.

Automated maritime telecommunications system (AMTS). An automatic, integrated and interconnected maritime communications system.

Automated mutual-assistance vessel rescue system (AMVER). An international system, operated by the U.S. Coast Guard, which provides aid to the development and coordination of search and rescue (SAR) efforts. Data is made available to recognized SAR agencies or vessels of any nation for reasons related to marine safety.

Bridge-to-bridge station. A radio station located on a ship's navigational bridge or main control station operating on a specified frequency which is used only for navigational communications, in the 156-162 MHz band.

Cargo ship safety radiotelegraphy certificate. A certificate issued after an inspection of a cargo ship radiotelegraph station which complies with the applicable Safety Convention radio requirements.

Cargo ship safety radiotelephony certificate. A certificate issued after inspection of a cargo ship radiotelephone station which complies with the applicable Safety Convention radio requirements.

Categories of ships. (1) When referenced in Part II of Title III of the Communications Act or the radio provisions of the Safety Convention, a ship is a *passenger ship* if it carries or is licensed or certificated to carry more than twelve passengers. A *cargo ship* is any ship not a passenger ship.

(2) A *commercial transport vessel* is any ship which is used primarily in commerce (i) for transporting persons or goods to or from any harbor(s) or port(s) or between places within a harbor or port area, or (ii) in connection with the construction, change in construction, servicing, maintenance, repair, loading, unloading, movement, piloting. or salvaging of any other ship or vessel.

(3) The term *passenger carrying vessel,* when used in reference to Part III, Title III of the Communications Act of the Great Lakes Radio Agreement, means any ship transporting more than six passengers for hire.

(4) *Power-driven vessel.* Any ship propelled by machinery.

(5) *Towing vessel.* Any commercial ship engaged in towing another ship astern, alongside or by pushing ahead.

(6) *Compulsory ship.* Any ship which is required to be equipped with radiotelecommunication equipment in order to comply with the radio or radio-navigation provisions of a treaty or statute to which the vessel is subject.

(7) *Voluntary ship.* Any ship which is not required by treaty or statute to be equipped with radiotelecommunications equipment.

Coast station. A land station in the maritime mobile service.

Commercial communications. Communications between coast stations and ship stations aboard commercial transport vessels, or between ship stations aboard commercial transport vessels, which relate directly to the purposes for which the ship is used including the piloting of vessels, movements of vessels, obtaining vessel supplies, and scheduling of repairs.

Day. (1) Where the word day is applied to the use of a specific frequency assignment or to a specific authorized transmitter power, its use means transmission on the frequency assignment or with the authorized transmitter power during that period of time included between one hour after local sunrise and one hour before local sunset.

(2) Where the word day occurs in reference to watch requirements, or to equipment testing, its use means the calendar day, from midnight to midnight, local time.

Digital selective calling (DSC). A synchronous system developed by the International Radio Consultative Committee (CCIR), used to establish contact with a station or group of stations automatically by means of radio. The operational and technical characteristics of this system are contained in CCIR Recommendation 493.

Direction finder (radio compass). Apparatus capable of receiving radio signals and taking bearings on these signals from which the true bearing and direction of the point of origin may be determined.

Distress signal. The distress signal is an internationally recognized radiotelegraph or radiotelephone transmission which indicates that a ship, aircraft, or other vehicle is threatened by grave and imminent danger and requests immediate assistance.

(1) In radiotelegraphy, the international distress signal consists of the group "three dots, three dashes, three dots", transmitted as a single signal in which the dashes are emphasized so as to be distinguished clearly from the dots.

(2) In radiotelephony, the international distress signal consists of the enunciation of the word "Mayday", pronounced as the French expression "m'aider". In case of distress, transmission of this particular signal is intended to ensure recognition of a radiotelephone distress call by stations of any nationality,

Distress traffic. All messages relative to the immediate assistance required by a ship, aircraft, or other vehicle in distress.

Emergency position indicating radiobeacon (EPIRB) station. A station in the maritime mobile service the emissions of which are intended to facilitate search and rescue operations.

Environmental communications. Broadcasts of information about the environmental conditions in which vessels operate, i.e., weather, sea conditions, time signals adequate for practical navigation, notices to mariners, and hazards to navigation.

Fleet radio station license. An authorization issued by the Commission for two or more ships having a common owner or operator.

Future global maritime distress and safety system (FGMDSS). An International Maritime Organization (IMO) worldwide coordinated maritime distress system designed to provide the rapid transfer of distress messages from vessels in distress to units best suited for giving or coordinating assistance. The system includes standardized equipment and operational procedures, unique identifiers for each station, and the integrated use of frequency bands and radio systems to ensure the transmission and reception of distress and safety calls and messages at short, medium and long ranges.

Great Lakes. This term, used in this part in reference to the Great Lakes Radio Agreement, means all of Lakes Ontario, Erie, Huron (including Georgian Bay), Michigan, Superior, their connecting and tributary waters and the St. Lawrence River as far east as the lower exit of the St. Lambert Lock as Montreal in the Province of Quebec, Canada, but does not include any connecting and tributary waters other than: the St. Marys River, the St. Clair River, Lake St. Clair, the Detroit River and the Welland Canal.

Harbor or port. Any place to which ships may resort for shelter, or to load or unload passengers or goods, or to obtain fuel, water, or supplies. This term applies to such places whether proclaimed public or not and whether natural or artificial.

Inland waters. This term, as used in reference to waters of the United States, its territories and possessions, means waters that lie landward of the boundary lines of inland waters as contained in 33 CFR Part 82, as well as waters within its land territory, such as rivers and lakes, over which the United States exercises sovereignty.

Marine utility station. A station in the maritime mobile service consisting of one or more handheld radiotelephone units licensed under a single authorization. Each unit is capable of operation while being hand-carried by an individual. The station operates under the rules applicable to ship stations when the unit is aboard a vessel, and under the rules applicable to private coast stations when the unit is on land.

Maritime control communications. Communications between private coast and ship stations or between ship stations licensed to a state or local governmental entity, which relate directly to the control of boating activities or assistance to ships.

Maritime mobile repeater station. A land station at a fixed location established for the automatic retransmission of signals to extend the range of communication of ship and coast stations.

Maritime mobile-satellite service. A mobile-satellite service in which mobile earth stations are located on board ships. Survival craft stations and EPIRB stations may also participate in this service.

Maritime mobile service. A mobile service between coast stations and ship stations, or between ship stations, or between associated on-board communication stations. Survival craft stations and EPIRB stations also participate in this service.

Maritime mobile service identities. An international system for the identification of radio stations in the maritime mobile service. The system is comprised of a series of nine digits which are transmitted over the radio path to uniquely identify ship stations, ship earth stations, coast stations, coast earth stations and groups of stations.

287

Maritime radiodetermination service. A maritime radiocommunication service for determining the position, velocity, and/or other characteristics of an object, or the obtaining of infor-mation relating to these parameters, by the propagation properties of radio waves.

Maritime support station. A station on land used in support of the maritime services to train personnel and to demonstrate, test and maintain equipment.

Navigational communications. Safety communications pertaining to the maneuvering of vessels or the directing of vessel movements. Such communications are primarily for the exchange of information between ship stations and secondary between ship stations and coast stations.

Noncommercial communications. Communication between coast stations and ship stations other than commercial transport ships, or between ship stations aboard other than commercial transport ships which pertain to the needs of the ship.

Non-selectable transponder. A transponder whose coded response is displayed on any conventional radar operating in the appropriate band.

On-board communication station. A low-powered mobile station in the maritime mobile service intended for use for internal communications on board a ship, or between a ship and its lifeboat and liferafts during lifeboat drills or operations, or for communication within a group of vessels being towed or pushed, as well as for line handling and mooring instructions.

On-board repeater. A radio station that receives and automatically retransmits signals between on-board communication stations.

Open sea. The water area of the open coast seaward of the ordinary low-water mark, or seaward of inland waters.

Operational fixed station. A fixed station, not open to public correspondence, operated by entities that provide their own radiocommunication facilities in the private land mobile, maritime or aviation services.

Passenger ship safety certificate. A certificate issued by the Commandant of the Coast Guard after inspection of a passenger ship which complies with the requirements of the Safety Convention.

Pilot. Pilot means a Federal pilot required by 46 U.S.C. 764, a state pilot required under the authority of 46 U.S.C. 211, or a registered pilot required by 46 U.S.C. 216.

Port operations communications. Communications in or near a port, in locks or in waterways between coast stations and ship stations or between ship stations, which relate to the operational handling, movement and safety of ships and in emergency to the safety of persons.

Portable ship station. A ship station which includes a single transmitter intended for use upon two or more ships.

Private coast station. A coast station, not open to public correspondence, which serves the operational, maritime control and business needs of ships.

Public coast station. A coast station that offers radio communication common carrier services to ship radio stations.

Public correspondence. Any telecommunication which the offices and stations must, by reason of their being at the disposal of the public, accept for transmission.

Radar beacon (RACON). A receiver-transmitter which, when triggered by a radar, automatically returns a distinctive signal which can appear on the display of the triggering radar, providing range, bearing and identification information.

Radioprinter operations. Communications by means of a direct printing radiotelegraphy system using any alphanumeric code, within specified bandwidth limitations, which is authorized for use between private coast stations and their associated ship stations on vessels of less than 1600 gross tons.

Safety communication. The transmission or reception of distress, alarm, urgency, or safety signals, or any communication preceded by one of these signals, or any form of radiocommunication which, if delayed in transmission or reception, may adversely affect the safety of life or property.

Safety signal. (1) The safety signal is the international radiotelegraph or radiotelephone signal which indicates that the station sending this signal is preparing to transmit a message concerning the safety of navigation or giving important meteorological warnings.

(2) In radiotelegraphy, the international safety signals consists of three repetitions of the group "TTT", sent before the call, with the letters of each group and the successive groups clearly separated from each other.

(3) In radiotelephony, the international safety signal consists of three oral repetitions of "Security", pronounced as the French word "Securite", sent before the call.

Selectable transponder. A transponder whose coded response may be inhibited or displayed on a radar on demand by the operator of that radar.

Selective calling. A means of calling in which signals are transmitted in accordance with a prearranged code to operate a particular automatic attention device at the station whose attention is sought.

Ship earth station. A mobile earth station in the maritime mobile-satellite service located on board ship.

Ship or vessel. Ship or vessel includes every description of watercraft or other artificial contrivance, except aircraft, capable of being used as a means of transportation on water whether or not it is actually afloat.

Ship radio station license. An authorization issued by the Commission to operate a radio station onboard a vessel.

Ship station. A mobile station in the maritime mobile service located on-board a vessel which is not permanently moored, other than a survival craft station.

Station. One or more transmitters or a combination of transmitters and receivers, including the accessory equipment, necessary at one location for carrying on radiocommunication services.

Survival craft station. A mobile station in the maritime or aeronautical mobile service intended solely for survival purposes and located on any lifeboat, liferaft or other survival equipment.

Urgency signal. (1) The urgency signal is the international radiotelegraph or radiotelephone signal which indicates that the calling station has a very urgent message to transmit concerning the safety of a ship, aircraft, or other vehicle, or of some person on board or within sight.

(2) In radiotelegraphy, the international urgency signal consists of three repetitions of the group "XXX", sent before the call, with the letters of each group and the successive groups clearly separated from each other.

(3) In radiotelephony, the international urgency signal consists of three oral repetitions of the group of words "PAN PAN", each word of the group pronounced as the French word "PANNE" and sent before the call.

Vessel traffic service (VTS). A U.S. Coast Guard traffic control service for ships in designated water areas to prevent collisions, groundings and environmental harm.

Watch. The act of listening on a designated frequency.

Subpart B - Applications and Licenses

§ 80.13 Station license required.

(a) All stations in the maritime services must be licensed by the FCC.

(b) One ship station license will be granted for operation of all maritime services transmitting equipment on board a vessel.

§ 80.15 Eligibility for station license.

(a) *General.* A station license cannot be granted to or held by a foreign government or its representative.

(b) *Public coast stations and Alaska public fixed stations.* A station license for a public coast station or an Alaska public fixed station cannot be granted to or held by:

(1) Any alien or the representative of any alien;

(c) *Private coast and marine utility stations.* The supplemental eligibility requirements for private coast and marine utility stations are contained in § 80.501(a).

(d) *Ship stations.* A ship station license may only be granted to:

(1) The owner or operator of the vessel;

(2) A subsidiary communications corporation of the owner or operator of to vessel;

(3) A State or local government subdivision; or

(4) Any agency of the U.S. Government subject to section 301 of the Communications Act.

(e) EPIRB stations. (1) Class C EPIRB stations will be authorized:

(i) For use on board vessels operating within 32 kilometers (approximately 20 miles) of shore and in the Great Lakes, or

(ii) On passenger and cargo vessels with survival craft as required or recommended by the U.S. Coast Guard.

(2) Class A or B EPIRB stations will be authorized for use on board the following types of vessels:

(i) Vessels authorized to carry survival craft; or

(ii) Vessels expected to travel in waters beyond the range of marine VHF distress coverage which is generally considered to be more than 32 kilometers (approximately 20 miles) offshore; or

(iii) Vessels required to be fitted with EPIRB's to comply with U.S. Coast Guard regulations.

(3) A 406.025 MHz EPIRBs may be used by any ship required by U.S. Coast Guard regulations to carry an EPIRB or by any ship that is equipped with a VHF ship radio station.

§ 80.17 Administrative classes of stations.

(a) Stations in the Maritime Mobile Service are licensed according to class of station as follows:

(1) *Public coast stations.*

(2) *Private coast stations.*

(3) *Maritime support stations.*

(4) *Ship stations.* The ship station license may include authority to operate other radio station classes aboard ship such as; radio-navigation, onboard, satellite, EPIRB, radio-telephone, radiotelegraph and survival craft.

(5) Marine utility stations.
(b) Stations on land in the Maritime Radiodetermination Service are licensed according to class of station as follows:
(1) Shore radiolocation stations.
(2) Shore radionavigation stations.
(c) Fixed stations in the Fixed Service associated with the maritime services are licensed as follows:
(1) Operational fixed stations.
(2) Alaska-public fixed stations.
(3) Alaska-private fixed stations.

§ 80.23 Filing of applications.

Rules about the filing of applications for radio station licenses are contained in this section.

(a) Each application must specify an address in the United States to be used by the Commission in serving documents or directing correspondence to the licensee.

(b) An original of each application must be filed.

(c) Each application must be filed with the Federal Communications Commission, Gettysburg, PA 17326 unless otherwise noted on the application form. Applications requiring fees as set forth at part 1, subpart G of this chapter must be filed in accordance with § 0.401(b) of the rules.

(d) One application for two or more new maritime utility stations may be submitted when the applicant and proposed area of operation for each station is the same.

(e) One application for transfer of control may be submitted for two or more stations subject to this part when the individual stations are clearly identified and the following elements are the same for all existing or requested station authorizations involved;
(1) Applicant;
(2) Specific details of basic request.

§ 80.25 License term.

(a) Licenses for stations in the maritime services will normally be issued for a term of five years from the date of original issuance, major modification, or renewal.

§80.29 Changes during license term.

(a) The following table indicates the required action for changes made during the license term:

Type of change	Required action
Mailing address	Written notice to the Commission
Name of licensee (without change in ownership, control or corporate structure).	Written notice to the Commission
Name of the vessel	Written notice to the Commission.

§ 80.31 Cancellation of license.

When a station subject to this part which is not a communication common carrier permanently discontinues operation, the licensee must return the station license to the Commission's office at 1270 Fairfield Rd., Gettysburg, PA 17325, for cancellation.

§ 80.43 Equipment acceptable for licensing.

Transmitters listed in § 80.203 must be type accepted for a particular use by the Commission based upon technical requirements contained in subparts E and F of this part.

§ 80.45 Frequencies.

When an application is submitted on FCC Form 503, the applicant must propose frequencies to be used by the station. The applicant must ensure that frequencies requested are consistent with the applicant's eligibility, the proposed class of station operation and the frequencies available for assignment as contained in subpart H of this part.

§ 80.47 Operation during emergency.

A station may be used for emergency communications when normal communication facilities are disrupted. The Commission may order the discontinuance of any such emergency communication service.

§ 80.53 Application for a portable ship station license.

(a) The Commission may grant a license permitting operation of a portable ship station aboard different vessels of the United States. Each application for a portable ship station must include a showing that:

(1) The station will be operated as an established class of station on board ship, and

(2) A station license for portable equipment is necessary to eliminate frequent application to operate a ship station on board different vessels.

§ 80.54 Automated Maritime Telecommunications System (AMTS) System Licensing.

AMTS licensees will be issued blanket authority for a system of coast stations and mobile units (subscribers). AMTS applicants will specify the maximum number of mobile units to be placed in operation during the license period.

§ 80.55 Application for a fleet station license.

(a) An applicant may apply for licenses for two or more radiotelephone stations aboard different vessels on the same application. Under these circumstances a fleet station license may be issued for operation of all radio stations aboard the vessels in the fleet.

(b) The fleet station license is issued on the following conditions:

(1) The licensee must keep a current list of vessel names and registration numbers authorized by the fleet license;

(2) The vessels do not engage in voyages to any foreign country;

(3) The vessels are not subject to the radio requirements of the Communications Act or the Safety Convention.

§ 80.56 Transfer of ship station license prohibited.

A ship station license may not be assigned. Whenever the vessel ownership is transferred, the previous authorization must be forwarded to the Commission for cancellation. The new owner must file for a new authorization.

§ 80.59 Compulsory ship stations.

(a) *Application for inspection and certification.* An application for inspection and certification must be submitted to the Engineer in Charge of the FCC District Office nearest the proposed place of inspection on one of the following forms at least three days before the proposed inspection date:

(1) FCC Form 801 must be used to apply for a ship radio inspection on board ships subject to Part II or III of Title III of the Communications Act, the Safety Convention or the Great Lakes Radio Agreement. In addition, FCC Form 801 must be used to apply for inspections of bridge-to-bridge radio stations on board vessels subject to the Vessel Bridge-to-Bridge Radiotelephone Act when they are additionally subject to any of the laws and treaties mentioned in the previous sentence.

(2) FCC Form 808 must be used to apply for a ship radio inspection on board ships subject to Part II of Title III of the Communications Act or the Great Lake Radio Agreement on a Sunday or national holiday or during other than established working hours on any other day.

(b) *Responsibilities.* Applicants for a ship radio inspection subject to Part II or III of Title III of the Communications Act, the Safety Convention, or the Great Lakes Radio Agreement must ensure that a licensed radio operator of the required class and endorsements, and sufficient personnel to lower and raise antennas and to launch any radio equipped survival craft are available on the ship at the time of inspection. The radio operator provided must be either a regularly assigned radio operator or a service representative.

(c) *Application for exemption.* FCC Form 820 must be used to apply for exemption from the radio provisions of Part II or III of Title III of the Communications Act, the Safety Convention, or the Great Lakes Radio Agreement, or for modification or renewal of an exemption previously granted. Applications for exemptions must be submitted to the Secretary, Federal Communications Commission, Washington, DC 20554. In cases of emergency, the Commission may consider an informal application which includes the full information normally furnished on the formal application.

(d) *Temporary waiver of annual inspection.* The Commission may grant a waiver of the annual inspection for a period not to exceed 30 days from the time of first arrival of a ship at a United States port directly from a foreign port for the sole purpose of enabling the vessel to proceed coastwise to another port in the United States where an inspection can be made.

(1) An informal application (such as a letter or telegram, or telephone call to be confirmed by letter) for waiver of inspection must be submitted by either the vessel owner, the vessel's operating agency, the ship station licensee or the master of the vessel. The application must be submitted not earlier than 3 days in advance of the vessel's arrival at the United States port. The application must be submitted to the Commission's Engineer in Charge of the FCC District Office nearest the port of arrival. The application must include:

(i) The ship's name and radio call sign;

(ii) The name of the first United States port of arrival directly from a foreign port;

(iii) The date of arrival;

(iv) The date and port at which annual inspection will be formally requested to be conducted;

(v) Reason for requesting waiver; and

(vi) A statement that the ship's compulsory radio equipment is operable.

Subpart C - Operating Requirements and Procedures

STATION REQUIREMENTS-GENERAL
§ 80.61 Commission inspection of stations.

All stations and required station records must be made available for inspection by authorized representatives of the Commission.

§ 80.63 Maintenance of transmitter power.

(a) The power of each radio transmitter must not be more than that necessary to carry on the service for which the station is licensed.

(b) Except for transmitters using single sideband and independent sideband emissions, each radio transmitter rated by the manufacturer for carrier power in excess of

GENERAL RADIOTELEPHONE OPERATOR LICENSE

100 watts must contain the instruments necessary to determine the transmitter power during its operation.

STATION REQUIREMENTS-LAND STATIONS
§ 80.67 General facilities requirements for coast stations.

(a) All coast stations licensed to transmit in the band 156-162 MHz must be able to transmit and receive on 156.800 MHz and at least one working frequency in the band.

(b) All coast stations that operate telephony on frequencies in the 1605-3500 kHz band must be able to transmit and receive using J3E emission on the frequency 2182 kHz and at least one working frequency in the band. In addition, each such public coast station must transmit and receive H3E emission on the frequency 2182 kHz.

§ 80.69 Facilities requirement for public coast stations using telephony.

Public coast stations using telephony must be provided with the following facilities.

(a) When the station is authorized to use frequencies in the 1605-3500 kHz band, equipment meeting the requirements of § 80.67(b) must be installed at each transmitting location.

(b) The transmitter power on the frequency 2182 kHz must not exceed 50 watts carrier power for normal operation. During distress, urgency and safety traffic, operation at maximum power is permitted.

§ 80.70 Special provisions relative to coast station VHF facilities.

(a) Coast stations which transmit on the same radio channel above 150 MHz must minimize interference by reducing radiated power, by decreasing antenna height or by installing directional antennas. Coast stations at locations separated by less than 241 kilometers (150 miles) which transmit on the same radio channel above 150 MHz must also consider a time-sharing arrangement. The Commission may order station changes if agreement cannot be reached between the involved licensees.

(b) Coast stations which transmit on a radio channel above 150 MHz and are located within interference range of any station within Canada or Mexico must minimize interference to the involved foreign station(s), and must notify the Commission of any station changes.

§ 80.72 Antenna requirements for coast stations.

All emissions of a coast station a marine-utility station operated on shore using telephony within the frequency band 30-200 MHz must be vertically polarized.

STATION REQUIREMENTS - SHIP STATIONS
§ 80.79 Inspection of ship station by a foreign government.

The governments or appropriate administrations of countries which a ship visits may require the license of the ship station or ship earth station to be produced for examination. When the license cannot be produced without delay or when irregularities are observed, governments or administrations may inspect the radio installations to satisfy themselves that the installation conforms to the conditions imposed by the Radio Regulations.

§ 80.80 Operating controls for ship stations.

(a) Each control point must be capable of:

(1) Starting and discontinuing operation of the station;

(2) Changing frequencies within the same sub-band;

(3) Changing from transmission to reception and vice versa.

(4) In the case of stations operating in the 156-162 MHz bands, reducing power output to one watt or less in accordance with § 80.215(e).1

(b) Each ship station using telegraphy must be capable of changing from telegraph transmission to telegraph reception and vice versa without manual switching.

(c) Each ship station using telephony must be capable of changing from transmission to reception and vice versa within two seconds excluding a change in operating radio channel.

(d) During its hours of service, each ship station must be capable of:

(1) Commencing operation within one minute;

(2) Discontinuing all emission within five seconds after emission is no longer desired.

(e) Each ship station using a multichannel installation for telegraphy (except equipment intended for use only in emergencies on frequencies below 515 kHz) must be capable of changing from one radio channel to another within:

(1) Five seconds if the channels are within the same sub-band; or

(2) Fifteen seconds if the channels are not within the same sub-band.

(f) Each ship station and marine-utility station using a multi-channel installation for telephony must be capable of changing from one radio channel to another within:

(1) Five seconds within the band 1605-3500 kHz; or

(2) Three seconds within the band 156-162 MHz.

(g)(1) Any telegraphy transmitter constructed since January 1, 1952, that operates

APPENDIX – PART 80

in the band 405-525 kHz with an output power in excess of 250 watts must be capable of reducing the output power to 150 watts or less.

(2) The requirement of paragraph (g)(1) of this section does not apply when there is available in the same station a transmitter capable of operation on the international calling frequency 500 kHz and at least one working frequency within the band 405-525 kHz, capable of being energized by a source of power other than an emergency power source and not capable of an output in excess of 100 watts when operated on such frequencies.

§ 80.81 Antenna requirements for ship stations.

All telephony emissions of a ship station or a marine utility station on board ship within the frequency band 30-200 MHz must be vertically polarized.

§ 80.83 Protection from potentially hazardous RF radiation.

Any license or renewal application for a ship earth station that will cause exposure to radiofrequency (RF) radiation in excess of the RF exposure guidelines specified in § 1.1307(b) of the Commission's Rules must comply with the environmental processing rules set forth in § 1.1301-1.1319 of this chapter.

OPERATING PROCEDURES - GENERAL
§ 80.86 International regulations applicable.

In addition to being regulated by these rules, the use and operation of stations subject to this part are governed by the Radio Regulations and the radio provisions of all other international agreements in force to which the United States is a party.

§ 80.87 Cooperative use of frequency assignments.

Each radio channel is available for use on a shared basis only and is not available for the exclusive use of any one station or station licensee. Station licensees must cooperate in the use of their respective frequency assignment in order to minimize interference and obtain the most effective use of the authorized radio channels.

§ 80.88 Secrecy of communication.

The station licensee, the master of the ship, the responsible radio operators and any person who may have knowledge of the radio communications transmitted or received by a fixed, land, or mobile station subject to this part, or of any radiocommunication service of such station, must observe the secrecy requirements of the Communications Act and the Radio Regulations. See sections 501, 502, and 705 of the Communications Act and Article 23 of the Radio Regulations.

§ 80.89 Unauthorized transmissions.

Stations must not:

(a) Engage in superfluous radiocommunication.

(b) Use telephony on 243 MHz.

(c) Use selective calling on 2182 kHz or 156.800 MHz.

(d) When using telephony, transmit signals or communications not addressed to a particular station or stations. This provision does not apply to the transmission of distress, alarm urgency, or safety signals or messages, or to test transmissions.

(e) When using telegraphy, transmit signals or communications not addressed to a particular station or stations, unless the transmission is preceded by CQ or CP or by distress, alarm, urgency, safety signals, or test transmissions.

(f) Transmit while on board vessels located on land. Vessels in the following situations are not considered to be on land for the purposes of this paragraph:

(1) Vessels which are aground due to a distress situation;

(2) Vessels in drydock undergoing repairs; and

(3) State or local government vessels which are involved in search and rescue operations including related training exercises.

(g) Transmit on frequencies or frequency bands not authorized on the current station license.

§ 80.90 Suspension of transmission.

Transmission must be suspended immediately upon detection of a transmitter malfunction and must remain suspended until the malfunction is corrected, except for transmission concerning the immediate safety of life or property, in which case transmission must be suspended as soon as the emergency is terminated.

§ 80.91 Order of priority of communications.

(a) The order of priority of radiotelegraph communications is as follows:

(1) Distress calls including the international distress signal for radiotelegraphy, the international radiotelegraph alarm signal, the international radiotelephone alarm signal, distress messages and distress traffic.

(2) Communications preceded by the international radiotelegraph urgency signal.

(3) Communications preceded by the international radiotelegraphy safety signal.

(4) Communications relative to radio direction-finding bearings.

(5) Communications relative to the navigation and safe movement of aircraft.

GENERAL RADIOTELEPHONE OPERATOR LICENSE

(6) Communications relative to the navigation, movements, and needs of ships, including weather observation messages destined for an official meteorological service.

(7) Government communications for which priority right has been claimed.

(8) Service communications relating to the working of the radiocommunication service or to communications previously transmitted.

(9) All other communications.

(b) The order of priority of radiotelephone communications is as follows:

(1) Distress calls including the international distress signal for radiotelephony, the international radiotelephone alarm signal, distress messages and distress traffic.

(2) Communications preceded by the international radiotelephone urgency signal, or known to the station operator to consist of one or more urgent messages concerning the safety of a person, aircraft or other mobile unit.

(3) Communications preceded by the international radiotelephone safety signal, or known to the station operator to consist of one or more messages concerning the safety of navigation or important meteorological warnings.

(4) Communications known by the station operator to consist of one or more messages relative to the navigation, movements and needs of ships, including weather observation messages destined for an official meteorological service.

(5) Government communications for which priority right has been claimed.

(6) All other communications.

§ 80.92 Prevention of interference.

(a) The station operator must determine that the frequency is not in use by monitoring the frequency before transmitting, except for transmission of signals of distress.

(b) When a radiocommunication causes interference to a communication which is already in progress, the interfering station must cease transmitting at the request of either party to the existing communication. As between nondistress traffic seeking to commence use of a frequency, the priority is established under § 80.91.

(c) Except in cases of distress, communications between ship stations or between ship and aircraft stations must not interfere with public coast stations. The ship or aircraft stations which cause interference must stop transmitting or change frequency upon the first request of the affected coast station.

§ 80.93 Hours of service.

(a) *All stations.* All stations whose hours of service are not continuous must not suspend operation before having concluded all communication required in connection with a distress call or distress traffic.

(b) *Public coast stations.* (1) Each public coast station whose hours of service are not continuous must not suspend operation before having concluded all communication involving messages or calls originating in or destined to mobile stations within range and mobile stations which have indicated their presence.

(2) Unless otherwise authorized by the Commission upon adequate showing of need, each public coast station authorized to operate on frequencies in the 3000-23,000 kHz band must maintain continuous hours of service.

(c) *Compulsory ship stations.* Compulsory ship stations whose service is not continuous may not suspend operation before concluding all traffic originating in or destined for public coast stations situated within their range and mobile stations which have indicated their presence.

(d) *Other than public coast or compulsory ship stations.* The hours of service of stations other than public coast or compulsory ship stations are determined by the station licensee.

§ 80.95 Message charges.

(a) Charges must not be made for service of:

(1) Any public coast station unless tariffs for the service are on file with the Commission;

(2) Any station other than a public coast station or an Alaska-public fixed station, except cooperatively shared stations covered by § 80.503;

(3) Distress calls and related traffic; and

(4) Navigation hazard warnings preceded by the SAFETY signal.

(b) The licensee of each ship station is responsible for the payment of all charges accruing to any other station(s) or facilities for the handling or forwarding of messages or communications transmitted by that station.

§ 80.101 Radiotelephone testing procedures.

This section is applicable to all stations using telephony except where otherwise specified.

(a) Station licensees must not cause harmful interference. When radiation is necessary or unavoidable, the testing procedure described below must be followed:

(1) The operator must not interfere with transmissions in progress.

(2) The testing station's call sign, followed by the word "test", must be announced on the radiochannel being used for the test.

(3) If any station responds "wait", the test must be suspended for a minimum of 30 seconds, then repeat the call sign followed by the word "test" and listen again for a response. To continue the test, the operator must use counts or phrases which do not conflict with normal operating signals, and must end with the station's call sign. Test signals must not exceed ten seconds, and must not be repeated until at least one minute has elapsed. On the frequency 2182 kHz or 156.800 MHz, the time between tests must be a minimum of five minutes.

(b) Testing of transmitters must be confined to single frequency channels on working frequencies. However, 2182 kHz and 156.800 MHz may be used to contact ship or coast stations as appropriate when signal reports are necessary. Short tests on 2182 kHz by vessels with DSB (A3) equipment for distress and safety purposes are permitted to evaluate the compatibility of that equipment with an A3J emission system. U.S. Coast Guard stations may be contacted on 2182 kHz or 156.800 MHz for test purposes only when tests are being conducted during inspections by Commission representatives, when qualified radio technicians are installing or repairing the station radiotelephone equipment, or when qualified ship's personnel conduct an operational check requested by the U.S. Coast Guard. In these cases the test must be identified as "PCC" or "technical".

(c) Survival craft transmitter tests must not be made within actuating range of automatic alarm receivers. Survival craft transmitters must not be tested on the frequency 500 kHz during the silence periods.

§ 80.102 Radiotelephone station identification.

This section applies to all stations using telephony which are subject to this part.

(a) Except as provided in paragraphs (d) and (e) of this section, stations must give the call sign in English. Identification must be made:

(1) At the beginning and end of each communication with any other station.

(2) At 15 minute intervals when transmission is sustained for more than 15 minutes. When public correspondence is being exchanged with a ship or aircraft station, the identification may be deferred until the completion of the communications.

(b) Private coast stations located at drawbridges and transmitting on the navigation frequency 156.650 MHz may identify by use of the name of the bridge in lieu of the call sign.

(c) Ship stations transmitting on any authorized VHF bridge-to-bridge channel may be identified by the name of the ship in lieu of the call sign.

(d) Ship stations operating in a vessel traffic service system or on a waterway under the control of a U.S. Government agency or a foreign authority, when communicating with such an agency or authority may be identified by the name of the ship in lieu of the call sign, or as directed by the agency or foreign authority.

(e) VHF public coast station may identify by means of the approximate geographic location of the station or the area it serves when it is the only VHF public coast station serving the location or there will be no conflict with the identification of any other station.

§ 80.104 Identification of radar transmissions not authorized.

This section applies to all maritime radar transmitters except radar beacon stations.

(a) Radar transmitters must not transmit station identification.

OPERATING PROCEDURES - LAND STATIONS
§ 80.105 General obligations of coast stations.

Each coast station or marine-utility station must acknowledge and receive all calls directed to it by ship or aircraft stations. Such stations are permitted to transmit safety communication to any ship or aircraft station.

§ 80.106 Intercommunication in the mobile service.

(a) Each public coast station must exchange radio communications with any ship or aircraft station at sea; and each station on shipboard or aircraft at sea must exchange radio communications with any other station on shipboard or aircraft at sea or with any public coast station.

(b) Each public coast station must acknowledge and receive all communications from mobile stations directed to it, transmit all communications delivered to it which are directed to mobile stations within range in accordance with their tariffs. Discrimination in service is prohibited.

§ 80.108 Transmission of traffic lists by coast stations.

(a) Each coast station is authorized to transmit lists of call signs in alphabetical order of all mobile stations for which they have traffic on hand. These traffic lists will be transmitted on the station's normal working frequencies at intervals of:

(1) In the case of telegraphy, at least two hours and not more than four hours during the working hours of the coast station.

GENERAL RADIOTELEPHONE OPERATOR LICENSE

(2) In the case of radiotelephony, at least one hour and not more than four hours during the working hours of the coast station.

(b) The announcement must be as brief as possible and must not be repeated more than twice. Coast stations may announce on a calling frequency that they are about to transmit call lists on a specific working frequency.

§ 80.110 Inspection and maintenance of tower markings and associated control equipment.

The licensee of any radio station which has an antenna structure required to be painted or illuminated pursuant to the provisions of section 303(q) of the Communications Act must operate and maintain the tower marking and associated control equipment in accordance with Part 17 of this chapter.

§ 80.111 Radiotelephone operating procedures for coast stations.

This section applies to all coast stations using telephony which are subject to this part.

(a) *Limitations on calling.* (1) Except when transmitting a general call to all stations for announcing or preceding the transmission of distress, urgency, or safety messages, a coast station must call the particular station(s) with which it intends to communicate.

(2) Coast stations must call ship stations by voice unless it is known that the particular ship station may be contacted by other means such as automatic actuation of a selective ringing or calling device.

(3) Coast stations may be authorized emission for selective calling on each working frequency.

(4) Calling a particular station must not continue for more than one minute in each instance. If the called station does not reply, that station must not again be called for two minutes. When a called station does not reply to a call sent intervals of two minutes, the calling must cease for fifteen minutes. However, if harmful interference will not be caused to other communications in progress, the call may be repeated after three minutes.

(5) A coast station must not attempt to communicate with a ship station that has specifically called another coast station until it becomes evident that the called station does not answer, or that communication between the ship station and the called station cannot be carried on because of unsatisfactory operating conditions.

(6) Calls to establish communication must be initiated on an available common working frequency when such a frequency exists and it is known that the called ship maintains a simultaneous watch on the common working frequency and the appropriate calling frequency(ies).

(b) *Time limitation on calling frequency.* Transmissions by coast stations on 2182 kHz or 156.800 MHz must be minimized and any one exchange of communications must not exceed one minute in duration.

(c) *Change to working frequency.* After establishing communications with another station by call and reply on 2182 kHz or 156.800 MHz coast stations must change to an authorized working channel for the transmission of messages.

(d) *Use of busy signal.* A coast station, when communicating with a ship station which transmits to the coast station on a radio channel which is a different channel from that used by the coast station for transmission, may transmit a "busy" signal whenever transmission from tile ship station is being received.

OPERATING PROCEDURES - SHIP STATIONS
§ 80.114 Authority of the master.

(a) The service of each ship station must at all times be under the ultimate control of the master, who must require that each operator or such station comply with the Radio Regulations in force and that the ship station is used in accordance with those regulations.

(b) These rules are waived when the vessel is under the control of the U.S. Government.

§ 80.115 Operational conditions for use of associated ship units.

(a) Associated ship units may be operated under a ship station authorization. Use of an associated ship unit is restricted as follows;

(1) It must only be operated on the safety and calling frequency 156.800 MHz or on commercial or noncommercial VHF intership frequencies appropriate to the class of ship station with which it is associated.

(2) Except for safety purposes, it must only be used to communicate with the ship station with which it is associated or with associated ship units of the same ship station. Such associated ship units may not be used from shore.

(3) It must be equipped to transmit on the frequency 156.800 MHz and at least one appropriate intership frequency.

(4) Calling must occur on the frequency 156.800 MHz unless calling and working on an intership frequency has been prearranged.

(5) Power is limited to one watt.

(6) The station must be identified by the call sign of the ship station with which

296

it is associated and an appropriate unit designator.

§ 80.116 Radiotelephone operating procedures for ship stations.

(a) *Calling coast stations.* (1) Use by ship stations of the frequency 2182 kHz for calling coast stations and for replying to calls from coast stations is authorized. However, such calls and replies should be on the appropriate ship-shore working frequency.

(2) Use by ship stations and marine utility stations of the frequency 156.800 MHz for calling coast stations and marine utility stations on shore, and for replying to calls from such stations, is authorized. However, such calls and replies should be made on the appropriate ship-shore working frequency.

(b) *Calling ship stations.* (1) Except when other operating procedure is used to expedite safety communication, ship stations, before transmitting on the intership working frequencies 2003, 2142, 2638, 2738, or 2830 kHz, must first establish communications with other ship stations by call and reply on 2182 kHz. Calls may be initiated on an intership working frequency when it is known that the called vessel maintains a simultaneous watch on the working frequency and on 2182 kHz.

(2) Except when other operating procedures are used to expedite safety communications, the frequency 156.800 MHz must be used for call and reply by ship stations and marine utility stations before establishing communication on one of the intership working frequencies. Calls may be initiated on an intership working frequency when it is known that the called vessel maintains a simultaneous watch on the working frequency and on 156.800 MHz.

(c) *Change to working frequency.* After establishing communication with another station by call and reply on 2182 kHz or 156.800 MHz stations on board ship must change to an authorized working frequency for the transmission of messages.

(d) *Limitations on calling.* Calling a particular station must not continue for more than 30 seconds in each instance. If the called station does not reply, the station must not again be called until after an interval of 2 minutes. When a called station called does not reply to a call sent three times at intervals of 2 minutes, the calling must cease and must not be renewed until after an interval of 15 minutes; however, if there is no reason to believe that harmful interference will be caused to other communications in progress, the call sent three times at intervals of 2 minutes may be repeated after a pause of not less than 3 minutes. In event of an emergency involving safety, the provisions of this paragraph do not apply.

(e) *Limitations on working.* Any one exchange of communications between any two ship stations on 2003, 2142, 2638, 2738, or 2830 kHz or between a ship station and a private coast station on 2738 or 2830 kHz must not exceed 3 minutes after the stations have established contact. Subsequent to such exchange of communications, the same two stations must not again use 2003, 2142, 2638, 2738, or 2830 kHz for communication with each other until 10 minutes have elapsed.

(f) *Transmission limitation on 2182 kHz and 156.800 MHz.* To facilitate the reception of distress calls, all transmissions on 2182 kHz and 156.800 MHz (channel 16) must be minimized and transmissions on 156.800 MHz must not exceed 1 minute.

(g) *Limitations on commercial communication.* On frequencies in the band 156-162 MHz, the exchange of commercial communication must be limited to the minimum practicable transmission time. In the conduct of ship-shore communication other than distress, stations on board ship must comply with instructions given by the private coast station or marine utility station on shore with which they are communicating.

(h) *2182 kHz silence periods.* To facilitate the reception of distress calls, transmission by ship or survival craft stations is prohibited on any frequency (including 2182 kHz) within the band 2173.5-2190.5 kHz during each 2182 kHz silence period.

SPECIAL PROCEDURES-SHIP STATIONS
§ 80.141 General provisions for ship stations.

(a) *Points of communication.* Ship stations and marine utility stations on board ships are authorized to communicate with any station in the maritime mobile service.

(b) *Service requirements for all ship stations.* (1) Each ship station must receive and acknowledge all communications which are addressed to the ship or to any person on board.

(2) Every ship, on meeting with any direct danger to the navigation of other ships such as ice, a derelict vessel, a tropical storm, subfreezing air temperatures associated with gale force winds causing severe icing on superstructures, or winds of force 10 or above on the Beaufort scale for which no storm warning has been received, must transmit related information to ships in the vicinity and to the authorities on land unless such action has already been taken by another station. All such radio messages must be preceded by the safety signal.

(3) A ship station may accept communications for retransmission to any other station in the maritime mobile service. Whenever such messages or communications have been received and acknowledged by a ship station for this purpose, that station must retransmit the message as soon as possible.

(c) *Service requirements for vessels.* Each ship station provided for compliance with Part II of Title III of the Communications Act must provide a public correspondence service on voyages of more than 24 hours for any person who requests the service.

(1) Compulsory radiotelegraph ships must provide this service during the hours the radio operator is normally on duty.

(2) Compulsory radiotelephone ships must provide this service for at least four hours daily. The hours must be prominently posted at the principal operating location of the station.

§ 80.143 Required frequencies for radiotelephony.

(a) Except for compulsory vessels, each ship radiotelephone station licensed to operate in the band 1605-3500 kHz must be able to receive and transmit J3E emission on the frequency 2182 kHz. Ship stations are additionally authorized to receive and transmit H3E emission for communications with foreign coast stations and with vessels of foreign registry. If the station is used for other than safety communications, it must be capable also of receiving and transmitting the J3E emission on at least two other frequencies in that band. However, ship stations which operate exclusively on the Mississippi River and its connecting waterways, and on high frequency bands above 3500 kHz, need be equipped with 2182 kHz and one other frequency within the band 1605-3500 kHz. Additionally, use of A3E emission is permitted for distress and safety purposes on 2182 kHz for portable survival craft equipment also having the capability to operate on 500 kHz and for transmitters authorized for use prior to January 1, 1972.

(b) Except as provided in paragraph (c) of this section, at least one VHF radiotelephone transmitter/receiver must be able to transmit and receive on the following frequencies:

(1) The distress, safety and calling frequency 156.800 MHz;

(2) The primary intership safety frequency 156.300 MHz;

(3) One or more working frequencies; and

(4) All other frequencies necessary for its service.

(c) Where a ship ordinarily has no requirement for VHF communications, handheld VHF equipment may be used solely to comply with the bridge-to-bridge navigational communication

SHIPBOARD GENERAL PURPOSE WATCHES
§ 80.146 Watch on 500 kHz

During their hours of service, ship stations using frequencies in the authorized bands between 405-525 kHz must, remain on watch on 500 kHz except when the operator is transmitting an 500 kHz or operating on another frequency. The provisions of this section do not relieve the ship from complying with the requirements for a safety watch as prescribed in § 80.304 and 80.305.

§ 80.147 Watch on 2182 khz.

Ship stations must maintain a watch on 2182 kHz as prescribed by § 80.304(b).

§ 80.148 Watch on 156.8 MHz (Channel 16).

Each VHF ship station, or if more than one VHF ship station is being operated from a vessel (for example, if a pilot is operating his radio equipment on board the vessel), then at least one VHF ship station per vessel must during its hours of service maintain a watch on 156.800 MHz whenever such station is not being used for exchanging communications. The watch is not required:

(a) Where a ship station is operating only with handheld bridge-to-bridge VHF radio equipment under § 80.143(c) of this part;

(b) For vessels subject to the Bridge-to-Bridge Act and participating in a Vessel Traffic Service (VTS) system when the watch is maintained on both the bridge-to-bridge frequency and a separately assigned VTS frequency; or

(c) For a station on board a voluntary vessel equipped with digital selective calling (DSC) equipment, maintaining a continuous DSC watch on 156.525 MHz whenever such station is not being used for exchanging communications, and while such station is within the VHF service area of a U.S. Coast Guard radio facility which is DSC equipped.

VIOLATIONS
§ 80.149 Answer to notice of violation.

(a) Any person receiving official notice of violation of the terms of the Communications Act, any legislative act, executive order, treaty to which the United States is a party, terms of a station or operator license, or the rules and regulations of the Federal Communications Commission must within 10 days from such receipt, send a written answer, in duplicate, to the office of the Commission originating the official notice. If an answer cannot be sent or an acknowledgment made within such 10-day period by reason of illness or other unavoidable circumstances, acknowledgment and answer must be made at the earliest practicable date with a satisfactory explanation of the delay. The answer to each notice must be complete in itself and must not be abbreviated by references to other communications or answers to other notices. The answer must contain a full explanation of the incident involved and must set forth the action taken to prevent a continuation or recurrence. If the notice relates to lack of attention to or improper operation of the station or to log or watch discrepancies, the answer must give the name and license number of the licensed operator on duty.

Subpart D - Operator Requirements

§ 80.151 Classification of operator licenses and endorsements.

(a) Commercial radio operator licenses issued by the Commission are classified in accordance with the Radio Regulations of the International Telecommunication Union.

(b) The following licenses are issued by the Commission. International classification, if different from the license name, is given in parentheses. The licenses and their alphanumeric designator are listed in descending order.

(1) T-1. First Class Radiotelegraph Operator's Certificate.

(2) T-2. Second Class Radiotelegraph Operator's Certificate.

(3) G. General Radiotelephone Operator License (radiotelephone operator's general certificate).

(4) T-3. Third Class Radiotelegraph Operator's Certificate (radiotelegraph operator's special certificate).

(5) MP. Marine Radio Operator Permit (radiotelephone operator's restricted certificate).

(6) RP. Restricted Radiotelephone Operator Permit (radiotelephone operator's restricted certificate).

(c) The following license endorsements are affixed by the Commission to provide special authorizations or restrictions. Applicable licenses are given in parentheses.

(1) Ship Radar endorsement (First and Second Class Radiotelegraph Operator's Certificate, General Radiotelephone Operator License).

(2) Six Months Service endorsement (First and Second Class Radiotelegraph Operator's Certificate).

(3) Restrictive endorsements; relating to physical handicaps, English language or literacy waivers, or other matters (all licenses).

SHIP STATION OPERATOR REQUIREMENTS
§ 80.155 Ship station operator requirements.

Except as provided in § 80.177 and 80.179, operation of transmitters of any ship station must be performed by a person holding a commercial radio operator license or permit of the class required below. The operator is responsible for the proper operation of the station.

§ 80.156 Control by operator.

The operator on board ships required to have a holder of a commercial operator license or permit on board may, if authorized by the station licensee or master, permit an unlicensed person to modulate the transmitting apparatus for all modes of communication except Morse code radiotelegraphy.

§ 80.157 Radio officer defined.

A radio officer means a person holding a first or second class radiotelegraph operator's certificate Issued by the Commission who is employed to operate a ship radio station in compliance with Part II of Title III of the Communications Act. Such a person is also required to be licensed as a *radio officer* by the U.S. Coast Guard when employed to operate a ship radiotelegraph station.

§ 80.159 Operator requirements of Title III of the Communications Act and the Safety Convention.

(a) Each telegraphy passenger ship equipped with a radiotelegraph station in accordance with Part II of Title III of the Communications Act must carry one radio officer holding a first or second class radiotelegraph operator's certificate and a second radio officer holding either a first or second class radiotelegraph operator's certificate. The holder of a second class radiotelegraph operator's certificate may not act as the chief radio officer.

(b) Each cargo ship equipped with a radiotelegraph station in accordance with Part II of Title III of the Communications Act and which has a radiotelegraph auto alarm must carry a radio officer holding a first or second class radiotelegraph operator's certificate who has had at least six months service as a radio officer on board U.S. ships. If the radiotelegraph station does not have an auto alarm, a second radio officer who holds a first or second class radiotelegraph operator's certificate must be carried.

(c) Each cargo ship equipped with a radiotelephone station in accordance with Part II of Title III of the Communications Act must carry a radio operator who meets the following requirements:

(1) Where the station power does not exceed 1500 watts peak envelope power, the operator must hold a marine radio operator permit or higher class license.

(2) Where the station power exceeds 1500 watts peak envelope power, the operator must hold a general radiotelephone radio operator license or higher class license.

GENERAL RADIOTELEPHONE OPERATOR LICENSE

(d) Each ship transporting more than six passengers for hire equipped with a radiotelephone station in accordance with Part III of Title III of the Communications Act must carry a radio operator who meets the following requirements:

(1) Where the station power does not exceed 250 watts carrier power or 1500 watts peak envelope power, the radio operator must hold a marine radio operator permit or higher class license.

(2) Where the station power exceeds 250 watts carrier power or 1500 watts peak envelope power, the radio operator must hold a general radiotelephone operator license or higher class license.

§ 80.161 Operator requirements of the Great Lakes Radio Agreement.

Each ship subject to the Great Lakes Radio Agreement must have on board an officer or member of the crew who holds a marine radio operator permit or higher class license.

§ 80.163 Operator requirements of the Bridge-to-Bridge Act.

Each ship subject to the Bridge-to-Bridge Act must have on board a radio operator who holds a restricted radiotelephone operator permit or higher class license.

§ 80.165 Operator requirements for voluntary stations.

Minimum operator license

Ship Morse telegraph	T-2.
Ship direct-printing telegraph	MP.
Ship telephone, more than 250 watts carrier power or 1,000 watts peak envelope power.	G.
Ship telephone, not more than 250 watts carrier power or 1,000 watts peak envelope power.	MP.
Ship telephone, not more than 100 watts carrier power or 400 watts peak envelope power:	
Above 30 MHz	None[1]
Below 30 MHz	RP.
Ship earth station	RP.

[1] RP required for international voyage.

GENERAL OPERATOR REQUIREMENTS
§ 80.167 Limitations on operators.

The operator of maritime radio equipment other than T-1, T-2, or G licensees, must not:

(a) Make equipment adjustments which may affect transmitter operation;

(b) Operate any transmitter which requires more than the use of simple external switches or manual frequency selection or transmitters whose frequency stability is not maintained by the transmitter itself.

§ 80.169 Operators required to adjust transmitters or radar.

(a) All adjustments of radio transmitters in any radiotelephone station or coincident with the installation, servicing, or maintenance of such equipment which may affect the proper operation of the station, must be performed by or under the immediate supervision and responsibility of a person holding a first or second class radiotelegraph operator's certificate or a general radiotelephone operator license.

(b) Only persons holding a first or second class radiotelegraph operator certificate must perform such functions at radiotelegraph stations transmitting Morse code.

(c) Only persons holding an operator certificate containing a ship radar endorsement must perform such functions on radar equipment.

§ 80.175 Availability of operator licenses.

All operator licenses required by this subpart must be readily available for inspection.

§ 80.177 When operator license is not required.

(a) No radio operator authorization is required to operate:

(1) A shore radar, a shore radiolocation, maritime support or shore radionavigation station;

(2) A survival craft station or an emergency position indicating radio beacon;

(3) A ship radar station if:

(i) The radar frequency is determined by a nontunable, pulse type magnetron or other fixed tuned device, and

(ii) The radar is capable of being operated exclusively by external controls;

(4) An on board station; or

(5) A ship station operating in the VHF band on board a ship voluntarily equipped with radio and sailing on a domestic voyage.

(b) No radio operator license is required to install a VHF transmitter in a ship station if the installation is made by, or under the supervision of, the licensee of the ship station and if modifications to the transmitter other than front panel controls are not made.

(c) No operator license is required to operate coast stations of 250 watts or less carrier power or 1500 watts or less peak envelope power operating on frequencies above 30 MHz, or marine utility stations.

(d) No radio operator license is required to install a radar station on a voluntarily equipped ship when a manual is included with the equipment that provides step-by-step instructions for the installation, calibration, and operation of the radar. The installation must be made by, or under the supervision of, the licensee of that ship station

APPENDIX – PART 80

and no modifications or adjustments other than to the front panel controls are to be made to the equipment.

Subpart E - General Technical Standards

§ 80.201 Scope.

This subpart gives the general technical requirements for the use of frequencies and equipment in the maritime services. These requirements include standards for equipment authorization, frequency tolerance, modulation, emission, power and bandwidth.

§ 80.203 Authorization of transmitters for licensing.

(a) Each transmitter authorized in a station in the maritime services after September 30, 1986, ...must be type accepted by the Commission for Part 80 operations. The procedures for type acceptance are contained in Part 2. Transmitters of a model type accepted or type approved before October 1, 1986 will be considered type accepted for use in ship or coast stations as appropriate.

(3) Programming of authorized channels must be performed only by a person holding a first or second class radiotelegraph operator's certificate or a general radiotelephone operator's license.

§ 80.213 Modulation requirements.

(a) Transmitters must meet the following modulation requirements:

(1) When double-sideband emission is used the peak modulation must be maintained between 75 and 100 percent;

(2) When phase or frequency modulation is used in the 156-162 MHz and 216-220 MHz bands the peak modulation must be maintained between 75 and 100 percent. A frequency deviation of ±5 kHz is defined as 100 percent peak modulation; and

(3) In single-sideband operation the upper sideband must be transmitted.

§ 80.215 Transmitter power.

(g) The carrier power of ship station radiotelephone transmitters, except portable transmitters, operating in the 156-162 MHz band must be at least 8 but not more than 25 watts. Transmitters that use 12-volt lead acid storage batteries as a primary power source must be measured with a primary voltage between 12.2 and 13.7 volts DC.

Subpart F - Equipment Authorization for Compulsory Ships

§ 80.251 Scope.

(a) This subpart gives the general technical requirements for type acceptance of equipment used on compulsory ships. Such equipment includes radiotelegraph transmitters, radiotelegraph auto alarms, automatic alarm-signal keying devices, survival craft radio equipment, watch receivers, and radar.

(b) The equipment described in this subpart must be type accepted.

Subpart G - Safety Watch Requirements and Procedures

COAST STATION SAFETY WATCHES

§ 80.301 Watch requirements.

(a) Each public coast station operating on telegraphy frequencies in the band 405-535 kHz must maintain a watch for classes AlA, A2B and H2B emissions by a licensed radiotelegraph operator on the frequency 500 kHz for three minutes twice each hour, beginning at x h.15 and x h.45 Coordinated Universal Time (UTC).

(b) Each public coast station licensed to operate in the band 1605-3500 kHz must monitor such frequency(s) as are used for working or, at the licensee's discretion, maintain a watch on 2182 kHz.

(c) Except for distress, urgency or safety messages, coast stations must not transmit on 2182 kHz during the silence periods for three minutes twice each hour beginning at x h.00 and x h.30 Coordinated Universal Time (UTC).

(d) Each public coast station must provide assistance for distress communications when requested by the Coast Guard.

§ 80.303 Watch on 156.800 MHz (Channel 16).

(a) During its hours of operation, each coast station operating in the 156-162 MHz band and serving rivers, bays and inland lakes except the Great Lakes, must maintain a safety watch on the frequency 156.800 MHz except when transmitting on 156.800 MHz.

SHIP STATION SAFETY WATCHES

§ 80.304 Watch requirement during silence periods.

(a) Each ship station operating on telegraphy frequencies in the band 405-535 kHz, must maintain. a watch on the frequency 500 kHz of three minutes twice each hour beginning at x h.15 and x h.45 Coordinated Universal Time (UTC) by a licensed radiotelegraph officer using either a loudspeaker or headphone.

(b) Each ship station operating on telephony on frequencies in the band 1605-3500 kHz must maintain a watch on the frequency 2182 kHz. This watch must be maintained at least twice each hour for 3 minutes commencing at x h.00 and x h.30 Coordinated Universal Time (UTC) using either a loudspeaker or headphone. Except for distress, urgency or safety messages, ship stations must not transmit during the silence periods on 2182 kHz.

301

§ 80.305 Watch requirements of the Communications Act and the Safety Convention.

(a) Each ship of the United States which is equipped with a radiotelegraph station for compliance with Part II of Title III of the Communications Act or chapter IV of the Safety Convention must:

(1) Keep a continuous and efficient watch on 500 kHz by means of radio officers while being navigated in the open sea outside a harbor or port. In lieu thereof, on a cargo ship equipped with a radiotelegraph auto alarm in proper operating condition, an efficient watch on 500 kHz must be maintained by means of a radio officer for at least 8 hours per day in the aggregate, i.e., for at least one-third of each day or portion of each day that the vessel is navigated in the open sea outside of a harbor or port.

(2) Keep a continuous and efficient watch on the radiotelephone distress frequency 2182 kHz from the principal radio operating position or the room from which the vessel is normally steered while being navigated in the open sea outside a harbor or port. A radiotelephone distress frequency watch receiver having a loudspeaker and a radiotelephone auto alarm facility must be used to keep the continuous watch on 2182 kHz if such watch is kept from the room from which the vessel is normally steered. After a determination by the master that conditions are such that maintenance of the listening watch would interfere with the safe navigation of the ship, the watch may be maintained by the use of the radiotelephone auto alarm facility alone.

(3) Keep a continuous and efficient watch on the VHF distress frequency 156.800 MHz from the room from which the vessel is normally steered while in the open sea outside a harbor or port.

§ 80.310 Watch required by voluntary vessels.

Voluntary vessels must maintain a watch on 156.800 MHz whenever the radio is operating and is not being used to communicate.

DISTRESS, ALARM, URGENCY AND SAFETY PROCEDURES

§ 80.311 Authority for distress transmission.

A mobile station in distress may use any means at its disposal to attract attention, make known its position and obtain help. A distress call and message, however, must be transmitted only on the authority of the master or person responsible for the mobile station. No person shall knowingly transmit, or cause to be transmitted, any false or fraudulent signal of distress or related communication.

§ 80.312 Priority of distress transmissions.

The distress call has absolute priority over all other transmissions. All stations which hear it must immediately cease any transmission capable of interfering with the distress traffic and must continue to listen on the frequency used for the emission of the distress call. This call must not be addressed to a particular station. Acknowledgement of receipt must not be given before the distress message which follows it is sent.

§ 80.313 Frequencies for use in distress.

The frequencies specified in the bands below are for use by mobile stations in distress. The conventional emission is shown. When a ship station cannot transmit on the designated frequency or the conventional emission, it may use any available frequency or emission. Frequencies for distress and safety calling using digital selective calling techniques are listed in § 80.359(b). Distress and safety NBDP frequencies are indicated by note 2 in § 80.361(b).

Frequency band	Emission	Carrier frequency
405-535 kHz	A2B	500 kHz
1605-3500 kHz	J3E	2182 kHz
4000-27,5000 kHz	A2B	8364 kHz
118-136 MHz	A3E	121.500 MHz
156-162 MHz	F3E, P0N	156.800 MHz
		156.750 MHz
243 MHz.	A3N	243.000 MHz

The maximum transmitter power obtainable may be used.

§ 80.314 Distress signals.

(a) The international radiotelegraphy distress signal consists of the group "three dots, three dashes, three dots" (...— ...), symbolized herein by SOS, transmitted as a single signal in which the dashes are slightly prolonged so as to be distinguished clearly from the dots.

(b) The international radiotelephone distress signal consists of the word MAYDAY, pronounced as the French expression "m'aider".

(c) These distress signals indicate that a mobile station is threatened by grave and imminent danger and requests immediate assistance.

§ 80.315 Distress calls.

(a) The radiotelegraph distress call consists of:

(1) The distress signal SOS, sent three times;
(2) The word DE;
(3) The call sign of the mobile station in distress, sent three times.

(b) The radiotelephone distress call consists of:

(1) The distress signal MAYDAY spoken three times;
(2) The words THIS IS;
(3) The call sign (or name, if no call sign assigned) of the mobile station in distress, spoken three times.

§ 80.316 Distress messages.

(a) The radiotelegraph distress message consists of:
(1) The distress signal SOS;
(2) The name of the mobile station in distress;
(3) Particulars of its position;
(4) The nature of the distress;
(5) The kind of assistance desired;
(6) Any other information which might facilitate rescue.

(b) The radiotelephone distress message consists of:
(1) The distress signal MAYDAY;
(2) The name of the mobile station in distress;
(3) Particulars of its position;
(4) The nature of the distress;
(5) The kind of assistance desired;
(6) Any other information which might facilitate rescue, for example, the length, color, and type of vessel, number of persons on board.

(c) As a general rule, a ship must signal its position in latitude and longitude, using figures for the degrees and minutes, together with one of the words NORTH or SOUTH and one of the words EAST or WEST. In radiotelegraphy, the signal .—.—.— must be used to separate the degrees from the minutes. When practicable, the true bearing and distance in nautical miles from a known geographical position may be given.

§ 80.327 Urgency signals.

(a) The urgency signal indicates that the calling station has a very urgent message to transmit concerning the safety of a ship, aircraft, or other vehicle, or the safety of a person. The urgency signal must be sent only on the authority of the master or person responsible for the mobile station.

(b) In radiotelegraphy, the urgency signal consists of three repetitions of the group XXX, sent with the individual letters of each group, and the successive groups clearly separated from each other. It must be transmitted before the call.

(c) In radiotelephony, the urgency signal consists of three oral repetitions of the group of words PAN PAN transmitted before the call.

(d) The urgency signal has priority over all other communications except distress. All mobile and land stations which hear it must not interfere with the transmission of the message which follows the urgency signal.

§ 80.329 Safety signals.

(a) The safety signal indicates that the station is about to transmit a message concerning the safety of navigation or giving important meteorological warnings.

(b) In radiotelegraphy, the safety signal consists of three repetitions of the group TTT, sent with the individual letters of each group, and the successive groups clearly separated from each other. It must be sent before the call.

(c) In radiotelephony, the safety signal consists of the word SECURITE, pronounced as in French, spoken three times and transmitted before the call.

(d) The safety signal and call must be sent on one of the international distress frequencies (500 kHz or 8364 kHz radiotelegraph; 2182 kHz or 156.8 MHz radiotelephone). Stations which cannot transmit on a distress frequency may use any other available frequency on which attention might be attracted.

§ 80.331 Bridge-to-bridge communication procedure.

(a) Vessels subject to the Bridge-to-Bridge Act transmitting on the designated navigational frequency must conduct communications in a format similar to those given below:
(1) This is the (name of vessel). My position is (give readily identifiable position, course and speed) about to (describe contemplated action). Out.
(2) Vessel off (give a readily identifiable position). This is (name of vessel) off (give a readily identifiable position). I plan to (give proposed course of action). Over.
(3) (Coast station), this is (vessel's name) off (give readily identifiable position). I plan to (give proposed course of action). Over.

(b) Vessels acknowledging receipt must answer (Name of vessel calling). This is (Name of vessel answering). Received your call, and follow with an indication of their intentions. Communications must terminate when each ship is satisfied that the other no longer poses a threat to its safety and is ended with "Out".

(c) Use of power greater than 1 watt in a bridge-to-bridge station shall be limited to the following three situations:
(1) Emergency.
(2) Failure of the vessel being called to respond to a second call at low power.
(3) A broadcast call as in paragraph (a)(1) of this section in a blind situation, e.g., rounding a bend in a river.

GENERAL RADIOTELEPHONE OPERATOR LICENSE

Subpart H - Frequencies

RADIOTELEGRAPHY

§ 80.351 Scope.

[This section] describes the carrier frequencies and general uses of radiotelegraphy with respect to the following:
- Distress, urgency, safety, call and reply.
- Working.
- Digital selective calling (DSC).
- Narrow-band direct-printing (NB-DP).
- Facsimile.

[This section also contains all of the Radiotelephony, Radiodetermination, Ship Earth Station, Maritime frequencies assigned to Aircraft, Operational Fixed Station, Vessel Traffic Service (VTS), Automated System, Alaska Fixed Station, Maritime Support Station and Developmental Station frequencies.]

Subpart I - Station Documents

§ 80.401 Station documents requirement

Licensees of ship and land radio stations are required to have various current station documents such as station licenses, operator authorizations, station logs, safety certificates, Part 80 (Rules and Regulations), maritime mobile call sign and coast/ship station lists ...and other station equipment records.

§ 80.403 Availability of documents.

Station documents must be readily available to the licensed operator(s) on duty during the hours of service of the station and to authorized Commission employees upon request.

§ 80.405 Station license.

(a) *Requirement.* Stations must have an authorization granted by the Federal Communications Commission.

(b) *Application.* Application for authorizations in the maritime services must be submitted on the prescribed forms in accordance with subpart B of this part.

(c) *Posting.* The current station authorization or a clearly legible copy must be posted at the principal control point of each station. If a copy is posted, it must indicate the location of the original. When the station license cannot be posted, as in the case of a marine utility station operating at temporary unspecified locations, it must be kept where it will be readily available for inspection. The licensee of a station on board a ship subject to Part II or III of Title III of the Communications Act or the Safety Convention must retain the most recently expired ship station license in the station records until the first Commission inspection after the expiration date.

§ 80.409 Station logs.

(a) *General requirements.* Logs must be established and properly maintained as follows:

(1) The log must be kept in an orderly manner. The required information for the particular class or category of station must be readily available. Key letters or abbreviations may be used if their proper meaning or explanation is contained elsewhere in the same log.

(2) Erasures, obliterations or willful destruction within the retention period are prohibited. Corrections may be made only by the person originating the entry by striking out the error, initialing the correction and indicating the date of correction.

(3) Ship station logs must identify the vessel name, country of registry, and official number of the vessel.

(4) The station licensee and the radio operator in charge of the station are responsible for the maintenance of station logs.

(b) *Availability and retention.* Station logs must be made available to authorized Commission employees upon request and retained as follows:

(1) Logs must be retained by the licensee for a period of one year from the date of entry, and when applicable for such additional periods as required by the following paragraphs:

(i) Logs relating to a distress situation or disaster must be retained for three years from the date of entry.

(ii) If the Commission has notified the licensee of an investigation, the related logs must be retained until the licensee is specifically authorized in writing to destroy them.

(iii) Logs relating to any claim or complaint of which the station licensee has notice must be retained until the claim or complaint has been satisfied or barred by statute limiting the time for filing suits upon such claims.

(2) Logs containing entries required by paragraphs (e) and (f) of this section must be kept at the principal radiotelephone operating location while the vessel is being navigated. All entries in their original form must be retained on board the vessel for at least 30 days from the date of entry.

(3) Ship radiotelegraph logs must be kept in the principal radiotelegraph operating room during the voyage.

Subpart J - Public Coast Stations

STATIONS ON LAND

§ 80.451 Supplemental eligibility requirements.

A public coast station license may be granted to any person meeting the citizenship provisions of § 80.15(b).

§ 80.453 Scope of communications.

Public coast stations provide ship/shore radiotelephone and radiotelegraph services.

(a) Public coast stations are authorized to communicate:

(1) With any ship or aircraft station operating in the maritime mobile service, for the transmission or reception of safety communication;

(2) With any land station to exchange safety communications to or from a ship or aircraft station;

(3) With Government and non-Government ship and aircraft stations to exchange public correspondence.

Subpart K - Private Coast Stations and Marine Utility Stations

§ 80.501 Supplemental eligibility requirements.

(a) A private coast station or a marine utility station may be granted only to a person who is:

(1) Regularly engaged in the operation, docking, direction, construction, repair, servicing or management of one or more commercial transport vessels or United States, state or local government vessels; or is

(2) Responsible for the operation, control, maintenance or development of a harbor, port or waterway used by commercial transport vessels; or is

(3) Engaged in furnishing a ship arrival and departure service, and will employ the station only for the purpose of obtaining the information essential to that service; or is

(4) A corporation proposing to furnish a nonprofit radio communication service to its parent corporation, to another subsidiary of the same parent, or to its own subsidiary where the party to be served performs any of the eligibility activities described in this section; or is

(5) A nonprofit corporation or association, organized to furnish a maritime mobile service solely to persons who operate one or more commercial transport vessels; or is

(6) Responsible for the operation of bridges, structures or other installations that area part of, or directly related to, a harbor, port or waterway when the operation of such facilities requires radio communications with vessels for safety or navigation; or is

(7) A person controlling public moorage facilities; or is

(8) A person servicing or supplying vessels other than commercial transport vessels; or is

(9) An organized yacht club with moorage facilities; or is

(10) A nonprofit organization providing noncommercial communications to vessels other than commercial transport vessels.

(b) Each application for station authorization for a private coast station or a marine utility station must be accompanied by a statement indicating eligibility under paragraph (a) of this section.

Subpart L - Operational Fixed Stations

§ 80.555 Scope of communication.

An operational fixed station provides control, repeater or relay functions for its associated coast station.

Subpart M - Stations in the Radiodetermination Service

§ 80.601 Scope of communications.

Stations on land in the Maritime Radiodetermination Service provide a radionavigation or radiolocation service for ships.

Subpart N - Maritime Support Stations

§ 80.653 Scope of communications.

(a) Maritime support stations are land stations authorized to operate at permanent locations or temporary unspecified locations.

(b) Maritime support stations are authorized to conduct the following operations:

(1) Training of personnel in maritime telecommunications;

(2) Transmissions necessary for the test and maintenance of maritime radio equipment at repair shops; and

(3) Transmissions necessary to test the technical performance of the licensee's public coast station(s) radiotelephone receiver(s); and

(4) Transmissions necessary for radar/racon equipment demonstration.

Subpart O - Alaska Fixed Stations

§ 80.701 Scope of service.

There are two classes of Alaska Fixed stations. Alaska-public fixed stations are common carriers, open to public correspondence, which operate on the paired duplex channels listed in Subpart H of this Part. Alaska-private fixed stations may operate on simplex frequencies listed in Subpart H of this Part to communicate with other Alaska private fixed stations or with ship stations, and on duplex frequencies listed in Subpart H of this Part when communicating with the Alaska public fixed stations. Alaska-private fixed stations must not charge for service, although third party traffic may be transmitted. Only Alaska public fixed stations are authorized to charge for communication services.

GENERAL RADIOTELEPHONE OPERATOR LICENSE

Subpart P - Standards for Computing Public Coast Station VHF Coverage

§ 80.751 Scope.

This subpart specifies receiver antenna terminal requirements in terms of power, and relates the power available at the receiver antenna terminals to transmitter power and antenna height and gain.

Subpart Q - Compulsory Radiotelegraph Installations for Vessels 1600 Gross Tons

STATIONS ON SHIPBOARD
§ 80.801 Applicability.

The radiotelegraph requirements of Part II of Title III of the Communications Act apply to all passenger ships irrespective of size and cargo ships of 1600 gross tons and upward. The Safety Convention applies to such ships on international voyages. These ships are required to carry a radiotelegraph installation complying with this subpart.

§ 80.802 Inspection of station.

(a) Every ship of the United States subject to Part II of Title III of the Communications Act or the radio provisions of the Safety Convention must have the required equipment inspected at least once every 12 months. If the ship is in compliance with the requirements of the Safety Convention, a Safety Certificate will be issued; if in compliance with the Communications Act, the license will be endorsed accordingly.

§ 80.807 Requirements of radiotelephone installation.

All radiotelephone installations in radiotelegraph equipped vessels must meet the following conditions:

(a) The radiotelephone transmitter must be capable of transmission of A3E or H3E emission on 2182 kHz and must be capable of transmitting clearly perceptible signals from ship to ship during daytime, under normal conditions over a range of 150 nautical miles when used with an antenna system in accordance with paragraph (c) of this section. The transmitter must:

(1) Have a duty cycle which allows for transmission of the radiotelephone alarm signal.

(2) Provide 25 watts carrier power for A3E emission or 60 watts peak power on H3E emission into an artificial antenna consisting of 10 ohms resistance and 200 picofarads capacitance or 50 ohms nominal to demonstrate compliance with the 150 nautical mile range requirement.

(b) The radiotelephone must receive A3E and H3E when connected to the system specified in paragraph (c) this section and must be preset to 2182 kHz.

(c) The antenna system must be as nondirectional and efficient as is practicable for the transmission and reception of radio ground waves over seawater.

§ 80.808 Requirements of reserve installation.

(a) All reserve radiotelegraph installations must comply with the following conditions, in additional to all other requirements:

(9) There must be readily available under normal load conditions a reserve power supply for the reserve installation which must be independent of the propelling power of the ship and of any other electrical system. The reserve power supply must simultaneously energize the reserve transmitter at its required antenna power and the reserve receiver for at least 6 hours continuously under normal working conditions, and energize the automatic-alarm-signal keying device continuously for a period of 1 hour.

§ 80.832 Tests of survival craft radio equipment.

(a) Except for emergency position indicating radio beacons and two-way radiotelephone equipment, inspections and tests of survival craft radio equipment must be conducted by the licensee at weekly intervals while the ship is at sea or, if a test or inspection has not been conducted within a week prior to its departure, within 24 hours prior to the ship's departure from a port. The inspection and tests must include operation of the transmitter connected to an artificial antenna and determination of the specific gravity or voltage under normal load of any batteries.

Subpart R - Compulsory Radiotelephone Installations for Vessels 300 Gross Tons

§ 80.851 Applicability.

The radiotelephone requirements of Part II of Title III of the Communications Act apply to cargo ships of 300 gross tons and upward but less than 1600 gross tons. The radiotelephone requirements of the Safety Convention apply to passenger ships irrespective of size and cargo ships of 300 gross tons and upward on international voyages. These ships are required to carry a radiotelephone installation complying with this Subpart.

§ 80.853 Radiotelephone station.

(a) The radiotelephone station is a radiotelephone installation and other equipment necessary for the proper operation of the installation.

(b) The radiotelephone station must be installed to insure safe and effective operation of the equipment and to facilitate repair.

APPENDIX – PART 80

Adequate protection must be provided against the effects of vibration, moisture, and temperature.

(c) The radiotelephone station and all necessary controls must be located at the level of the main wheelhouse or at least one deck above the ship's main deck.

(d) The principal operating position of the radiotelephone station must be in the room from which the ship is normally steered while at sea.

§ 80.855 Radiotelephone transmitter.

(a) The transmitter must be capable of transmission of H3E and J3E emission on 2182 kHz, and J3E emission on 2638 kHz and at least two other frequencies within the band 1605 to 3500 kHz available for ship-to-shore or ship-to-ship communication.

(b) The duty cycle of the transmitter must permit transmission of the international radiotelephone alarm signal.

(c) The transmitter must be capable of transmitting clearly perceptible signals from ship to ship during daytime under normal conditions over a range of 150 nautical miles.

(d) The transmitter complies with the range requirement specified in paragraph (c) of this section if:

(1) The transmitter is capable of being matched to actual ship station transmitting antenna meeting the requirements of § 80.863; and

(2) The output power is not less than 60 watts peak envelope power for H3E and J3E emission on the frequency 2182 kHz and for J3E emission on the frequency 2638 kHz into either an artificial antenna consisting of a series network of 10 ohms resistance and 200 picofarads capacitance, or an artificial antenna of 50 ohms nominal impedance. An individual demonstration of the power output capability of the transmitter, with the radiotelephone installation normally installed on board ship, may be required.

(e) The transmitter must provide visual indication whenever the transmitter is supplying power to the antenna.

(f) The transmitter must be protected from excessive currents and voltages.

(g) A durable nameplate must be mounted on the transmitter or made an integral part of it showing clearly the name of the transmitter manufacturer and the type or model of the transmitter.

(h) An artificial antenna must be provided to permit weekly checks of the automatic device for generating the radiotelephone alarm signal on frequencies other than the radiotelephone distress frequency.

§ 80.873 VHF radiotelephone transmitter.

(a) The transmitter must be capable of transmission of G3E emission on 156.300 MHz and 156.800 MHz, and on frequencies which have been specified for use in a system established to promote safety of navigation.

(b) The transmitter must be adjusted so that the transmission of speech normally produces peak modulation within the limits of 75 percent and 100 percent.

(c) The transmitter must deliver a carrier power between 8 and 25 watts into a 50 ohm effective resistance. Provision must be made for reducing the carrier power to a value between 0.1 and 1.0 watt.

Subpart S - Compulsory Radiotelephone Installations for Small Passenger Boats

§ 80.901 Applicability.

The provisions of Part III of Title III of the Communication Act require United States vessels which transport more than six passengers for hire while such vessels are being navigated on any tidewater within the jurisdiction of the United States adjacent or contiguous to the open sea, or in the open sea to carry a radiotelephone installation complying with this subpart. The provisions of Part III do not apply to vessels which are equipped with a radio installation for compliance with Part II of Title III of the Act, or for compliance with the Safety Convention, or to vessels navigating on the Great Lakes.

§ 80.903 Inspection of radiotelephone installation.

Every vessel subject to Part III of Title III of the Communications Act must have a detailed inspection by the Commission of the prescribed installation once every five years. If after inspection the Commission determines that all relevant provisions of Part III of Title III of the Communications Act, the rules of the Commission, and the station license are met a Communications Act Safety Radiotelephone Certificate will be issued. The effective date of this certificate is the date the installation is found to be in compliance, or not more than one business day later.

§ 80.905 Vessel radio equipment.

(a) Vessels subject to Part III of Title III of the Communications Act that operate in the waters described in § 80.901 of this section must, at a minimum, be equipped as follows:

(1) Vessels operated solely within the communications range of a VHF public coast station or U.S. Coast Guard station that maintains a watch on 156.800 MHz while the vessel is navigated must be equipped

GENERAL RADIOTELEPHONE OPERATOR LICENSE

with a VHF radiotelephone installation. Vessels in this category must not operate more than 20 nautical miles from land.

Subpart T - Radiotelephone Installation Required for Vessels on the Great Lakes

§ 80.951 Applicability.

The Agreement Between the United States of America and Canada for Promotion of Safety on the Great Lakes by Means of Radio, 1973, applies to vessels of all countries when navigated on the Great Lakes.

Subpart U - Radiotelephone Installations Required by the Bridge-To-Bridge Act

§ 80.1001 Applicability.

The Bridge-to-Bridge Act and the regulations of this part apply to the following vessels in the navigable waters of the United States:

(a) Every power-driven vessel of 300 gross tons and upward while navigating;

(b) Every vessel of 100 gross tons and upward carrying one or more passengers for hire while navigating;

(c) Every towing vessel of 26 feet (7.8 meters) or over in length, measured from end to end over the deck excluding sheer, while navigating; and

(d) Every dredge and floating plant engaged, in or near a channel or fairway, in operations likely to restrict or affect navigation of other vessels. An unmanned or intermittently manned floating plant under the control of a dredge shall not be required to have a separate radiotelephone capability.

Subpart V - Emergency Position Indicating Radiobeacons (EPIRBs)

§ 80.1051 Scope.

This subpart describes the technical and performance requirements for Classes A, B, C, and S, and Categories 1, 2, and 3 EPIRB stations.

§ 80.1053 Special requirements for Class A EPIRB stations.

(a) A Class A EPIRB station must meet the following:

(1) Float free of a sinking ship;

(2) Activate automatically when it floats free of a sinking ship;

(3) Have an antenna that deploys automatically when the EPIRB activates;

(4) Use A3X emission on a mandatory basis and A3E and NØN emissions on an optional basis on the frequencies 121.500 MHz and 243.000 MHz;

(b) Class A EPIRB's must have a manually activated test switch which must be held in position for test operation and when released return the EPIRB to its normal state.

(c) EPIRBs that meet the output power characteristics of this section must have a permanent label prominently displayed on the outer casting stating, "Meets FCC Rules for improved satellite detection."

(d) Vacuum tubes are not permitted in EPIRBs.

(e) EPIRBs must be powered by a battery contained within the transmitter case or in a battery holder that is rigidly attached to the transmitter case. The battery connector must be corrosion resistant and positive in action and must not rely for contact upon spring force alone. The useful life of the battery is the length of time that the battery can be stored under marine environmental conditions without the EPIRB transmitter peak effective radiated power falling below 75 milliwatts prior to 48 hours of continuous operation. The month and year of the battery's manufacture must be permanently marked on the battery and the month and year upon which 50 percent of its useful life will have expired must be permanently marked on both the battery and the outside of the transmitter. The batteries must be replaced if 50 percent of their useful life has expired or if the transmitter has been used in an emergency situation.

(f) The EPIRB must be waterproof and must not be accidentally activated by rain, seaspray, hose washdown spray or storage in high humidity conditions.

(g) Operating instructions understandable by untrained personnel must be permanently displayed on the equipment.

(h) The exterior of the equipment must have no sharp edges or projections. Means must be provided to fasten the EPIRB to a survival craft or person.

(i) The antenna must be deployable to its designed length and operating position in a foolproof manner. The antenna must be securely attached to the EPIRB and easy to de-ice. The antenna must be vertically polarized and omnidirectional.

Subpart W - Global Maritime Distress and Safety System (GMDSS) [Reserved]

This Subpart contains the rules applicable to the Global Maritime Distress and Safety System (GMDSS).

§ 80.1065 Applicability

(b) The regulations contained within this Subpart apply to all passenger ships regardless of size and cargo ships of 300 tons gross tonnage and upward.

APPENDIX – PART 80

(4) Ships constructed before February 1, 1995, must comply with all requirements of this Subpart as of February 1, 1999.

§ 80.1073 Radio operator requirements for ship stations.

(a) Ships must carry at least two persons holding GMDSS Radio Operator Licenses. This license qualifies personnel as GMDSS radio operators for the purposes of operating GMDSS radio installations, including basic equipment adjustments.

(1) One of the qualified GMDSS radio operators must be designated to have primary responsibility for radiocommunications during distress incidents.

(2) A second qualified GMDSS radio operator must be designated as backup for distress and safety radiocommunications.

§ 80.1074 Radio maintenance personnel for at-sea maintenance.

(a) Ships that elect the at-sea option for maintenance of GMDSS equipment must carry at least one person who qualifies as a GMDSS radio maintainer for the maintenance and repair of equipment specified in this Subpart. This person may be, but need not be, the person designated as GMDSS radio operator as specified in § 80.1073.

(b) [On an interim basis until GMDSS examinations are available] The following licenses qualify personnel as GMDSS radio maintainers to perform at-sea maintenance of equipment specified in this Subpart.

(1) T-1: First Class Radiotelegraph Operator's License;

(2) T-2: Second Class Radiotelegraph Operator's License;

(3) G: General Radiotelephone Operator's License.

Subpart X - Voluntary Radio Installations

GENERAL

§ 80.1151 Voluntary radio operations

Voluntary ships must meet the rules applicable to the particular mode of operation as contained in the following subparts of this part and as modified by § 80.1153:

§ 80.1153 Station log and radio watches.

(a) Licensees of voluntary ships are not required to operate the ship radio station or to maintain radio station logs.

(b) When a ship radio station of a voluntary ship is being operated, appropriate general purpose watches must be maintained in accordance with § 80.146 (500 kHz), § 80.147 (2182 kHz) and § 80.148 (156.800 MHz - Channel 16).

VOLUNTARY TELEGRAPHY
§ 80.1155 Radioprinter.

Radioprinter operations provide a record of communications between authorized maritime mobile stations.

(a) Supplementary eligibility requirements. Ships must be less than 1600 gross tons.

(b) Scope of communication.

(1) Ship radioprinter communications may be conducted with an associated private coast station.

(2) Ships authorized to communicate by radioprinter with a common private coast station may also conduct intership radioprinter operations.

(3) Only those communications which are associated with the business and operational needs of the ship are authorized.

§ 80.1165 Assignment and use of frequencies.

Frequencies for general radiotelephone purposes are available to ships in three radio frequency bands. Use of specific frequencies must meet the Commission's rules concerning the scope of service and the class of station with which communications are intended. The three frequency bands are:

(a) *156-158 MHz (VHF/FM Radiotelephone)*. Certain frequencies within this band are public correspondence frequencies and they must be used as working channels when communicating with public coast stations. Other working frequencies within the band are categorized by type of communications for which use is authorized when communicating with a private coast station or between ships. Subpart H of this Part lists the frequencies and types of communications for which they are available.

(b) *1600-4000 kHz (SSB Radiotelephone)*. Specific frequencies within this band are authorized for single sideband (SSB) communications with public and private coast stations or between ships. The specific frequencies are listed in Subpart H of this Part.

(c) *4000-23000 kHz (SSB Radiotelephone)*. Specific frequencies within this band are authorized for SSB communications with public and private coast stations. The specific frequencies are listed in Subpart H of this Part.

GENERAL RADIOTELEPHONE OPERATOR LICENSE

Federal Communications Commission - Rules and Regulations – Excerpts from
47 C.F.R. – PART 87 – STATIONS IN THE AVIATION SERVICES

Subpart A - General Information

§87.5 Definitions.

Aeronautical advisory station (unicom). An aeronautical station used for advisory and civil defense communications primarily with private aircraft stations.

Aeronautical enroute station. An aeronautical station which communicates with aircraft stations in flight status or with other aeronautical enroute stations.

Aeronautical fixed service. A radiocommunication service between specified fixed points provided primarily for the safety of air navigation and for the regular, efficient and economical operation of air transport. A station in this service is an aeronautical fixed station.

Aeronautical Mobile Off-Route (OR) Service. An aeronautical mobile service intended for communications, including those relating to flight coordination, primarily outside national or international civil air routes. (RR)

Aeronautical Mobile Route (R) Service. An aeronautical mobile service reserved for communications relating to safety and regularity of flight, primarily along national or international civil air routes. (RR)

Aeronautical Mobile-Satellite Off-Route (OR) Service. An aeronautical mobile-satellite service intended for communications, including those relating to flight coordination, primarily outside national and international civil air routes. (RR)

Aeronautical Mobile-Satellite Route (R) Service. An aeronautical mobile-satellite service reserved for communications relating to safety and regularity of flights, primarily along national or international civil air routes. (RR)

Aeronautical mobile service. A mobile service between aeronautical stations and aircraft stations, or between aircraft stations, in which survival craft stations may also participate; emergency position-indicating radio-beacon stations may also participate *in this service* on designated distress and emergency frequencies.

Aeronautical multicom station. An aeronautical station used to provide communications to conduct the activities being performed by, or directed from, private aircraft.

Aeronautical radionavigation service. A radionavigation service intended for the benefit and for the safe operation of aircraft.

Aeronautical search and rescue station. An aeronautical station for communication with aircraft and other aeronautical search and rescue stations pertaining to search and rescue activities with aircraft.

Aeronautical station. A land station in the aeronautical mobile service. In certain instances an aeronautical station may be located, for example, on board ship or on a platform at sea.

Aeronautical utility mobile station. A mobile station used on airports for communications relating to vehicular ground traffic.

Air carrier aircraft station. A mobile station on board an aircraft which is engaged in, or essential to, the transportation of passengers or cargo for hire.

Aircraft station. A mobile station in the aeronautical mobile service other than a survival craft station, located on board an aircraft.

Airport. An area of land or water that is used or intended to be used for the landing and takeoff of aircraft, and includes its buildings and facilities, if any.

Airport control tower (control tower) station. An aeronautical station providing communication between a control tower and aircraft.

Automatic weather observation station. A land station located at an airport and used to automatically transmit weather information to aircraft.

Aviation service organization. Any business firm which maintain facilities at an airport for the purposes of one or more of the following general aviation activities: (a) Aircraft fueling; (b) aircraft services (e.g. parking, storage, tie-downs); (c) aircraft maintenance or sales; (d) electronics equipment maintenance or sales; (e) aircraft rental, air taxi service or flight instructions; and (f) baggage and cargo handling, and other passenger or freight services.

Aviation services. Radio-communication services for the operation of aircraft. These services include aeronautical fixed service, aeronautical mobile service, aeronautical radiodetermination service, and secondarily, the handling of public correspondence on frequencies in the maritime mobile and maritime mobile satellite services to and from aircraft.

Aviation support station. An aeronautical station used to coordinate aviation services with aircraft and to communicate with aircraft engaged in unique or specialized activities. (See Subpart K)

Civil Air Patrol station. A station used exclusively for communications of the Civil Air Patrol.

Emergency locator transmitter(ELT). A transmitter of an aircraft or a survival craft actuated manually or automatically that is used as an alerting and locating aid for survival purposes.

Emergency locator transmitter (ELT) test station. A land station used for testing ELTs or for training in the use of ELTS.

Expendable Launch Vehicle (ELV). A booster rocket that can be used only once to launch a payload, such as a missile or space vehicle.

Flight test aircraft station. An aircraft station used in the testing of aircraft or their major components.

Glide path station. A radionavigation land station which provides vertical guidance to aircraft during approach to landing.

Instrument landing system (ILS). A radionavigation system which provides aircraft with horizontal and vertical guidance just before and during landing and, at certain fixed points, indicates the distance to the reference point of landing.

Instrument landing system glide path. A system of vertical guidance embodied in the instrument landing system which indicates the vertical deviation of the aircraft from its optimum path of descent.

Instrument landing system localizer. A system of horizontal guidance embodied in the instrument landing system which indicates the horizontal deviation of the aircraft from its optimum path of descent along the axis of the runway or along some other path when used as an offset.

Land station. A station in the mobile service not intended to be used while in motion.

Localizer station. A radionavigation land station which provides horizontal guidance to aircraft with respect to a runway center line.

Marker beacon station. A radionavigation land station in the aeronautical radionavigation service which employs a marker beacon. A marker beacon is a transmitter which radiates vertically a distinctive pattern for providing position information to aircraft.

Mean power (of a radio transmitter). The average power supplied to the antenna transmission line by a transmitter during an interval of time sufficiently long compared with the lowest frequency encountered in the modulation taken under normal operating conditions.

Microwave landing system. An instrument landing system operating in the microwave spectrum that provides lateral and vertical guidance to aircraft having compatible avionics equipment.

Mobile service. A radiocommunication service between mobile and land stations, or between mobile stations. A mobile station is intended to be used while in motion or during halts at unspecified points.

Operational fixed station. A fixed station, not open to public correspondence, operated by and for the sole use of persons operating their own radiocommunication facilities in the public safety, industrial, land transportation, marine, or aviation services.

Peak envelope power (of a radio transmitter). The average power supplied to the antenna transmission line by a transmitter during one radio frequency cycle at the crest of the modulation envelope taken under normal operating conditions.

Private aircraft station. A mobile station on board an aircraft not operated as an air carrier. A station on board an air carrier aircraft weighing less than 12,500 pounds maximum certified takeoff gross weight may be licensed as a private aircraft station.

Racon station. A radionavigation land station which employs a racon. A racon (radar beacon) is a transmitter-receiver associated with a fixed navigational mark, which when triggered by a radar, automatically returns a distinctive signal which can appear on the display of the triggering radar, providing range, bearing and identification information.

Radar. A radiodetermination system based upon the comparison of reference signals with radio signals reflected, or retransmitted, from the position to be determined.

Radio altimeter. Radionavigation equipment, on board an aircraft or spacecraft, used to determine the height of the aircraft or spacecraft above the Earth's surface or another surface.

Radiobeacon station. A station in the radionavigation service the emissions of which are intended to enable a mobile station to determine its bearing or direction in relation to the radiobeacon station.

Radiodetermination service. A radiocommunication service which uses radiodetermination. Radiodetermination is the determination of the position, velocity and/or other characteristics of an object, or the obtaining of information relating to these parameters, by means of the propagation of radio waves. A station in this service is called a radiodetermination station.

Radiolocation service. A radiodetermination service for the purpose of radiolocation. Radiolocation is the use of radiodetermination for purposes other than those of radionavigation.

Radionavigation land test stations. A radionavigation land station which is used to transmit information essential to the testing and calibration of aircraft navigational aids, receiving equipment, and interrogators at predetermined surface locations. The Maintenance Test Facility (MTF) is used primarily to permit maintenance testing by aircraft radio service personnel. The Operational Test Facility (OTF) is used primarily to permit the pilot to check a radionavigation system aboard the aircraft prior to takeoff.

Radionavigation service. A radiodetermination service for the purpose of radionavigation. Radionavigation is the use of radiodetermination for the purpose of navigation, including obstruction warning.

Re-usable launch vehicle (RLV). A booster rocket that can be recovered after launch, refurbished and relaunched.

Surveillance radar station. A radionavigation land station in the aeronautical radionavigation service employing radar to display the presence of aircraft within its range.

Survival craft station. A mobile station in the maritime or aeronautical mobile service intended solely for survival purposes and located on any lifeboat, life raft **or** other survival equipment.

VHF Omni directional range station (VOR). A radionavigation land station in the aeronautical radionavigation service providing direct indication of the bearing (omni-bearing) of that station from an aircraft.

Subpart B - Applications and Licenses

§ 87.17 Scope.

This subpart contains the procedures and requirements for the filing of applications for radio station licenses in the aviation services.

§ 87.19 Basic eligibility.

(a) *General.* Foreign governments or their representatives cannot hold station licenses.

(b) *Aeronautical enroute and aeronautical fixed stations.* The following persons cannot hold an aeronautical enroute or an aeronautical fixed station license.

(1) Any alien or the representative of any alien;

(2) Any corporation organized under the laws of any foreign government;

(3) Any corporation of which any officer or director is an alien;

(4) Any corporation of which more than one-fifth of the capital stock is owned of record or voted by aliens or their representatives or by a foreign government or its representative, or by a corporation organized under the laws of a foreign country; or

(5) Any corporation directly or indirectly controlled by any other corporation of which more than one-fourth of the capital stock is owned of record or voted by aliens, their representatives, or by a foreign government or its representatives, or by any corporation organized under the laws of a foreign country, if the Commission finds that the public interest will be served by the refusal or revocation of such license.

§ 87.21 Standard forms to be used.

Applications must be submitted on prescribed forms which may be obtained from the Commission in Washington, DC 20554 or from any of its field offices.

§ 87.25 Filing of applications.

Rules about the filing of applications for aviation radio station licenses are contained in this section.

§ 87.27 License term.

Licenses for regular stations will normally be issued for five years.

§ 87.33 Transfer of aircraft station license prohibited.

An aircraft station license cannot be assigned. If the aircraft ownership is transferred, the previous license must be returned to the Commission. The new owner must file for a new license.

§ 87.35 Cancellation of license.

When a station permanently discontinues operation, the license must be returned to the Commission, Gettysburg, PA 17326.

§ 87.39 Equipment acceptable for licensing.

Transmitters listed in this part must be type accepted for a particular use by the Commission based upon technical requirements contained in Subpart D of this part.

§ 87.41 Frequencies.

(a) *Applicant responsibilities.* The applicant must propose frequencies to be used by the station consistent with the applicant's eligibility, the proposed operation and the frequencies available for assignment. Applicants must cooperate in the selection and use of frequencies in order to *minimize* interference and obtain the most effective use of stations. See Subpart E and the appropriate Subpart applicable to the class of station being considered.

(b) *Licensing limitations.* Frequencies are available for assignment to stations on a shared basis only and will not be assigned for the exclusive use of any licensee. The use of any assigned frequency may be restricted to one or more geographical areas.

(c) *Government frequencies.* Frequencies allocated exclusively to federal government radio stations may be licensed. The applicant for a government frequency must provide a satisfactory showing that such assignment is required for inter-communication with government stations or required for coordination with activities of the federal government. The Commission will coordinate with the appropriate government agency before a government frequency is assigned.

(d) *Assigned frequency.* The frequency coinciding with the center of an authorized bandwidth of emission must be specified as the assigned frequency. For single sideband emission, the carrier frequency must also be specified.

§ 87.43 Operation during emergency.

A station may be used for emergency communications in a manner other than that specified in the station license or in the operating rules when normal communication facilities are disrupted. The Commission may order the discontinuance of any such emergency service.

Subpart C - Operating Requirements and Procedures

OPERATING REQUIREMENTS

§ 87.69 Maintenance tests.

The licensee may make routine maintenance tests on equipment other than emergency locator transmitters if there is no interference with the communications of any other station. Procedures for conducting tests on emergency locator transmitters are contained in Subpart F.

§ 87.71 Frequency measurements.

A licensed operator must measure the operating frequencies of all landbased transmitters at the following times:

(a) When the transmitter is originally installed;

(b) When any change or adjustment is made in the transmitter which may affect an operating frequency; or

(c) When an operating frequency has shifted beyond tolerance.

§ 87.73 Transmitter adjustments and tests.

A general radiotelephone operator must directly supervise and be responsible for all transmitter adjustments or tests during installation, servicing or maintenance of a radio station. A general radiotelephone operator must be responsible for the proper functioning of the station equipment.

§ 87.75 Maintenance of tower marking and control equipment.

Section 303(q) of the Communications Act of 1934, as amended, requires some antenna structures to be painted or illuminated. The licensee of any radio station which has such an antenna structure must operate and maintain the tower marking and associated control equipment in accordance with Part 17.

§ 87.77 Availability for inspections.

The licensee must make the station and its records available for inspection upon request.

§ 87.79 Answer to notice of violation.

(a) Any person who receives an official notice of violation of the Communications Act, any legislative act, executive order, treaty to which the U.S. is a party, terms of a station or operator license, or the Commission's rules must send a written answer, in duplicate, to the office which originated the notice, within 10 days of receipt. If the licensee cannot acknowledge within the allotted period due to unavoidable circumstances, an answer must be given at the earliest practicable date with a satisfactory explanation of the delay.

(b) The answer to each notice must be complete in itself. The answer must contain a full explanation of the incident involved and must give the action taken to prevent a recurrence of the violation. If the notice relates to operator errors, the answer must give the name and license number of the operator on duty.

RADIO OPERATOR REQUIREMENTS

§ 87.87 Classification of operator licenses and endorsements.

(a) Commercial radio operator licenses issued by the Commission are classified in accordance with the Radio Regulations of the International Telecommunication Union.

(b) The following licenses are issued by the Commission. International classification, if different from the license name, is given in parentheses. The licenses and their alphanumeric designator are listed in descending order.

(1) T-1 First Class Radiotelegraph Operator's Certificate

(2) T-2 Second Class Radiotelegraph Operator's Certificate

(3) G General Radiotelephone Operator License (radiotelephone operator's general certificate)

(4) T-3 Third Class Radiotelegraph Operator's Certificate (radiotelegraph operator's special certificate)

GENERAL RADIOTELEPHONE OPERATOR LICENSE

(5) MP Marine Radio Operator Permit (radiotelephone operator's restricted certificate)

(6) RP Restricted Radiotelephone Operator Permit (radiotelephone-operator's restricted certificate)

§ 87.89 Minimum operator requirements.

(a) A station operator must hold a commercial radio operator license or permit, except as listed in paragraph (d).

(b) The minimum operator license or permit required for operation of each specific classification is:

MINIMUM OPERATOR LICENSE OR PERMIT

Land stations, all classes
- All frequencies except VHF telephony transmitters providing domestic service .... RP

Aircraft stations, all classes
- Frequencies below 30 MHz allocated exclusively to aeronautical mobile services .... RP
- Frequencies below 30 MHz not allocated exclusively to aeronautical mobile services ... MP or higher
- Frequencies above 30 MHz not allocated exclusively to aeronautical mobile services and assigned for international use .... MP or higher
- Frequencies above 30 MHz not assigned for international use... none
- Frequencies not used solely for telephone or exceeding 250 watts carrier power or 1000 watts peak envelope power .... G or higher

(c) The operator of a telephony station must directly supervise and be responsible for any other person who transmits from the station, and must ensure that such communications are in accordance with the station license

(d) No operator license is required to:

(1) Operate an aircraft radar set, radio altimeter, transponder or other aircraft automatic radionavigation transmitter by flight personnel;

(2) Test an emergency locator transmitter or a survival craft station used solely for survival purposes;

(3) Operate an aeronautical enroute station which automatically transmits digital communications to aircraft stations;

(4) Operate a VHF telephony transmitter providing domestic service or used on domestic flights.

§ 87.91 Operation of transmitter controls.

The holder of a marine radio operator permit or a restricted radiotelephone operator permit must perform only transmitter operations which are controlled by external switches. These operators must not perform any internal adjustment of transmitter frequency determining elements. Further, the stability of the transmitter frequencies at a station operated by these operators must be maintained by the transmitter itself. When using an aircraft radio station on maritime mobile service frequencies the carrier power of the transmitter must not exceed 250 watts (emission A3E) or 1000 watts (emission R3E, H3E, or J3E).

OPERATING PROCEDURES

§ 87.103 Posting station license.

(a) *Stations at fixed locations.* The license or a photocopy must be posted or retained in the station's permanent records.

(b) *Aircraft radio stations.* The license must be either posted in the aircraft or kept with the aircraft registration certificate. If a single authorization covers a fleet of aircraft, a copy of the license must be either posted in each aircraft or kept with each aircraft registration certificate.

(c) *Aeronautical mobile stations.* The license must be retained as a permanent part of the station records.

§ 87.105 Availability of operator permit or license.

All operator permits or licenses must be readily available for inspection.

§ 87.107 Station identification.

(a) *Aircraft station.* Identify by one of the following means:

(1) Aircraft radio station call sign.

(2) Assigned FCC control number (assigned to ultralight aircraft).

(3) The type of aircraft followed by the characters of the registration marking ("N" number) of the aircraft, omitting the prefix letter "N". When communication is initiated by a ground station, an aircraft station may use the type of aircraft followed by the last three characters of the registration marking.

(4) The FAA assigned radiotelephony designator of the aircraft operating organization followed by the flight identification number.

(5) An aircraft identification approved by the FAA for use by aircraft stations participating in an organized flying activity of short duration.

(b) *Land and fixed stations.* Identify by means of radio station call sign, its location, its assigned FAA identifier, the name of the city area or airport which it serves, or any additional identification required. An aeronautical enroute station which is part of a multistation network may also be identified by the location of its control point.

(c) *Survival craft station.* Identify by transmitting a reference to its parent aircraft. No identification is required when distress signals are transmitted automatically. Transmissions other than distress or emergency signals, such as equipment testing or adjustment, must be identified by the call sign or by the registration marking of the parent aircraft followed by a single digit other than 0 or 1.

(d) *Exempted station.* The following types of stations are exempted from the use of a call sign: Airborne weather radar, radio altimeter, air traffic control transponder, distance measuring equipment, collision avoidance equipment, racon, radio relay, radionavigation land test station (MTF), and automatically controlled aeronautical enroute stations.

§ 87.109 Station logs.

A station at a fixed location in the international aeronautical mobile service must maintain a written or automatic log in accordance with Paragraph 3.5, Volume II, Annex 10 of the ICAO Convention.

§ 87.111 Suspension or discontinuance of operation.

The licensee of any airport control tower station or radionavigation land station must notify the nearest FAA regional office upon the temporary suspension or permanent discontinuance of the station. The FAA center must be notified again when service resumes.

Subpart D - Technical Requirements

This subpart contains the power, frequency stability, bandwidth, emission, modulation and acceptability of transmitters for licensing requirements .

Subpart E - Frequencies

This subpart contains class of station symbols and a frequency table which lists assignable frequencies.

Subpart F - Aircraft Stations

This subpart covers communications limitations of domestic and foreign aircraft stations and requirements for public correspondence. Aircraft stations must limit their communications to the necessities of safe, efficient, and economic operation of aircraft and the protection of life and property.

Subpart G - Aeronautical Advisory Stations (Unicoms)

This subpart covers the guidelines for unicom communications which must be limited to the necessities of safe and expeditious operation of aircraft such as condition of runways, types of fuel available, wind conditions, weather information, dispatching, or other necessary information.

Subpart H - Aeronautical Multicom Stations

The communications of an aeronautical multicom station (multicom) must pertain to activities of a temporary, seasonal or emergency nature involving aircraft in flight. Communications are limited to directing or coordinating ground activities from the air or aerial activities from the ground.

Subpart I - Aeronautical Enroute and Aeronautical Fixed Stations

Aeronautical enroute stations provide operational control communications to aircraft along domestic or international air routes. Operational control communications include the safe, efficient and economical operation of aircraft, such as fuel, weather, position reports, aircraft performance, and essential services and supplies. Public correspondence is prohibited. Aeronautical fixed stations provide non-public point-to-point communications service pertaining to safety, regularity and economy of flight.

Subpart J - Flight Test Stations

The use of flight test stations is restricted to the transmission of necessary information or instructions relating directly to tests of aircraft or components thereof.

Subpart K - Aviation Support Stations

Aviation support stations are used for the following types of operations:

(a) Pilot training;

(b) Coordination of soaring activities between gliders, tow aircraft and land stations;

(c) Coordination of activities between free balloons or lighter-than-air aircraft and ground stations;

(d) Coordination between aircraft and aviation service organizations located on an airport concerning the safe and efficient portal-to-portal transit of the aircraft, such as the types of fuel and ground services available, and

(e) Promotion of safety of life and property.

Subpart L - Aeronautical Utility Mobile Stations

Aeronautical utility mobile stations provide communications for vehicles operating on an airport movement area. An airport movement area is defined as the runways, taxiways and other areas utilized for taxiing, takeoff and landing of aircraft, exclusive of loading ramp and parking areas.

GENERAL RADIOTELEPHONE OPERATOR LICENSE

Subpart M - Aeronautical Search and Rescue Stations

Aeronautical search and rescue land and mobile stations must be used only for communications with aircraft and other aeronautical search and rescue stations engaged in search and rescue activities.

Subpart N - Emergency Communications

This Subpart provides the rules governing operation of stations in the Aviation Services during any national or local emergency situation constituting a threat to national security or safety of life and property.

Subpart O - Airport Control Tower Stations

Airport control tower stations (control towers) and control tower remote communications outlet stations (RCOS) must limit their communications to the necessities of safe and expeditious operations of aircraft operating on or in the vicinity of the airport. Control towers and RCOs provide air traffic control services to aircraft landing, taking off and taxing on the airport as well as aircraft transmitting the airport traffic area.

Subpart P - Operational Fixed Stations

An operational fixed station provides control, repeater or relay functions for its associated aeronautical station.

Subpart Q - Stations in the Radiodetermination Service

Stations in the aeronautical radiodetermination service provide radionavigation and radiolocation services which must be limited to aeronautical navigation, including obstruction warning.

Subpart R - Civil Air Patrol Stations

Civil Air Patrol land and mobile stations must be used only for training, operational and emergency activities of the Civil Air Patrol. They may communicate with other land and, mobile stations of the Civil Air Patrol. When engaged in training or on actual missions in support of the U.S. Air Force, Civil Air Patrol stations may communicate with U.S. Air Force stations on the frequencies specified in Subpart E.

Subpart S - Automatic Weather Observation Stations

Automatic weather observation stations provide up-to-date weather information including the time of the latest weather sequence, altimeter setting, wind speed and direction, dewpoint, temperature, visibility and other pertinent data needed at airports having neither a full-time control tower nor a full-time FAA Flight Service Station.

APPENDIX – QUESTION POOL FORMULAS

SUMMARY OF QUESTION POOL FORMULAS

Question	Formula	Where:
3B6	**VHF Range to Horizon** $$D = 1.415 \times \sqrt{H}$$	D = Distance in **Miles** H = Antenna Height in **feet**
3C6-11	**Counter Readout Error** $$\text{Readout Error} = f \times a$$	f = Frequency in **MHz** being measured a = Counter accuracy in **parts per million**
3D5-7	**Resonance** $$X_L = X_C$$ $$2\pi f_r L = \frac{1}{2\pi f_r C}$$ $$f_r = \frac{1}{2\pi \sqrt{LC}}$$	X_L = Inductive reactance in **ohms** X_C = Capacitive reactance in **ohms** f_r = Resonant frequency in **hertz** L = Inductance in **henrys** C = Capacitance in **farads** π = 3.14
3D25-31,99	**True Power** For AC: $P_T = P_A \times P_F$ $P_A = E \times I$ $PF = \cos \phi$ For DC: $P_T = E \times I$	P_T = True power in **watts** P_A = Apparent power in **watts** ϕ = Phase angle in **degrees** PF = Power factor E = Applied voltage in **volts** I = Circuit current in **amps**
3D49-76	**Time Constants** $$\tau = RC$$	τ = Time Constant in **seconds** R = Resistance in **ohms** C = Capacitance in **farads**
	$$\tau = \frac{L}{R}$$	τ = Time Constant in **seconds** R = Resistance in **ohms** L = Inductance in **henries**

Time Constant Curve

		RC		RL	
Time Constant τ	% of Change in τ	% of Final Q or V on C When Charging	% of Initial Q or V on C When Discharging	% of Final I When Increasing	% of Initial I When Decreasing
1	63.2	63.2	36.8	63.2	36.8
2	23.3	86.5	13.5	86.5	13.5
3	8.5	95.0	5.0	95.0	5.0
4	3.2	98.2	1.8	98.2	1.8
5	1.1	99.3	0.7	99.3	0.7

Question	Formula	Where:
3D59	**Capacitors** **Resistors** Series: $C_T = \dfrac{C_1 \times C_2}{C_1 + C_2}$ $R_T = R_1 + R_2$ Parallel: $C_T = C_1 + C_2$ $R_T = \dfrac{R_1 \times R_2}{R_1 + R_2}$	C = Capacitance in **farads** R = Resistance in **ohms**
3D77	**Inductive Reactance** $$X_L = 2\pi f L$$	X_L = Inductive reactance in **ohms** L = Inductance in **henries** f = Frequency in **hertz** π = 3.14

SUMMARY OF QUESTION POOL FORMULAS (Continued)

Question	Formula	Where:
3D77-96	**Rectangular Coordinates** $$Z = R \pm jX$$ **Polar Coordinates** $$Z = Z \underline{/\pm\Theta}$$ **Conversion:** $Z = R \pm jX$ to $Z \underline{/\pm\Theta}$ $Z = \sqrt{R^2+X^2} \underline{/\arctan \frac{X}{R}}$ $Z \underline{/\pm\Theta}$ to $Z = R \pm jX$ $R = Z \cos\Theta$ $\pm jX = Z \sin\Theta$	Z = Impedance in **ohms** R = Resistance in **ohms** $+jX$ = Inductive reactance X_L in **ohms** $-jX$ = Capacitive reactance X_C in **ohms** Θ = Phase angle in **degrees** arctan = angle whose tangent is
3D77-96	**Do Addition and Subtraction in Rectangular Coordinates** $Z_1 = R_1 + jX_1$ $Z_2 = R_2 - jX_2$ **Addition** $Z_T = (R_1 + jX_1) + (R_2 - jX_2)$ $Z_T = (R_1 + R_2) + j(X_1 - X_2)$ **Subtraction** $Z_T = (R_1 + jX_1) - (R_2 - jX_2)$ $Z_T = (R_1 - R_2) + j(X_1 + X_2)$	**Do Multiplication and Division in Polars Coordinates** $Z_1 = Z_1 \underline{/\Theta_1}$ $Z_2 = Z_2 \underline{/\Theta_2}$ **Multiply** $Z_1 Z_2 = Z_1 \times Z_2 \underline{/\Theta_1 + \Theta_2}$ **Divide** $\dfrac{Z_1}{Z_2} = \dfrac{Z_1 \underline{/\Theta_1}}{Z_2 \underline{/\Theta_2}} = \dfrac{Z_1}{Z_2} \underline{/\Theta_1 - \Theta_2}$
3D80,86,89	**Capacitive Reactance** $$X_C = \frac{1}{2\pi fC}$$ $$X_C = \frac{10^6}{2\pi fC}$$ **Impedance in Series** $$Z_T = Z_1 + Z_2$$ **Impedance in Parallel** $$Z_T = \frac{Z_1 \times Z_2}{Z_1 + Z_2}$$	X_C = Capacitive reactance in **ohms** C = Capacitance in **farads** f = frequency in **hertz** π = 3.14 f = frequency in **MHz** C = Capacitance in **picofarads** Z = Impedance in **ohms** π = 3.14
3F53	**Transistor Amplifier Load Resistor** $$R_L = \frac{(V_{CC})^2}{2P_O}$$	R_L = Load resistance in **ohms** V_{CC} = Collector voltage in **volts** P_O = Amplifier's power output in **watts**
3G13-16	**Deviation Ratio** $$\text{Deviation Ratio} = \frac{\text{Maximum Carrier Frequency Deviation (in kHz)}}{\text{Maximum Modulation Frequency (in kHz)}}$$	
3G17-21	**Modulation Index** $$\text{Modulation Index} = \frac{\text{Deviation of FM Signal (in Hz)}}{\text{Modulating Frequency (in Hz)}}$$	

APPENDIX – QUESTION POOL FORMULAS

SUMMARY OF QUESTION POOL FORMULAS (Continued)

Question	Formula	Where:
3G48-52, 89-90	**RMS, Peak and Peak-to-Peak Voltage** $V_{RMS} = 0.707\ V_{PK}$ $V_{PK} = 1.414\ V_{RMS}$ $V_{PP} = 2\ V_{PK}$ $V_{PK} = \dfrac{V_{PP}}{2}$	V_{PP} = Peak-to-peak voltage in **volts** V_{PK} = Peak voltage in **volts** V_{RMS} = Root-mean-square voltage in **volts**
3G53-57	**Peak Envelope Power** $PEP = P_{DC} \times \text{Efficiency}$ **Amplifier Class / Efficiency** C — 80% B — 60% AB — 50% A — <50%	PEP = Peak envelope power in **watts** P_{DC} = Input DC power in **watts**
3G70-80	**Bandwidth for Digital** $BW = \text{baud rate} + (1.2 \times f_S)$	BW = Necessary bandwidth in **hertz** f_S = frequency shift in **hertz** Baud rate = Digital signal rate in **bauds**
3G70-80	**Bandwidth for CW** $BW_{CW} = \text{baud rate} \times \text{wpm} \times \text{fading factor}$	BW_{CW} = Necessary bandwidth in **hertz** wpm = Morse code signal rate in **words per minute** Fading factor = Constant of 5 for **CW**
3H31-34	**Physical Length vs Electric Length** $L = \dfrac{984\ \lambda V}{f}$ **Feedline / Velocity Factor** Coax — 0.66 Parallel — 0.82 Twin-Lead — 0.80	L = Physical length in **feet** λ = Electrical length in **wavelength** V = Velocity factor of feedline f = Frequency in **MHz**
3H60-63	**Antenna Bandwidth** $\text{Bandwidth} = \dfrac{203}{(\sqrt{10})^x}$ where $X = \dfrac{A_G}{10}$	Bandwidth = Antenna bandwidth in **degrees** A_G = Antenna gain in **dB**
3H100	**Modulated Antenna Current** % increase in modulation $= \left(\sqrt{1 + \dfrac{m^2}{2}} - 1\right) \times 100$	m = A value between 0 and 1 representing the % modulation, e.g., 1 = 100%

GENERAL RADIOTELEPHONE OPERATOR LICENSE

FEDERAL COMMUNICATIONS COMMISSION
Restricted Radiotelephone Operator Permit Application

FCC 753

Approved by OMB
3060-0049
Expires 3/31/91

- FEE PAYMENT REQUIRED. SEE ACCOMPANYING FEE FILING GUIDE.
- Before completing this form, see other side.
- Complete Parts 1 and 2—Print or Type.
- If you need a temporary Permit, complete Part 3.
- No examination is required.

I Certify that:
- I can keep at least a rough written log.
- I am familiar with the provisions of applicable treaties, laws, and rules and regulations governing the radio station which I will operate.
- I can speak and hear.
- I am legally eligible for employment in the United States.
- I need this permit because I intend to engage in international flights or voyages (Aviation and Marine Services only).
- The statements made on this application and any attachments are true to the best of my knowledge.

Estimated Average Burden Hr. Per Resp.: 6 min.

① Name—Last: **JONES** First: **JAMES** M.I.: **B.**
② Address Number & Street: **126 MAIN STREET**
  City: **CUSHING** State: **OK** ZIP Code: **74023**
③ Date of Birth Use numerals: MONTH **08** / DAY **12** / YEAR **1945**

④ Signature: *James B. Jones*   Date: **4/12/94**

Willful false statements made on this form are punishable by fine and/or imprisonment (U.S. Code, Title 18, Section 1001), and/or revocation of any station license or construction permit (U.S. Code, Title 47, Section 312(a)(1)), and/or forfeiture (U.S. Code, Title 47, Section 503).

FCC Form 753—Part 1
February 1991

DO NOT DETACH   **Fee Code: PARR**

Federal Communications Commission
1270 Fairfield Road, Gettysburg PA 17325-7245
Official Business
Penalty for Private Use—$300

UNITED STATES OF AMERICA
Federal Communications Commission
Restricted Radiotelephone Operator Permit

JAMES B. JONES

is authorized to operate any radio station which may be operated by a person holding this class of license. This permit is issued in conformity with Paragraphs 3454 and 3945 of the Radio Regulations, Geneva 1987, and is valid for the lifetime of the holder unless suspended by the FCC.

Not Valid without FCC Seal

James B. Jones
FCC Form 753–Part 2

⑤ Print or Type Your Full Name

**JAMES B. JONES
126 MAIN ST.
CUSHING, OK 74023**

POSTAGE AND FEES PAID
FEDERAL COMMUNICATIONS COMMISSION
FCC 615

⑥ Signature (Keep your signature within the box)
⑦ Print *Your* Name and Address Above
DO NOT ADDRESS TO THE F.C.C.

FCC Form 753—**Part 2**
February 1991

FEDERAL COMMUNICATIONS COMMISSION
Temporary Restricted Radiotelephone Operator Permit

FCC Form 753—Part 3
February 1991

Approved by OMB
3060-0049
Expires 3/31/91

If you need a temporary Restricted Radiotelephone Operator Permit while your application is being processed, do the following:
- Complete Parts 1 & 2 of this form and mail to the FCC.
- Complete this part of the form and keep it.

If you have done the above, you now hold a temporary Restricted Radiotelephone Operator Permit. This is your temporary permit. DO NOT MAIL IT TO THE FCC.

This permit is valid for 60 days from the date Parts 1 and 2 of this form were mailed to the FCC.

You must obey all applicable laws, treaties, and regulations.

**DO NOT MAIL THIS PART 3 OF THE FORM
IT IS YOUR TEMPORARY PERMIT**

Read, Fill in the Blanks, and Sign:

Name: **JAMES B. JONES**
Date FCC Form 753, Parts 1 & 2 mailed to FCC: **APRIL 12, 1994**

I Certify that:
- The above information is true.
- I have completed and signed FCC Form 753, Parts 1 and 2, and mailed it to the FCC.
- I have never had a license suspended or revoked by the FCC.

Signature: *James B. Jones*   Date: **4/12/94**

If you cannot certify to all of the above, you are not eligible for a temporary permit.

FCC Form 753 – Use if you are eligible for employment in the United States.

APPENDIX – FORMS

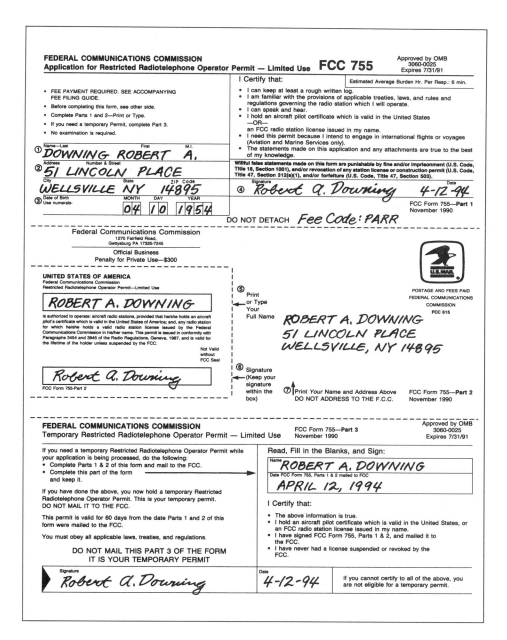

FCC Form 755 – Use if you *are not* eligible for employment in the United States.

321

GENERAL RADIOTELEPHONE OPERATOR LICENSE

Approved by OMB
3060-0069
Expires 9/30/96
See reverse for Public
Burden Estimate.

FEDERAL COMMUNICATIONS COMMISSION

FOR FCC USE ONLY

APPLICATION FOR COMMERCIAL RADIO OPERATOR LICENSE
(OTHER THAN RESTRICTED RADIOTELEPHONE OPERATOR PERMIT)

PLEASE READ INSTRUCTIONS CAREFULLY BEFORE COMPLETING THIS FORM

1. Applicant Name (Last Name) (Suffix): **WILTON**   (First Name): **HERMAN**   (Middle Initial): **T.**
2. (a) Number and Street address: **962 SHERWOOD COURT**
   (b) Street address continuation: **APT 3**
3. City: **SARATOGA**   4. State: **CA**   5. ZIP Code: **95070**
6. Date of Birth: Month **05** Day **09** Year **52**
7. Fee Type Code: **PACQ**
8. Fee Multiple: **0 0 0 1**
9. Fee Due: **$105.00**
   FCC USE ONLY

Check "YES" or "NO" to each of the following questions and provide the information requested.   YES   NO

10. Are you legally eligible for employment in the United States? (All U.S. citizens are considered, for the purpose of this application, to be legally eligible for employment in the U.S.)   [✓] [ ]

11. Do you have a speech impediment, blindness, acute deafness or any other disability which will impair or handicap you in properly using the license for which you are applying? If "YES", attach details.   [ ] [✓]

12. Have you held an FCC Commercial Radio Operator License(s) during the past five years? If "YES", list the license(s) held below. Renewal applicants must list data from expiring permit or certificate.   [ ] [✓]

Class	Endorsement	License Number	Date Issued	Place Issued

13. I hereby apply for: (Check appropriate boxes)
   [✓] New
   [ ] Renewal
   [ ] Duplicate/Replacement

   [✓] General Radiotelephone Operator License
   [ ] GMDSS Radio Operator License
   [ ] GMDSS Radio Maintainer License
   [ ] Marine Radio Operator Permit
   [ ] Radiotelegraph 1st Class
   [ ] Radiotelegraph 2nd Class
   [ ] Radiotelegraph 3rd Class

   [ ] Ship Radar Endorsement
   [ ] Six Months Service Endorsement

By checking "yes", the applicant certifies that, in the case of an individual applicant, he or she is not subject to a denial of federal benefits, that includes FCC benefits, pursuant to Section 5301 of the Anti-Drug Abuse Act of 1988, 21 USC, Section 862.

[✓] YES   [ ] NO

Failure to check "YES" may be cause for dismissal of your application.

I CERTIFY that I am the above-named applicant and that all statements made on this application and any attachment hereto are true and complete to the best of my knowledge.

Willful false statements made on this form are punishable by fine and/or imprisonment (U.S. Code, Title 18, Section 1001), and/or revocation of any station license or construction permit (U.S. Code, Title 47, Section 312(a)(1)), and/or forfeiture (U.S. Code, Title 47, Section 503).

Signature of Applicant: *Herman T. Wilton*   Date Signed: **8-12-94**

Telephone number where you can be reached during daytime hours (optional):
Area Code (**408**) Number **555-1212**

NOTICE TO INDIVIDUALS REQUIRED BY THE PRIVACY ACT OF 1974 AND THE PAPERWORK REDUCTION ACT OF 1980
The solicitation of personal information requested in this form is authorized by the Communications Act of 1934, as amended. The Commission will use the information provided in this form to determine whether grant of this application is in the public interest. In reaching that determination, or for law enforcement purposes, it may become necessary to refer personal information contained in this form to another government agency. In addition, all information provided in this form, as well as the form itself, will be available for public inspection. If information requested on the form is not provided, processing of the application may be delayed or the application may be returned without action pursuant to Commission Rules. The foregoing notice is required by the Privacy Act of 1974, 5 U.S.C. Section 522a(e)(3).

FCC 756
October 1993

FCC Form 756 – Use for all new licenses, permits and endorsements, and for renewal, replacement and duplicate licenses.

Glossary

AM – Amplitude Modulation: An audio signal such as a voice modulates a carrier wave. This creates a signal which, when transmitted, carries information. Intelligence is determined by varying the intensity of the radio wave.

Auto Alarm: Also called an automatic alarm. A device which monitors the distress frequencies and alerts personnel when traffic is received. Its fundamental purpose is to stand watch when the radio operator is not on duty.

Authorized bandwidth: The maximum permitted band of frequencies as specified in the FCC authorization to be occupied by an emission. Includes a total of the frequency departure above and below the carrier frequency.

Authorized frequency: The radio frequency assigned to a station by the FCC and specified in the station license or other instrument of authorization.

Authorized power: Unless transmitting distress calls, the minimum amount of output power necessary to carry on the telecommunications for which the station is licensed. (See § 80.63(a))

Avionics: The electronics including the radio system aboard an aircraft which must be maintained by a General Radiotelephone Operator.

Base Station: A fixed land station used to communicate with mobile radio stations installed in motor vehicles.

Break: Phrase spoken just before a brief pause in radiotelephone conversation to allow the other station to acknowledge your transmission.

Bridge-to-Bridge Station: A VHF radio station located on a ship's navigational bridge used for navigational purposes.

Calling Frequency: Agreed-upon frequency which stations use to initially call one another. Upon contact, both stations switch over to another "working" frequency so others may use the calling frequency.

Call Sign: A unique identifier issued to a radio station by the FCC as an aid to enforcement of radio regulations. A call sign identifies the country of origin and individual station.

Cargo Vessel: Any ship not licensed to carry more than 12 passengers.

Carrier: An alternating-current wave radiated from a transmitter of constant frequency with no modulation present. A carrier is a radio wave intended to "carry" the modulation or information.

Clear (or Out): Word used in radiotelephony to indicate that a transmission has ended and no response is expected.

Coast Station: Land-based radio station for the maritime services. Class-I stations provide long distance communications to ships at sea. Class-II coast stations provide regional service. A public coast station is open to public correspondence; a limited coast station may not transmit telecommunications for the public.

Commercial Operator Licensing Examination Manager (COLEM): Entity approved by the FCC to provide commercial radio license exams.

Communications Act of 1934: The basic document for controlling telecommunications in the United States.

Communications Priority: The order of priority of communications in the mobile service is: (1) Distress messages, (2) Communications preceded by the urgency signal, (3) Communications preceded by the safety signal, (4) Radio direction finding commu-

GENERAL RADIOTELEPHONE OPERATOR LICENSE

nications, (5) Navigation and safe movement of aircraft, (6) Navigation, movement of ships, and weather observations, (7) Government radiotelegrams, (8) Government communications for which priority has been requested, (8) Communications relating to previously exchanged and; (9) All other communications.

Compulsory Ship: Ship required by international law to carry radio equipment, licensed radio operators and to keep logs.

Distress: Requiring immediate assistance. Distress traffic in radio communications receives the highest priority because distress calls indicate imminent disaster. MAYDAY is the radiotelephone distress signal; SOS in radiotelegraphy.

Distress Frequency: An internationally recognized frequency set aside for distress traffic such as 2182 kHz (single sideband), 156.8 MHz (FM) and 500 kHz (telegraph.) 121.5 MHz (AM double sideband-full carrier) is the aircraft distress channel.

Downlink: A one-way wide angle radio beam from a communications satellite in earth orbit to a station on the surface of the earth. See uplink, transponder

Duplex: Two-way communications with both stations transmitting and receiving on different frequencies.

Earth Station: A station in the earth-space service located on the surface of the earth or on a ship or aircraft.

Effective Radiated Power (ERP): The output of a transmitter taking antenna gain into consideration.

Emergency Locator Transmitter (ELT): A battery-operated automatic transmitter in the aviation service used as a locating device for survival purposes.

Emergency Position Indicating Radio Beacon (EPIRB): A battery-operated radio transmitter in the maritime mobile service that activates upon the sinking of a ship to facilitate search and rescue operations.

Emission Designator: A series of three alpha-numeric characters used to identify radio signal properties. For example: A1A, manual radiotelegraphy. (A=Double-sideband amplitude modulation, full carrier; 1=One channel of digital modulation; A=Morse code for manual reception.)

Facsimile: The exchange of scanned "still" images via wire or radio circuits. A fax signal is modulated into a range of audio tones for transmission and demodulated by the receiving facsimile machine.

FCC – Federal Communications Commission: The official telecommunications agency in the United States. Among its duties is the allocation and regulation of radio frequency assignments within a framework of international agreements.

FM – Frequency Modulation: The process of varying a radio signal to convey intelligence by changing the transmitting frequency.

Frequency Allocation: A radio frequency or band of frequencies internationally or nationally assigned by an authorized body.

Frequency Deviation: VHF-FM transmitters in the maritime service are determined to be operating properly (100% modulation) when the maximum amount by which the carrier frequency changes either side of center frequency is plus-or-minus ($\pm$) 5 kHz.

General Radiotelephone Operator License (GROL): License issued by the FCC to individuals qualified to service, maintain, repair and operate radiotelephony communications equipment.

GHz – Gigahertz: term for one billion cycles per second. Also 1,000 megahertz (MHz).

Global Maritime Distress and Safety System (GMDSS): An automated ship-to-shore distress alerting system that uses satellites and advanced terrestrial communications systems. It is coordinated throughout the world by the IMO, the International Maritime Organization. It picks up radio distress messages and relays them to the proper authorities.

Ground: A connection with the earth to establish ground potential. A common connection in an electrical or electronic circuit. The area directly below an antenna. With Marconi antennas, the ground becomes one-half of the antenna.

GLOSSARY

Great Lakes Radio Agreement: A rule applying to all ships in the Great Lakes region. Technical requirements are stated for radio equipment and operators.

Harmful Interference: Any emission or radiation which interrupts or degrades a radio communications service operating in accordance with the rules. Operators must never deliberately interfere with any radio signal.

Harmonics: Spurious signals that show up at integer multiples of the main frequency.

Hertz: A measure of frequency equal to one cycle per second identified as hertz with lower case h.

HF – High Frequency: The radio frequency band that occupies 3 MHz to 30 MHz.

ILS – Instrument Landing System: An electronic aircraft guidance system using a radio beam to direct a pilot along a glide path.

Interference: The presence of unwanted atmospheric noise or man-made signals that obstructs or inhibits the reception of radio communications.

International Fixed Public Radio Service: A point-to-point radio communications service open to public correspondence.

International Phonetic Alphabet: Worldwide method of substituting words for individual letters to increase understanding.

International Radiotelegraph Alarm Signal: A signal consisting of a series of twelve dashes sent in one minute. This signal activates automatic devices to inform the operator that traffic is being received on a distress frequency.

International Radiotelephone Alarm Signal: A signal consisting of two sinusoidal audio tones transmitted alternately. This signal activates automatic devices to inform the operator that traffic is being received on a distress frequency.

International Telecommunication Union – ITU: The worldwide governing body controlling wire and radio telecommunications.

Ionosphere: An electrically charged frequency sensitive portion of the upper atmosphere that has the ability to reflect radio waves back to earth.

Kilohertz (kHz): One thousand cycles per second.

LF – Low Frequency: This band occupies 30 kHz to 300 kHz.

Licensee: The entity to which a radio station is licensed by the Federal Communications Commission.

License term: Ship and aircraft stations are licensed for five years. Most commercial radio operators are licensed for five years except Restricted Permits and General Radiotelephone Operator Licenses are issued for the life of the operator.

Line-of-Sight: The most effective mode of VHF communications is direct from the antenna to the receiver.

Log: Diary of radio communications kept by the station operator. It must contain frequencies used, any technical problems encountered, what action has been taken to correct technical problems, and if any distress traffic has been intercepted. The log is the written report of the station's performance and activities.

Loran-C: Acronym for LOng RAnge Navigation. Loran-C is a system of radio transmitters broadcasting low-frequency pulses to allow ships to determine their positions. The difference in time it takes for pulses from different transmitters to reach a ship allows Loran-C to determine the ship's location very accurately.

Marconi Antenna: A one-quarter wavelength antenna fed at one end and operated against a good RF grounding system.

Marine Radio Operator Permit (MROP): A permit earned by passing a 24 question examination on regulations, operating techniques and practices in the maritime services. The MROP is granted by passing Element 1 – Radio Law. MROP holders may not make internal adjustments to radio transmitting equipment.

Maritime: Relating to navigation or commerce on the seas.

Maritime Mobile Radio Service: A two-way mobile communications service between ships, or ships and coast stations.

325

GENERAL RADIOTELEPHONE OPERATOR LICENSE

Maritime Mobile Repeater Station: A land station at a fixed location established for the automatic retransmission of signals to extend the range of communication of ship and coast stations.

Master: A person licensed to command a merchant ship.

Mayday: A word spoken three times spoken during radiotelephone distress messages.

MF – Medium Frequency: This radio frequency band occupies 300 kHz to 3,000kHz (or 0.3 MHz to 3 MHz.)

Megahertz (MHz): One million cycles per second. Also 1,000 kilohertz (kHz).

Microwaves: Radio waves generally beginning at 1,000 MHz or 1 GHz. Most microwave activity is in the 1 to 50 GHz range.

Modem: A modulator/demodulator that converts digital signals to audio tones for transmission over wirelines or via radio wave. The process is reversed at the receiver.

Modulation: The process of modifying a radio wave so that information may be transmitted. The desired signal is superimposed onto a higher "carrier" radio frequency. A radio wave has three basic properties that can be varied: amplitude, frequency and phase.

Multihop – Multipath: Radio waves can bounce back and forth between the earth and the ionosphere or follow more than one route to a receiving point.

Nautical Mile: The fundamental unit of distance used in navigation. One nautical mile = 1.15 statute miles (or 6,080 feet). One knot is one nautical mile per hour.

Navigational communications: Safety communications exchanged between ships and/or coast stations concerning the maneuvering of vessels.

Omega: A radio navigation system relying on eight land transmitters throughout the world. Ships carry special receivers to listen to the transmitters and determine from the information carried exactly where the ship is at all times. Omega relies on phase differences between received signals.

Omni-directional: Performing equally well in all directions.

Over: Word spoken in radiotelephone conversation to indicate that it is the other station's turn to speak.

Overmodulation: Driving a transmitter over its designed parameters causes adjacent frequency interference. Can be caused by shouting into a microphone. Peak modulation should not exceed 100%.

PAN: The internationally recognized radiotelephone urgency signal. The words "PAN PAN" are spoken three times in succession to indicate that an urgent message will follow.

Part 13: The rules issued by the FCC governing commercial radio operators.

Part 23: The group of rules issued by the FCC governing stations in the international fixed public radio communication services.

Part 73: The group of rules issued by the FCC governing radio broadcasting services.

Part 80: The group of rules issued by the FCC governing stations in the maritime services.

Part 87: The group of rules issued by the FCC governing stations in the aviation services.

Passenger Ship: Any ship carrying more than twelve paying passengers. (Six passengers when used in reference to the Great Lakes Radio Agreement.)

Peak Envelope Power (PEP): Method of measuring the output power of a single sideband signal since no carrier is transmitted.

Phonetic Alphabet: A system of substituting easily understood words for corresponding letters.

Power-driven Vessel: Any ship propelled by machinery.

Propagation, Radio Wave: The method of radio wave travel which may be along the earth's surface, directly through space or reflected from the upper atmosphere.

GLOSSARY

Public Correspondence: Any third-party telecommunication (message, image or voice traffic) that must be accepted for transmission.

Radar: Acronym for RAdio Detection And Ranging. A method of tracking objects by analyzing reflected microwave radio signals or echoes. Doppler radar is used to measure speed.

Radio Operator: The FCC licensed operator in charge of the station who is responsible for its proper use and operation.

Radioprinter: A means of exchanging alphanumeric codes by direct printing.

Radio Services: Radio operations are classified into services according to the nature and purpose of the transmission.

Radiotelephony: Method of transmitting voice over radio waves.

Restricted Radiotelephone Operator Permit: Permit allowing certain radio privileges in the aviation, broadcast and maritime services. No examination is required.

Roger: A word in radiotelephone conversation to indicate that you have received and understood all of the other station's transmission.

Safety communications: A radio transmission indicating that a station is about to transmit an important navigation or weather warning.

Safety Convention: International agreement which spells out certain safety requirements for on-board radio equipment and operators.

Security: The word "SECURITY" is spoken three times prior to the transmission of a safety message.

Secrecy of Communications: Other than broadcasts to the general public, persons may not divulge the content of telecommunications nor use the information obtained to benefit anyone other than the intended recipient.

Selective Calling: A coded transmission to a particular radio station. Other stations do not hear it.

Selectivity: The ability of a radio receiver to separate the desired signal from other signals.

Separation: A method of minimizing mutual interference by spacing stations using the same frequency at required distance intervals.

Sensitivity: The ability of a radio receiver to respond to weak input signals.

Ship Earth Station: A mobile satellite station located on board a vessel.

Silent period: The three-minute duration of time during a continuous watch on a distress frequency when a maritime radio operator must not transmit.

Simplex: Two-way communications with both stations transmitting and receiving on the same frequency. Only one station may transmit at a time.

Single Sideband (SSB): Method of transmitting radiotelephony where one sideband is filtered out and the carrier suppressed or reduced. SSB is more efficient than double sideband signals since it takes up less radio spectrum. Emission: J3E

SOS: The radiotelegraphy distress signal sent three times followed by DE ("this is") and the call sign of the station in danger.

Station Authorization: Any construction permit, license or special temporary authorization issued by the FCC for activating a radio station.

Statute Mile: 5,280 feet. Unit of distance commonly used on land in the United States. One statute mile equals 0.8684 nautical miles.

Sunspot Cycle: The height, thickness and intensity of the ionosphere from which radio waves are reflected vary according to a cycle of approximately 11 years.

Survival Craft Station: A mobile station on a lifeboat, life raft or other survival equipment aboard a ship or aircraft intended for emergency purposes

Telecommunication: The transfer of sound, images or other intelligence by electromagnetic means.

Telegraphy: The process of sending and receiving information through the use of Morse code.

Telephony: The process of exchanging information through the use of speech transmissions.

Teleprinter: A mechanical typewriter–like device that prints text sent over wire or radio circuits. In a radioteleprinter, a modem converts the audio output of a receiver into electrical impulses to drive the individual keys. See modem.

Traffic: Radio messages exchanged between stations.

Translator, Broadcast: A relay station used to improve the reception of weak television and FM broadcast signals in remote locations. Translator equipment rebroadcast the input signal on another frequency or channel.

Transmission Line: The conduit by which radio frequency energy is transferred from the transmitter to the antenna.

Transponder: A device in an orbiting satellite that receives uplink (transmitted) signals from earth and downlinks (retransmits) them back to earth. A transponder is a wide coverage space repeater. See uplink, downlink.

Type Acceptance: Radio equipment that has met FCC specifications. All transmitters in the Fixed, Aviation and Maritime Services must be "type accepted." Type acceptance is based on data submitted by the manufacturer. "Type Approval" is granted after tests are made by FCC technical personnel.

UHF – Ultra High Frequency: This radio frequency band occupies 300 MHz to 3,000 MHz (or 0.3 GHz to 3 GHz).

Universal Coordinated Time – (UTC): Sometimes called Greenwich Mean Time (GMT), the time appearing at the zero meridian near Greenwich, England. UTC is the standard for time throughout the world.

Uplink: The ground-based frequency on which an orbiting satellite receives its radio signals from earth. See downlink, transponder.

Upper Sideband: The information carrying portion of the signal just above the amplitude modulated (AM) carrier frequency which is reduced or eliminated before transmission.

Urgency Communication: Urgent message concerning the safety of a ship, aircraft, other vehicle or person. Urgency traffic has slightly lower priority than distress traffic.

Vertical Polarization: Standard method of orienting maritime antennas operating at frequencies above 30 MHz: perpendicular to the ground or water. Polarization is determined by the direction of the electric component of the electromagnetic field. Vertically oriented antennas produce vertically polarized waves.

Vessel Traffic Service (VTS): Traffic management service operated by the U.S. Coast Guard in certain water areas to prevent ship collisions, groundings and environmental harm.

VHF – Very High Frequency: This radio frequency band occupies 30 MHz to 300 MHz. (or 3,000 kHz to 30,000 kHz).

Violation Notices: Notification from the FCC of a rule infraction. A written response must be made within 10 days containing a full explanation and action taken to prevent reoccurrence.

VLF – Very Low Frequency: This band occupies 3 kHz to 30 kHz.

Voluntary Ship: A ship not required to carry radio equipment, licensed radio operators or keep logs. When radio equipment is carried, however, appropriate listening watches must be maintained on 2182 kHz.

WWV: The precise standard frequency and time transmissions of the National Bureau of Standards.

Watch: The act of keeping close observation on distress frequencies for any distress messages.

Wilco: Phrase spoken in radiotelephony to acknowledge that a message has been received and that the receiving station will comply. "Wilco" is short for "will comply."

Working Frequency: A frequency establishing for conducting communications after first being established on a Calling Frequency.

Index

A

Aeronautical stations, 310
Aircraft stations, 11, 310
    identification, 314
    operating procedures, 314
Alarm signals, automatic, 52, 56, 62, 68, 88
Amplifiers, 184-186
    intermediate frequency, 198-199
    power amplifier stage, 227
    radio frequency, RF, 198
Amplitude modulation, 4
    compandored, ACSB, 233-234
    double-sideband, 194, 216
    single-sideband, 194, 217
Antenna
    bandwidth, 238
    beamwidth, 251-253
    dipole, 243
    dummy, 259, 268
    efficiency, 241-242
    electrical length, 264-266
    folded dipole, 242-243
    gain, 237-238
    half-wave, 246, 259, 267
    impedance, 254-256
    isotropic radiator, 248-250
    loading coil, 247-248
    loop, 261
    Marconi, 260, 264, 266-268
    maritime requirements, 37, 46, 47
    mobile, 247-248
    parabolic dish, 251
    parasitic elements, 239-240, 257, 260, 262
    radiation resistance, 257
    resonant frequency, 258, 262
    stacked, 258-259, 262
    trap, 38-239
    vertical, 258, 261
    yagi, 240-241
Application "filing" fees, 19
Applications, Commercial radio operator
    completed samples, 320-322
    where to get assistance, 23
    where to obtain forms, 17-18
    where to send, 19-20
Armstrong, Edwin H., 6
Associated ship unit, 41, 57, 74, 286
Automated maritime telecommunications system,
    AMTS, 286, 290
Automated Mutual-Assistance Vessel
    Rescue System, AMVER, 48, 286
Auxiliary broadcast stations, 11
    remote pickup, 12
Aviation Services, 10, 310

B

Bipolar transistors, 168-169

Bridge-to-bridge station, 48, 56, 308
    definition, 286
    power limitations, 55
    purpose, 60, 303

C

Capacitance, 127
Capture effect, 106-107
Cargo ship, definition, 286
Citizens Band, CB, 12
Civil Air Patrol, CAP, 11, 311, 316
Coast station, 11, 39, 40, 73, 286, 292, 304
    antenna requirements, 292
    operating procedures, 296
    traffic lists, 295
Commercial Operator Examination Manager, COLEM
    addresses, 275
    definition, 270
    qualifications, 283
    records, 283
    responsibilities, 29, 33, 282-283
Commercial Radio Operator
    classification, 12-17, 299
    requirements, aircraft stations, 314
    requirements, Bridge-to-Bridge Act, 300
    requirements, GMDSS, 27-28, 309
    requirements, Great Lakes, 300
    requirements, Safety Convention, 299
    requirements, voluntary ship stations, 300
Communications Act of 1934, 6
Compandoring, 233
Compulsory ship station, 52, 61, 286, 291, 306
    authorized equipment, 301
    safety watches, 301-302
    small passenger boats, 307
Cross modulation, 105-106
Crystal-lattice filter, 174-175

D

de Forest, Lee, 3
Desensing, receiver, 105, 111
Detection, signal, 195-196
Digital codes, 229
Digital selective calling, 61, 286
Diodes
    junction, 166, 171
    point contact, 167
    tunnel, 165
    varactor, 165-166
    zener, 164
Dip meter, 100-102
Direction finder, 287
Distress
    call, 1, 36, 45, 59, 78, 302
    frequency, 42, 58, 66, 302
    message traffic, 35, 55, 72, 287, 303
    radiotelegraph, radiotelephone signal, 287, 302
Dynamic range, 208-209

E
Earth station, ship, 36
Effective radiated power, 130-133
Emergency locator-transmitter, ELT, 311
Emergency position-indicating radio beacon, EPIRB
 class A, 48, 75, 308
 definition, 55, 287
 license requirements, 13
 survival craft, 67
Endorsement
 aircraft Radiotelegraph, 9
 broadcast, 9
 restrictive, 278
 ship Radar, 9, 278
Examination
 credit, 27, 32, 279
 elements, 28, 281
 fees, 19, 24, 270
 qualifying for a license, 281
 question pools, 283
 preparing, 282
 procedures, 282
 requirements, 27

F
Facsimile, 80, 214-215
FCC Form 753, 755, 756, 17-18, 274
 ordering forms, 18
Federal Communications Commission, FCC, 6
 field offices, 276
Federal Radio Commission, FRC, 5
Fees, License, 21
 application, 19
 examination, 19
 fee filing guide, 24
 regulatory, 19
 type codes, 20-22, 24
 where to send, 19
Field-effect transistors, FET, 175-176
Filter
 constant-K, 189
 m-derived, 190
First Class Radiotelephone License, 9
Fixed public service, 284
Flip-flop circuit, 199
Formulas, mathematical, 317-319
Frequency
 bands, 113
 boosters, 11
 broadcast translators, 11
 conversion, 196
 cooperative use, 293
 required radiotelephony, 298
 tolerance, 284
Frequency modulation, 6, 7
 deviation ratio, 217-218
 frequency discriminator, 196
 reactance modulator, 193, 216
Frequency standard, 97
 marker generator, 97, 202, 203
 counter, 98-100, 102, 203

G
General Mobile Radio Service, GMRS, 12
General Radio Telephone Operator License, GROL
 classification, 277
 examinations, 27
 fees, 21
 license term, 12
 question pools, 30
 requirements, 14, 65
Glide-path station, 311
Global Maritime Distress and Safety System
 applicability, 308
 definition, 34, 277, 287
 license requirements, 16, 309
GMDSS Radio Maintainer's License, GMDSS/M
 classification, 277
 examinations, 27, 282
 examination credit, 16
 requirements, 16
GMDSS Radio Operator's License, GMDSS/O
 classification, 277
 examinations, 27, 282
 requirements, 16

H
Handicapped applicants, 274
Harmonics, 236-237, 264
High frequency, HF, 11, 74
Hoover, Herbert, 4, 5

I
Identification, station, 40, 63, 68, 295
Ignition noise, 109-110, 118
Impedance, network, 147-158
 input/output, 211
 match, 161
Inspection, see Ship inspection
Instrument landing system, ILS, 311
 ILS localizer, 86, 119, 120, 311
Interference prevention, 294
Intermodulation interference, 103
International Fixed Service, 10
International Radiotelegraph Convention, 6
International Telecommunication Union, ITU, 1, 7, 277
 ITU Regions, 8

K
Knots, 87

L
Land mobile radio stations, 12
Land stations, Part 80, 292
License, Commercial Radio Operator
 application forms, 279
 classifications, 277
 duplicate, renewal, replacement, 20, 279-280
 eligibility, 278
 examination credit, 279
 examination requirements, 27, 281
 fraudulent, 279

INDEX

holding more than one, 279
new, 19, 279
operator's responsibility, 280
when required, 11, 269, 289
License, station, 304
Light-emitting diodes, LED, 172-173
Logic
   AND, NAND, OR, NOR, NOT, 201
   positive, negative, 202
   probe, 109
Logs, station, 34, 45, 46, 53, 58, 62, 304
Loran-C, 75

M

Marconi
   antenna, 260, 264, 266-268
   company, 3
   Guglielmo, 1
Marine Radio Operator Permit, MROP, 10, 34
   classification, 277
   examinations, 27
   requirements, 14, 16
   fees, 21
Marine utility station, 287
Maritime Services, 10
Maritime mobile repeater station, 35, 67, 287
Maritime mobile-satellite service, 287
Maritime mobile service, 287
Maritime radiodetermination service, 288
Maritime support station, 288, 305
Marker beacon station, 311
Marker generator, see frequency standard
Master, ship's, 41
Mayday, 1, 59
Modulation
   definition, 193
   double sideband, 194, 216
   index, 218-219
   pulse, 228-229
   requirements, 71, 301
   single-sideband, 194, 217
   spread-spectrum, 234-235
Morse Code, International, 1, 9, 277
   examinations, 31, 282
Multivibrator, 200

N

Name change on license, 21, 23
National Television System Committee, NTSC, 7
Nautical miles, 68, 85, 88-89
Navigational communications, 50, 288

O

Omega radio navigation, 74
Op amp circuit, 177-178, 204-205, 209-210
Oscillator circuits, 190-192
Oscilloscope, 102-103

P

Packet radio, 229
Passenger ship, definition, 286
Passenger ship safety certificate, 288
Part 13 Commercial Radio Operator Rules, 277
Part 23 International Fixed Public Radiocommunications Services, 284
Part 73 Radio Broadcast Services, 285
Part 80 Maritime Radio Services, 286
Part 87 Aviation Radio Services, 310
Peak envelope power, PEP, 226, 311
Phase angle, 129-130
Phase-locked Loop, PLL, 178-179
Phonetic alphabet, 53, 62, 73, 77
Photoconductivity, 133-135
Pi-network, 187-188
Pi-L, 187-189
Polarization
   circular, 221, 253
   horizontal, 220
   vertical, 221
Posting license, 23
Power consumption, 159-164
Power, transmitter, 41, 47, 62, 69, 74, 291
Priority of communications, 51, 73, 75, 293
Private coast station, 288
Proof-of-Passing Certificate, PPC, 24, 273
Propagation, radio wave
   electromagnetic waves, 219-221
   ionospheric, 93
   knife edge, 94-95, 257
   selective fading, 90-91
   transequatorial, 93
   velocity, 116
   VHF, 92, 94
Public correspondence, 288
   messages charges, 294

Q

Question coding, 31
Question pool, 29, 32, 277, 283
Question pool, Element 1, 33
Question pool, Element 3, 79
Question set, 277

R

Radar beacon, RACON, 288, 311
Radar endorsement, 9, 16
   examination, 27, 282
   fees, 21-22
Radar range, 95, 96, 311
Radio Act of 1912, 3, 4
Radio Act of 1927, 5
Radio Corporation of America, RCA, 3
Radiolocation service, 312
Radio officer, defined, 299
Radioprinter operations, 288
Radiotelegraph, definition, 284
Radiotelegraph Operator Certificates
   examinations, 27
   fees, 21
   First Class, T-1, 9, 15, 277
   license terms, 16
   photograph requirements, 18
   Second Class, T-2, 9, 15, 22, 277
   six-months service endorsement, 17, 278
   Third Class, T-3, 9, 14, 277
Radiotelephone, definition, 284

Radiotelephone Operator License
  First Class, 9, 10
  Second Class, 9, 10
Radioteletype, RTTY, 230
Reactance, Inductive/Capacitance, 122-123
Regulatory "User" fees, 19
Renewals, license
  fees, 21, 22
Replacement licenses, 21, 22
Resonance, 122-123, 124
Restricted Radiotelephone Operator Permit, RP, 9
  classification, 277
  fees, 21
  license term, 12
  requirements, 13, 27, 53
Root-mean-square voltage, RMS, 224-225

S
Safety communication, 288
Safety convention, 49
Safety signals, 288, 303
Safety transmission, 36, 59, 68
Satellite stations, 12, 81
Second Class Radiotelephone, 9
Secrecy of communications, 293
Selective calling, 288
Selectivity, 206-208
Ship earth station, 289
Ship inspection, 37, 47, 71, 291
Ship radar endorsement, see radar endorsement
Ship radar station, 13
Ship station, 11, 289
    operating controls, 292
    operating procedures, 296
    station requirements, 292
Silent periods, 56, 64, 72, 76, 301
Sine wave, 222
Single sideband
    amplitude compandored, ACSB, 233-234
    balanced modulator, 194
Six-months service endorsement, 17, 21, 22, 27
Skin effect, 125
Soldering, 85, 87, 88, 89, 113, 119, 121
Spectrum analyzer, 107-108
Spread spectrum communications, 234-235
Square wave, 223
Standing-wave ratio, SWR, 263, 266-267
Statute miles, see nautical miles
Super high frequency, SHF, 112
Survival craft station, 66, 67, 289, 312

T
Telegraphy examinations, see Morse code
Television signals, 215-216
Testing, transmitter, 51, 63, 65, 70, 313
    auto alarm, 68
    EPIRBs, 75
    radiotelephone procedures, 294
    survival craft, 66

Time constant, 135-143
Time, UTC/GMT, 86, 88
Titanic disaster, 2
Transfer of station license, 291, 312
Transistors, 170-172
    CMOS, 180-181
    TTL, 179-180
Translators, FM, TV, 11
Transmission line
    electrical length, 244
    loss, 116
    physical length, 244-245
    power loss, 263
    shielding, 118
    velocity factor, 243
Transmitters, Maritime Service, 301
    modulation requirements, 301
Traveling-wave tube, TWT, 181

U
Ultra-high frequency, UHF, 76, 89
    insulation, 112
Unauthorized transmissions, 293
Urgency signal, 36, 59, 64, 289, 303

V
VEC System, 25
Very high frequency, VHF stations, 7, 62, 70
Video
    blanking, 83
    levels, 84
    voltage, 83
Violations, 298
Voltage
    drop, 212
    half-wave dipole, 257
    peak-to-peak, 226, 235-236
    regulation, 159-160, 182-183
    root-mean-square, RMS, 224-225
Voluntarily-equipped ship stations
    definition, 286
    operator requirements, 44, 52, 63, 300
    watch requirements, 302, 309

W
Watch requirements,
    coast stations, 301
    compulsory ship stations, 301-302
    voluntary ship stations, 48, 302
Waveguides, 88, 116, 263, 265-268
Waves, sine, sawtooth, square, 222-224
Wilson, Woodrow, 3
Wireless Ship Act of 1910, 2
World Administrative Radio Conference, WARC, 7

Z
Zener diodes, 164
Zenith, 4